WEAPONS IN SPACE

WEAPONS IN SPACE

Edited by Franklin A. Long,
Donald Hafner, and Jeffrey Boutwell

W·W·NORTON & COMPANY
New York · London

Copyright © 1986 by the American Academy of Arts and Sciences
All rights reserved.
Published simultaneously in Canada by Penguin Books Canada Ltd, 2801 John Street, Markham, Ontario L3R 1B4
Printed in the United States of America.

First Edition

Library of Congress Cataloging-in-Publication Data
Main entry under title:
Weapons in space.
   Includes index.
   1. Strategic Defense Initiative—Addresses, essays, lectures. 2. Ballistic missile defenses—Addresses, essays, lectures. 3. Space weapons—Addresses, essays, lectures. I. Long, Franklin A., 1910–   . II. Hafner, Donald. III. Boutwell, Jeffrey D.
UG743.W43   1986   358′.1754   85–28549

ISBN 0-393-01989-6
ISBN 0-393-95527-3 (pbk.)

W. W. Norton & Company, Inc., 500 Fifth Avenue, New York, N.Y. 10110
W. W. Norton & Company, Ltd., 37 Great Russell Street, London WC1B 3NU

1 2 3 4 5 6 7 8 9 0

Contents

5   Acknowledgments

F.A. Long
7   Foreword

13   Glossary of Acronyms

I. Concepts and Technologies

Herbert F. York
17   Nuclear Deterrence and the Military Uses of Space

Alexander Flax
33   Ballistic Missile Defense: Concepts and History

Hans A. Bethe, Jeffrey Boutwell, Richard L. Garwin
53   BMD Technologies and Concepts in the 1980s

Gerold Yonas
73   The Strategic Defense Initiative

Donald L. Hafner
91   Assessing the President's Vision: The Fletcher, Miller, and Hoffman Panels

2   Contents

Charles A. Zraket
109 Strategic Defense: A Systems Perspective

Paul Stares
127 U.S. and Soviet Military Space Programs: A Comparative Assessment

Kurt Gottfried and Richard Ned Lebow
147 Anti-Satellite Weapons: Weighing the Risks

Ashton B. Carter
171 The Relationship of ASAT and BMD Systems

II. Implications for Security

Abram Chayes, Antonia Handler Chayes, Eliot Spitzer
193 Space Weapons: The Legal Context

John C. Toomay
219 The Case for Ballistic Missile Defense

George Rathjens and Jack Ruina
239 BMD and Strategic Instability

David Holloway
257 The Strategic Defense Initiative and the Soviet Union

Christoph Bertram
279 Strategic Defense and the Western Alliance

Jeffrey Boutwell and F. A. Long
297 The SDI and U.S. Security

Contents

Appendix A: New BMD Technologies
313 Hans A. Bethe and Richard L. Garwin

Appendix B: Relevant Documents
351 President Reagan's Address to the Nation, p. 351;
     1972 ABM Treaty, p. 355

371 Contributors

375 Index

Acknowledgments

SINCERE THANKS ARE DUE to a number of individuals and organizations for helping make this book possible. The Carnegie Corporation and its president, David Hamburg, provided the funds and encouragement necessary for the American Academy of Arts and Sciences to put together its Weapons in Space working group. The initial decision to proceed with the study came out of a May 1983 planning conference at the Academy, for which we thank Rita Goodman and the Johnson Foundation.

The editors and authors of this volume were greatly aided by an advisory group that helped shape the study in its early stages. Members of this group included Ruth Adams, Harvey Brooks, Herman Feshbach, Spurgeon Keeny, Judith Reppy, Walter Slocombe, and Jasper Welch, Jr. Thanks are also due to many members of the Academy's Committee on International Security Studies for their advice and support.

Our task of molding an extensive amount of information and analysis into a coherent volume was facilitated by superb editorial help. Stephen Graubard, Editor of *Daedalus*, provided thoughtful criticism and wise counsel throughout the study. Elizabeth Lilla, Manuscript Editor of *Daedalus*, expertly shepherded the papers from rough draft to finished manuscript. We are also indebted to Barbara Flanagan, Benjamin Loeb, and Daniel Snodderly for editorial help along the way, to Catherine Girrier and Katherine Gunness for invaluable research help, and to Executive Officer John Voss and the staff of the American Academy for providing essential in-house support.

Finally, we would like to thank our authors for their patience, forebearance, and, most of all, for their contributions to this study.

F.A.L.
J.B.
D.L.H.

F.A. Long

Foreword

IN HIS 1983 "Star Wars" speech, President Reagan set a major new goal for the United States in its efforts to avoid nuclear war. The U.S. military, with the aid of the nation's scientists, was to study the science and technology of defense against nuclear weapons so that America could ultimately develop and deploy a defense so totally effective that nuclear weapons would be rendered "impotent and obsolete." As a first step, the Department of Defense established a large new program, the Strategic Defense Initiative (SDI), to manage and support a greatly expanded research effort into nationwide defenses.

The fundamental question underlying this effort is how best to avoid nuclear destruction. The Soviet Union, confronted with the U.S. nuclear arsenal, faces the same problem, and arms negotiations between the two nations have long been based on a perceived mutual interest in avoiding war. The immense destructiveness of nuclear weapons contributes directly to the continuing mutual deterrence against their use. This deterrence stems from the recognition by the two superpowers that even a surprise first strike will not prevent the nation being attacked from launching a devastating retaliatory strike: destruction will be mutual. What makes the "Star Wars" proposal so dramatic is that it calls for the U.S. to move away from retaliatory nuclear deterrence (or mutual assured destruction, as it is sometimes called) by building fully effective defenses against nuclear weapons. This change in strategy by the U.S., of course, leaves open the question of Soviet military and political responses.

This is not the first time that the U.S. and USSR have each been actively involved in developing and deploying defenses against nuclear weapons. The first was in the mid-1960s, and the difference between then and now is instructive. Then, as now, there were strong advocates for deploying defenses against nuclear-armed ballistic missiles. Then, as now, both nations were engaged in vigorous research and development programs for defensive systems. However, there were serious doubts in the U.S. concerning the strategic implications of the proposed defenses. And detailed cost analyses conducted by the Pentagon concluded that deploying missile defenses was not cost-effective against obvious possible Soviet responses. In the end, although both nations deployed limited defensive systems, the two nations agreed to negotiate to ban such systems. The result was the bilateral Anti-Ballistic Missile Treaty of 1972, which set severe mutual restrictions on the development, testing, and deployment of ABM systems. The treaty remains in effect, and is a factor in current discussions about the SDI.

Although defense against nuclear attack is one of the items being considered in the current arms-control talks between the U.S. and the Soviet Union, their negotiating positions appear to be far apart. The aim of the Soviets seems to be to stop the expanded U.S. research program, and the aim of the American negotiators seems to be to maintain the SDI as a major U.S. effort. The most likely outcome is agreement to disagree: this would leave the U.S. free to proceed with the Strategic Defense Initiative and related programs, and the Soviet Union free to respond as it sees fit. The likelihood and character of Soviet responses will, of course, greatly influence the feasibility and costs of the U.S. program, and as Paul Nitze recently noted, the projected defenses will need to be both cost-effective and survivable, even in the face of major Soviet responses.

A complicating aspect of the SDI is that many of the most interesting new technologies are designed to be deployed in space, on earth-orbiting satellites—some as sensors, others as weapons. However, many of these new technologies are only in the very early stages of study; hence, it will be many years before enough is known to permit a decision to deploy them, and even then the ultimate success of the overall defense system will be far from certain. This uncertainty will stem partly from doubts about systems capabilities and reliability, and partly from concern about possible Soviet countermeasures such as reducing the vulnerability and increasing the numbers and penetrability of their nuclear weapons.

Until there is high confidence that the new defensive systems are impregnable, the U.S. will be quite unwilling to abandon its reliance on retaliatory deterrence based on large offensive forces. This raises the prospect of a complex offensive and defensive arms competition between the U.S. and Soviet Union that could continue for years, if not decades. A combination of large military expenditures and great political uncertainty will have an unsettling effect on American domestic politics and U.S. relations with its allies, and one can assume that debates on strategic defenses will continue and intensify.

The contributors to this volume have attempted to deal constructively with the principal issues raised by the SDI. They give a broad picture of the weapons that are being proposed for use in space, and assess their opportunities and problems. They offer arguments for and against U.S. deployment of defensive weapons systems, and in so doing raise political as well as military and technical concerns.

Following the introductory chapter by Herbert York, which examines U.S. and Soviet military space activities in the postwar period, there is a general consideration of the problem of defense against land- and sea-based ballistic missiles. ICBMs and SLBMs represent the major current threat to the U.S., and the Strategic Defense Initiative is therefore concentrating its research on ballistic missile defense (BMD). Since the travel times for the warheads launched by these missiles is in the range of ten to thirty minutes, a major problem confronting a defense system is how to respond reliably and promptly to a major missile attack. Thus any BMD system that is developed will have to be both technically feasible and coordinated by an effective and reliable command-and-control system. Analyses of previous BMD efforts and the current SDI are provided by Alexander Flax, Gerold Yonas, Donald Hafner, and Charles Zraket.

Because Soviet missiles are deployed either in underground silos deep within the USSR or on submarines in the open ocean, it is essential for the defense to use space-based systems, i.e., orbiting satellites, for launch detection, tracking, and eventual attack on the missile or its warheads. It is highly desirable to destroy missiles at launch before they have deployed their multiple warheads, but because the earth is round, the launch sites distant, and the time for attack short, a successful attack at launch must come from space, preferably using kill mechanisms that travel at or near the speed of light. Major research and development, however, will be required to identify likely kill mechanisms and determine their effectiveness, their probable costs, and their vulnerability to countermeasures. An extended analysis of these prospects and problems is provided by Hans A. Bethe and Richard L. Garwin in the technical appendix to this collection; a paper by Bethe, Garwin, and Jeffrey Boutwell summarizes these analyses from a somewhat less technical point of view.

Three articles in this collection deal with the anti-satellite weapons (ASATs) that may be deployed by the U.S. and the Soviet Union, and that may seriously complicate development of an effective defense against nuclear weapons. Weapons for use against satellites can be ground-based, air-based, or space-based and their task is relatively uncomplicated, given that satellites are relatively fragile and travel in predictable orbits. The Soviets have tested and deployed a comparatively simple space-based ASAT, and the U.S, which in the 1960s deployed a ground-based system and then abandoned it, is now developing more advanced ASAT systems. Although the objectives and difficulties of space-based systems for ASAT and for BMD are quite different, there is or can be considerable technical overlap between the two. This relationship of ASAT and BMD weapons systems could influence decisions to deploy either one, and may also complicate negotiations to ban or limit weapons for either purpose. The wide variety of ASAT issues are treated by Paul Stares, Kurt Gottfried and Ned Lebow, and Ashton B. Carter.

The second half of *Weapons in Space* begins with an evaluation of how U.S. and Soviet BMD and ASAT efforts could effect the 1972 ABM treaty and other international legal agreements. Particular attention is paid to the SDI program and its implications for the ABM treaty in the article by Abram Chayes, Antonia Handler Chayes, and Eliot Spitzer.

This is followed by analyses of the military utility and political implications of different levels of BMD deployments. An important point in the debate over strategic defenses is that the Soviets can take a number of military countermeasures that will greatly complicate the development of a fully effective defense: it can deploy additional numbers of ballistic missiles, shorten their launch time, and equip their missile systems with various types of penetration aids, including decoys and chaff. All these possibilities have been studied extensively, and appear both feasible and relatively inexpensive. In contrast, there are many uncertainties regarding the technical feasibility, cost, and survivability of the BMD concepts being studied by the SDI. It is difficult, therefore, to escape the conclusion that the stated goal of obtaining a defense so complete as to make nuclear weapons "impotent and obsolete" is almost surely unobtainable. This is doubly so as effective defenses would also have to be deployed to protect against manned bombers, cruise missiles, and other delivery systems.

It is at least partly in recognition of this that a number of suggestions have been proposed for U.S. deployment of limited defenses that would protect missile silos and other important hardened sites in the U.S.. It is possible that deployment of defenses for some of these partial objectives would be cost-effective. However, their purpose would clearly be to bolster deterrence by retaliation, not to replace it. Discussion of these topics can be found in the articles by Maj. Gen. John Toomay, and George Rathjens and Jack Ruina.

So dramatic a change as moving from deterrence by threat of retaliation to a defense so complete as to make nuclear weapons "impotent and obsolete" is certain to have large political implications. The reactions of the Soviet Union are particularly important, as David Holloway points out in his article, as they will affect not only the feasibility and costs of strategic defenses, but also the likelihood of agreements on mutual arms reductions, and perhaps even the probability of nuclear war.

Reactions of America's allies to the SDI program are covered in the essay by Christoph Bertram. While considerable Allied apprehension about the implications of the new U.S. defense effort clearly exists, there are also significant voices of support for the SDI. It is probably too soon to expect any Allied consensus about the implications of U.S. deployment of strategic defenses. President Reagan's recognition in his 1983 speech of the importance of the NATO alliance, and his pledge to "continue to honor our commitments," was reassuring, but it will be some years before it will be clear how this will be reflected in new and continuing alliance policies.

The concluding article of *Weapons in Space* examines the pros and cons of the debate over the SDI, and considers some of the policy alternatives that the U.S. might consider. A theme that underlies many of these alternatives is the desirability, perhaps even the necessity, of exploring with the Soviet Union further measures for avoiding war, and the possibility of devising mutually useful ways to employ defense to enhance each nation's security.

GLOSSARY OF ACRONYMS

ABM	anti-ballistic missile
ACDA	Arms Control and Disarmament Agency
ALCM	air-launched cruise missile
AOA	airborne optical adjunct
ASAT	anti-satellite
ASW	anti-submarine warfare
ATBM	anti-tactical ballistic missile
ATP	Acquisition Tracking and Pointing [system]
AWACS	airborne warning and control system
BAMBI	ballistic missile boost intercept [system]
BMD	ballistic missile defense
BPI	boost-phase intercept
BSTS	Boost Surveillance and Tracking System
C^3I	command, control, communications, and intelligence
DARPA	Defense Advanced Research Projects Agency
DEW	directed energy weapon
DOD	Department of Defense
DOE	Department of Energy
DSAT	defensive satellite
DTST	Defensive Technologies Study Team [Fletcher panel]
ELINT	electronic intelligence
EMP	electromagnetic pulse
FOBS	fractional orbital bombardment system
FOFA	follow-on-force attack
GLCM	ground-launched cruise missile
GEO	geosynchronous orbit
GPS	Global Positioning System [Navstar]
HADS	high-altitude defense system
HF	hydrogen fluoride
HKV	homing kill vehicle
HOE	Homing Overlay Experiment
HSD	hard site defense
ICBM	intercontinental ballistic missile

INF	intermediate-range nuclear forces
IR	infrared
LEO	low earth orbit
MAD	mutual assured destruction
MaRV	maneuvering reentry vehicle
MIRV	multiple independently targetable reentry vehicle
MSR	missile site radar
NCA	National Command Authority
NTM	national technical means [of verification]
PAR	perimeter acquisition radar
R&D	research and development
RDT&E	research, development, testing, and engineering
RORSAT	radar ocean reconnaissance satellite
RV	reentry vehicle
SALT	Strategic Arms Limitation Talks
SAM	surface-to-air missile
SATKA	surveillance, acquisition, tracking, kill assessment
SCC	Standing Consultative Commission
SDI	Strategic Defense Initiative
SDS	satellite data system
SLBM	submarine-launched ballistic missile
SLCM	sea-launched cruise missile
SSTS	Space Surveillance and Tracking System
START	Strategic Arms Reduction Talks
TNF	tactical nuclear forces
UV	ultraviolet

I. Concepts and Technologies

Herbert F. York

Nuclear Deterrence and the Military Uses of Space

ON MARCH 23, 1983 PRESIDENT REAGAN made a speech that was almost as unexpected as it was portentous. In it, he called on American science to create a total defense against ballistic missiles, a defense of such qualitative difference that it could support a radical change in our nuclear strategy. Nuclear strategy would no longer be based on offense, but rather on defense; no longer on assured destruction, but on assured survival; or, as some have put it, no longer on death, but on life.

The official name of the program is the Strategic Defense Initiative (SDI), but because the technical ideas underlying it involve deploying defensive weapons in space, the program has come to be referred to—even by most of its supporters—as "Star Wars."

Although the *strategy* involved in the SDI represents a radical departure from past thinking, the *technology* that underlies it is the direct and logical result of many decades of research and development. Indeed, proposals for using satellites to supplement earth-based military capabilities were first put forth seriously in the mid-1940s. Ten years later, both the United States and the Soviet Union initiated programs designed to implement certain of these ideas. For the next quarter of a century, the development and deployment of these military space assets, as they came to be called, proceeded in a roughly balanced and relatively non-threatening manner. Recent developments—including anti-satellite weapons, the direct use of space assets in warfare, as well as the president's Strategic Defense

Initiative—are bringing about a new, more costly, and more dangerous phase in this ongoing process.

THE EVOLVING MILITARY USES OF SPACE

As early as 1946, a group of scientists and engineers at the RAND Corporation published a report entitled "Preliminary Design of an Experimental World-Circling Spaceship." The report was remarkable for its accuracy and prescience. Most of it is devoted to the question of how to build adequate launching systems, and while its estimates in this regard are not perfect, they are very good. More important, the report also devoted some attention to the possible military applications of satellites:

... The military importance of establishing vehicles in satellite orbits arises largely from the circumstances that defenses against airborne attack are rapidly improving. ... Under these circumstances, a considerable premium is put on high missile velocity, to increase the difficulty of interception.

This being so, we can assume that an air offensive of the future will be carried out largely or altogether by high-speed pilotless missiles ... [accordingly] precise observations of the position of the missile can be made from the satellite, and a final control impulse applied to bring the missile down on its intended target. This scheme, while it involves considerable complexity in instrumentation, seems entirely feasible.

Alternatively, the satellite itself can be considered as the missile. After observation of its trajectory, a control impulse can be applied in such a direction, and at such a time that the satellite is brought down on its target.[1]

We have here the germ of two important ideas that were realized only decades later. The first finds its modern expression in the current plan to use navigation satellites of the global positioning system (GPS) to provide prompt location and steering information to a variety of weapons systems, including aircraft and missiles, thus increasing their accuracy and lethality. The second is the notion of an orbiting weapons platform, an idea implemented by the Soviets in their fractional orbital bombardment system (FOBS) in the early 1960s.

Other proposed military uses for satellites advanced in the 1946 report included locating targets, observing weather conditions over enemy territory, and using satellites to relay military communications. The report specified that satellites in geosynchronous orbits—36,000 kilometers above the earth—would be best for this last purpose.

Since no rockets suitable for launching a satellite into orbit existed at that time, it was soon concluded that the potential benefits of satellites would not by themselves justify the very large costs of the development and construction of the necessary rockets. For the next several years, then, work on satellites proceeded on a research basis only. As a result of a series of unexpected events, however—including, especially, the first Soviet A-bomb test, the Korean War, and the discovery of the existence of a substantial Soviet program dedicated to the development of long-range rockets—the U.S. in 1954 decided to establish several parallel programs aimed at developing huge rockets (Atlas, Titan, Thor, and Jupiter) specifically designed to power large and very long-range ballistic missiles. These rockets, of course, proved to be just right for the launching of artificial satellites, and as a result the idea of building such devices took on new meaning. By this time, satellite designers, still concentrated primarily at RAND, had begun to focus on those goals—overhead reconnaissance and early warning—that they felt were most promising for near-term practical use. In short order, a RAND study entitled "Project Feedback" was published on March 1, 1954, delineating these possibilities, followed a year later by a U.S. Air Force contract calling for the development of reconnaissance satellites.

All of these developments were, of course, proceeding on a highly classified basis. At about the same time, but in the public sphere, plans were being elaborated to declare July 1957–58 an "International Geophysical Year." One of the highlights of the year was to be the launching of artificial satellites that would perform scientific observations of our planet. In 1954 and 1955, work began on two additional satellites: a much smaller scientific satellite called "Project Orbiter," and the Navy's "Project Vanguard" (an alternative to "Orbiter"). However, before either of these programs could place a satellite in orbit, the Soviets, to the great shock and surprise of the general public and the Congress, launched the first Sputnik on October 4, 1957.

The launch of Sputnik prompted the U.S. government to undertake a thorough, and often frantic, review of the state of "high-technology" in America. Out of this came a total revamping and expansion of the then much smaller, but very prominent, civilian space program and the creation of a new agency, NASA, to oversee it. The still very secret military satellite program, on the other hand, experienced only minor changes in either its substance or organization. Like many other "high-tech" programs of the time, however, its priority and funding were eventually substantially increased.

This brief review of the first steps towards the military use of space exposes a fact that is commonly missed or glossed over in the folk-historic views of the origins and nature of our space programs: our space programs, *from the beginning*, have been primarily of a military, not a civilian or scientific nature. Public attention may have focused on Mercury, Gemini, and Apollo and the corresponding Soviet projects, but the fact remains that both the U.S. and Soviet space programs overall have always been driven more by military considerations and requirements than civilian and scientific ones.

Moreover, the military space program has always included a component dedicated to warfighting, as well as components devoted to general intelligence, administration, and logistical support. Our strategic missiles, for example, are currently primed to attack some 10,000 military and industrial targets, mostly in the Soviet Union. For all practical purposes, every one of these targets was discovered and located by reconnaissance satellites and was given precise coordinates by means of data produced by geodetic satellites. Our missile submarines derive their initial coordinates from data provided by navigational satellites. In the future, the use of these global positioning system (GPS) satellites will markedly improve the effectiveness of aircraft and missiles by correcting their trajectories while in flight. Mapping satellites are currently providing the input data for the super-accurate "tercom" (terrain comparison) system that will steer our cruise missiles. And we are already in the early stages of supplying data gleaned by satellites directly to battlefield commanders.

AMBIGUITY OF MILITARY SPACE MISSIONS

Setting aside for the moment consideration of ASATs and the SDI, are the various elements of today's military space progam stabilizing or destabilizing? Are they benign or aggressive? Are they intended to reinforce deterrence, or is their primary purpose to improve our fighting capability should deterrence fail? The answers to these questions are by no means simple or straightforward.

Let us first consider overhead reconnaissance in the broadest sense. Reconnaissance is used for several major purposes, each of which affects the stability of the relationship between the superpowers. One purpose is *target discovery and location*—the "spotting" function of the 1946 RAND report. Another is *general intelligence* to evaluate the state of preparedness of an enemy. A third is *verification of arms-control agreements*. This last purpose receives most of the attention in those political circles primarily interested in arms control, detente, and diplomatic approaches to solving the world's problems. But it is the first two purposes that provide the main rationale for spending billions of dollars annually to operate the existing system and to develop new ones. General intelligence-gathering and the verification of treaty compliance are, by and large, stabilizing. Target "spotting" is not. On the one hand, the discovery by satellites in 1961 that there was in fact no "missile gap" between the U.S. and the Soviet Union prevented us from over-reacting to Sputnik even more than we did. In this case, satellite intelligence contributed to stability. On the other hand, these same satellites are also capable of providing data that is sufficiently complete and accurate to make a successful preemptive first strike against land-based military forces seem possible.

Early-warning satellites are similarly ambiguous in their function. They are most important for their ability to give reliable warning of a missile attack twice as quickly as conventional earth-based radars, thus making it more difficult for the enemy to achieve a successful first strike. For this reason, early-warning satellites are said to be stabilizing. At the same time, these same characteristics are important to the launching of a retaliatory strike before absorbing a Soviet attack (launch-on-warning), a tactic advanced by some strategists as the best and cheapest answer to a surprise first strike on our land-based force. By making early warning more accurate, these satellites make launch-on-warning more thinkable.

The data produced by earth-mapping satellites gives modern cruise missiles their exceptional accuracy. Air-launched cruise missiles have as their primary purpose extending the life of the bomber leg of the strategic triad; they are therefore commonly said to contribute to strategic stability because they reinforce deterrence in a fundamental way. Other types of cruise missiles, such as the ground-launched cruise missiles deployed in Europe, add very little to deterrence itself; they are primarily warfighting devices designed to destroy specific military targets in the event deterrence fails.

Similarly, the GPS, the Soviet radar ocean reconnaissance satellite (RORSAT), and other systems designed to supply critical information rapidly to tactical commanders, are clearly much more sharply oriented towards warfare than towards deterrence. Indeed, these systems belong to a special class of space weaponry that one American official has described as the "most important class of weapons in space," yet is "almost totally neglected by the arms-control community."[2]

Space, then, long militarized and weaponized, is becoming steadily more so. Up to now, however, these developments have taken place in a roughly balanced fashion, in which the actions of one superpower have not generally been construed by the other as immediately menacing or even particularly hostile. The reason for this is not that satellites have made no direct contribution to warfighting, but simply that they have so far done so only by a slow accumulation of data during peacetime. This situation is obviously about to change: new systems—GPS, RORSAT, and the direct use of space-based assets by tactical commanders—will soon be supplying essential data to warfighting systems in real time. Such a move will inevitably be seen as hostile and dangerous by both sides; a clear example of this is the U.S. Navy's strong reaction to the Soviet RORSAT, whose principal purpose is to locate and target U.S. surface ships.

In what ways, then, would the deployment of a serious anti-satellite (ASAT) capability fit into this existing situation?

ASATS AND MILITARY SATELLITES

The pressure to acquire an anti-satellite capability is a natural response to the superpowers' increased use of space-based assets for military purposes. Indeed, the very first anti-satellite system was deployed by the United States during the Kennedy administration as a direct and immediate response to two seemingly ominous events in the Soviet Union. One of these was the development of an orbital bombardment system, that is, a system in which bombs would be deployed on satellites in low orbits and either exploded in space or brought down and detonated at specific points on the ground. The other was the explosion of an extraordinarily large nuclear device in 1961 at Novaya Zemlya in central Asia. While its estimated size was "only" fifty-eight megatons, American scientists believed the Soviets could have easily increased the yield to one hundred megatons or more. Many analysts concluded that bombs of this type could, if exploded at an altitude of one hundred miles or so, set fires on the ground over an area of perhaps a hundred thousand square miles—the area of a typical midwestern American state.

As a natural response to the threat posed by the combination of these two developments, Secretary of Defense McNamara ordered the deployment on Johnston Island of an ASAT system that used a Thor rocket to deliver a nuclear warhead to an altitude sufficient for intercepting bombs in orbit. While the Soviets did in fact eventually deploy a so-called fractional orbital bombardment system (FOBS), and although both the Soviet FOBS and the U.S. anti-satellite system remained deployed for more than a decade, each was eventually decommissioned.

Despite this instance of provocation and response, U.S. authorities during this period declined to authorize the development and deployment of a general purpose anti-satellite system. The rationale was simple: the U.S. relied heavily on satellites for intelligence and support purposes, and it was in our direct interest to maintain the idea of space as a special sanctuary for such systems. If we wanted no Soviet ASAT, we were willing to pay the necessary price: no American ASAT.

The Soviets, however, did not practice similar restraint. As is discussed more fully elsewhere in this volume, they initiated testing of a general-purpose ASAT in 1968.[3] After almost a decade of further restraint on our part, the Carter administration in 1977 finally responded to this Soviet ASAT project with a three-pronged approach. The first was the initiation of an ASAT development program of our own. The second was an increased effort to find ways of defending satellites against attack, the so-called DSAT program. Last was the initiation of negotiations with the Soviets designed to eliminate ASATs altogether, if possible, and hence to render the first two responses unnecessary. The Soviet reply was carefully ambiguous, indicating a preference for a "rules of the road" approach (i.e., regulating satellite use) over elimination of ASATs. Indeed, the Soviets made it clear that they regarded ASATs as a natural response to threats from space, mentioning in this context "third countries" (i.e., China) and "threats against our sovereignty from space." As it now stands, development of a general-purpose ASAT is going forward on both sides, despite the resumption in March 1985 of U.S.-Soviet negotiations on a broad range of weapons.

The main argument of those who favor ASAT arms control centers on the special uncertainties and dangers that the introduction of general-purpose ASATs will give rise to. So far, warfare, at least in the direct sense, has not extended to space; indeed, we have treaties now in force that were specifically designed to maintain outer space as a zone of peace. The introduction of ASATs, its critics contend, would destroy all that. Satellites whose purposes are largely benign and stabilizing—that is, those we depend on for intelligence, warning, communications, and the like—would be placed in peril and subject to destruction at any time. These critics foresee not only needlessly heightened fears and tensions, but increasing expenditures of money that could otherwise go to more useful military programs or neglected civil needs.

ASAT supporters argue that space will soon harbor even more satellites that contribute directly to an enemy's warfighting capability, and that it makes no sense, at this late date, to confer on space special sanctuary status. American officials cite the existence of Soviet ocean reconnaissance satellites (RORSAT) and contend that other systems, presumably parallel to those that we are developing, are in prospect.[4] They point to the difficulties of verifying any ban on ASAT deployment, and argue that new non-nuclear ABM systems (such as the one tested by the U.S. Army in 1984 at Kwajalein Atoll), to say nothing of powerful ground-based lasers and the like, make a treaty prohibiting ASATs too late and impractical.

Given these diametrically opposed positions, how is the American body politic likely to react? One basic fact to consider is that only one-third of the Senate is necessary to block any treaty, and no arms-control treaty that is openly opposed by the armed forces can make it over this hurdle. Although our military leadership is currently split on this issue, a growing majority appears to belong to the camp that regards ASATs as a natural response to a coming threat. Given that the most influential members of the office of the secretary of defense have similar sentiments, it is unlikely that a treaty totally banning ASATs will be negotiated in the near future.

On the other hand, a treaty establishing clear-cut "rules of the road" could probably be negotiated, even in the present circumstances. The world has had such rules governing behavior at sea for centuries. The oceans are full of fighting machines, yet these co-exist easily with civilian ships during peacetime. A similar arrangement could surely be worked out for space, one that would give all parties the same kind of assurance of safe passage for military as well as civilian satellites during peacetime that is now the case for all nations' warships now at sea. Such a policy could go a long way in guaranteeing the survival of intelligence and warning systems, as well as other systems that add to both international stability and military readiness in peacetime (in the event of a major war, of course, all space assets would be in grave danger). As with all high-technology weapons systems, especially nuclear systems, it is those components that play a major role in deterring war that should command primary attention, not those that would be important in combat, should deterrence fail. Of course, a total ban on ASATs would be the best way to reassure the continuing role of our space assets in deterring war, but a "rules of the road" agreement, if carefully crafted, could accomplish this almost as well, and from a practical political perspective seems much more feasible.

SPACE AND STRATEGIC DEFENSE

The president's new Strategic Defense Initiative was introduced into a world in which the militarization of space by the United States and the Soviet Union was already solidly underway, and in which it was widely recognized that this process ought to be limited or contained before it escalated beyond our control. Indeed, the technological side of the SDI is based, to a large extent, on concepts already deployed or currently under development in military space programs. Approximately 90 percent of the 1985 SDI budget is devoted to continuing the development of devices and the exploration of ideas that were already programmed *before* the SDI was introduced in March 1983.

For all practical purposes, the technological side of the SDI consists of those ideas, devices, and systems discussed in the report of the Fletcher panel, one of three committees set up by President Reagan to investigate strategic defense.[5] This report called for the creation of a multilayered defense system made up of a variety of components designed to detect, target, and destroy attacking ballistic missiles and their nuclear warheads all along their trajectories from their initial launch to their final targets. As noted above, most of the systems involved were based on concepts either already employed in our military space program or in various stages of development for possible future space applications. Many others were taken from terminal ballistic missile R&D programs established over a quarter of a century ago. And some came from unrelated programs, particularly from the research laboratories of our defense establishment. Among these are the so-called "Third Generation Weapons," including laser and particle beams and kinetic energy weapons that are considered especially useful in a defensive mode against Soviet missiles in boost phase, immediately after they have been launched.

There is, of course, much debate over whether the SDI will result in a total defensive system possessing the characteristics and qualities necessary to bring about the strategic revolution the president called for. We can here summarize this debate by examining two quotations from expert studies of the issue. On the one side, the Fletcher panel noted that: "By taking an optimistic view of newly emerging technologies, we concluded that a robust BMD system can be made to work eventually. The ultimate effectiveness, complexity, and degree of technical risk in this system will depend not only on the technology itself, but also on the extent to which the Soviet Union either agrees to mutual defense arrangements and offense limitations, or embarks on new strategic directions in response to our initiative."[6]

On the other side, in a report written by Drell, Farley, and Holloway, we find the statement: "We do not now know how to build a strategic defense of our Society that can render nuclear weapons impotent and obsolete as called for by President Reagan. Nor can one foresee the ability to achieve the President's goal against an unconstrained, responsive threat."[7] Many other studies by knowledgeable experts have reached very similar conclusions.[8]

Note that experts on both sides agree that the present technology cannot produce a system adequate to the president's goals; the disagreement is over whether the technology that will become available in the next ten or twenty years will be adequate to provide a truly useful level of defense, and whether the best way to generate such technology is to proceed along the lines now being proposed in the Department of Defense. The argument is actually over the future course of technology: to what degree we can extrapolate from existing know-how to estimate our capabilities in the distant future.

SPACE WEAPONS AND NUCLEAR DETERRENCE

During a press interview conducted soon after his "Star Wars" speech, President Reagan was asked about his new initiative. At one point he replied:

To look down on an endless future with both of us sitting here with these horrible missiles aimed at each other and the only thing preventing a holocaust is just so long as no one pulls the trigger . . . this is unthinkable.[9]

In giving this answer, the president, perhaps unwittingly, joined a host of others who believe that peace through a balance of terror is untenable in the long run, that deterrence through the threat of mutual assured destruction (MAD) cannot be the basis for an enduring peace, that the threat to kill hundreds of millions of civilians as punishment for some unacceptable political act by their government is immoral in the deepest sense of the word. These ideas are not new; all previous U.S. presidents have held similar ideas, and as far as we can tell, so have all Soviet leaders, at least since Stalin. The problem is—and has been for forty years—that no viable alternative presents itself. Supporters of nuclear deterrence have often pointed to the fact that, whatever the long-term prospects may be, deterrence has so far "worked"; that peace in Europe has been preserved for more than forty years, a time during which any number of crises might otherwise have precipitated a major war.

The introduction of nuclear explosives and other high-technology weapons only a few decades ago resulted in a radical change in the vulnerability of both the United States and the Soviet Union. Before that, the oceans that had from our origins isolated us from the rest of the world still protected us from both invasion and attack. And in 1941, when the Nazi war machine attacked the Soviet Union, the great size of that country enabled it to absorb the full weight of attack long enough to rally, muster outside help, and turn the tide of the invasion. Today, each of the superpowers could, if it chose to, literally annihilate the other in less than an hour, and the only recourse either nation has to forestall such a horror is to threaten to wreak revenge in kind. That is the stark meaning of Reagan's obviously heartfelt remark that the only thing preventing a nuclear holocaust is the fact that no one is pulling the trigger. Nor is this stalemate a static situation; things are in fact becoming steadily worse. The thoroughness of the potential destruction is increasing, the chance of survival for individuals and for civilization itself is decreasing. The combination of technical improvements in accuracy and multiple warheads (MIRVs) is making the possibility of a first strike seem, at least to some, more thinkable. To guard against this, some strategists advocate a policy of launch-on-warning, though such a strategy carries with it the possibility of accidental war. And last but not least, the ability to wreak this sort of total devastation is slowly but steadily spreading to more countries.

In sum, for forty years now our military power, as measured by our ability to wreak death and destruction on an enemy in an even shorter time, has been steadily increasing. At the same time, our national security, as measured by the combination of the ability of others to wreak death and destruction on us and our inability to do anything about it, has been steadily decreasing. This has been, and remains, our mutual and fundamental security dilemma.

Spiritual leaders such as the Pope and the American and French Catholic bishops have perceived the issue not very differently from our political leaders. While they have all condemned the use of the threat of nuclear annihilation as the principal means of maintaining peace, they have in various ways also said that this can be tolerated so long as we have no alternative way of avoiding large-scale war. More important, they have added that the current situation is tolerable only so long as we are actively and seriously seeking alternatives. Not everyone may agree with the way these spiritual leaders have stated their case, but nearly everyone believes that we should indeed be seriously seeking alternative ways of maintaining stability and assuring peace.

Our lives, the lives of our progeny, and the continuation of our civilization do depend on "nobody pulling the trigger." That, truly, "is unthinkable," and necessarily evokes extreme proposals for a solution. Some people suggest panaceas: general and complete disarmament, unilateral nuclear disarmament, the abolition of sovereignty, the creation of a super-state, and, not so long ago, preemptive wars (America against Russia, Russia against China). Others have solved the problem in their own minds by wishing it away; some in this group suggest that the Russians aren't nearly so bad as our leaders say, and if we would only stop provoking them all would be well. Still others suggest that if our leaders would only "stand tall in the saddle" our enemies would be cowed into decent behavior.

Where does the SDI stand in this regard? Is it a technological panacea, mere wishful thinking, or is it a grand idea whose time has come?

From the beginning of the nuclear age until the Reagan administration, all American presidents have placed great emphasis on searching for and exploiting political means for maintaining peace and stability and have played down the possibility of an active defense against a nuclear attack. In addition to maintaining an adequate level of nuclear deterrence, they forged and strengthened alliances, they pursued detente and peaceful coexistence by a variety of diplomatic means, and they sought to negotiate agreements that would first limit and eventually reverse the nuclear arms race among the nuclear powers and prevent its spread to other nations. This has not been a partisan matter: every president from Truman to Carter—four Democrats and three Republicans—placed the same trust in political measures, including arms negotiations, as the favored alternative to the naked threat of nuclear revenge as the means for averting large-scale war. In addition, all of the postwar presidents also supported research and development on a variety of active defense schemes. In the 1940s and early 1950s, primary attention was focused on the construction of air defenses intended to keep long-range bombers away from our shores. In the 1960s, the primary focus was on the development of anti-ballistic missile systems. While the prospects for such systems have waxed and waned in accordance with the technological facts as well as the temper of the times, at no point in the thirty years prior to President Reagan did a president place greater emphasis on developing technological means of defense against nuclear attack than on promoting political means for maintaining international stability and world peace.

It is this balance between the search for technical solutions and the search for political solutions that President Reagan seems intent on changing. We now have the president's challenge to scientists to build a defense so effective that it might eventually "make nuclear weapons impotent and obsolete," a challenge accompanied by an extraordinary national program administered by a specially created new organization. We also have a long list of statements by the president and some of his closest advisers and political associates concerning the perfidy of the Soviet leaders, and the consequent difficulty of making any useful agreements about important matters with them. While this has not prevented the two countries from resuming arms-control negotiations in March 1985, the complexity of combining negotiations on strategic offensive *and* defensive systems makes the outcome of these talks very problematic.

Is there reason to believe this new approach would be better than the one favored by other presidents? Is there reason to hope that we can, at last, find security through technology? The unhappy fact is that, thus far, whenever we (or others) have sought to solve our national security dilemma by technical means, we have in the end only made matters worse. Back in the 1940s and 1950s, the answer to active air defenses was the introduction of ballistic missiles. Ballistic missiles overflew the air defense, making the attack more certain and much quicker. In response to ABM defenses, efforts in the early 1960s to put multiple warheads on missiles (MIRV) were accelerated. This increase in warheads could saturate and swamp any conceivable defense, at the same time making it possible to increase greatly the number of individual targets that could be attacked. MIRVs, however, make the world more dangerous because, when combined with increased accuracy, they make a first strike seem more promising, thus undermining stability in certain crisis conditions.

* * *

It is easy to understand why we look to high technology for solutions to our military problems. The United States, after all, is the world leader in this area; technology truly is one of our strongest suits. Those who support various extreme "high-tech" ideas, such as the SDI, can and do point to any of a number of remarkable and dramatic past successes. Radar did tip the balance in the Battle of Britain, and nuclear bombs did end World War II earlier than would otherwise have been the case. In 1961, a president issued an order to go to the moon within the decade and we did precisely that. On an everyday level, our homes and lives are filled with "technological miracles," including televisons that pick up signals relayed by satellites from around the world and home computers as powerful as the largest of their institutional ancestors only a generation ago. The computer revolution, after thirty years of surpassing most short-term predictions, seems destined to offer equally rapid rates of change in the years ahead.

Against this background, the Fletcher panel's optimistic extrapolations of current capabilities are not prima facie unreasonable. On the other hand, we must note that most of these frequently cited technological triumphs involved a struggle of man against nature; the SDI, while it does require some new inventions, also involves a struggle of man against man—of measure against countermeasure—and that is quite a different matter.

We are truly on the horns of a dilemma. Is the SDI, as some think, an instance of exceedingly expensive technological exuberance sold privately to an uninformed leadership by a tiny in-group of especially privileged advisors? If it is, then it is a perfect example of what President Eisenhower had in mind when he warned that "in holding scientific research and discovery in respect, as we should, we must also be alert to the equal and opposite danger that public policy could itself become the captive of a scientific technological elite."

Or is the SDI, as others maintain, a good idea whose time has come? Is it a possible means for replacing assured destruction by assured survival, even, perhaps, for eventually making nuclear weapons obsolete? The issues involved in the debate over strategic defenses are many and complex, and the debate itself will surely go on for many years. What will be vitally important during this debate is that the American public realize that the SDI is not proceeding in a vacuum, but is enmeshed in a web of issues concerning defenses against all types of nuclear weapons, as well as current and projected military uses of space, anti-satellite weapons, offensive nuclear forces, and strategic stability between the superpowers.

ENDNOTES

[1]*Preliminary Design of an Experimental World-Circling Spaceship*, (Santa Monica: RAND Corporation, 1946); see especially the section on the possible military applications of satellites by Louis Ridenour, "Significance of a Satellite Vehicle."

[2] Dr. Robert Cooper (director of the Defense Department Advanced Research Projects Agency), *The San Diego Union,* June 17, 1984, page C4.

[3] See the contribution to this volume by Paul Stares, "U.S. and Soviet Military Space Programs: A Comparative Assessment," on p. 127.

[4] Speaking about the ocean reconnaissance systems specifically, Dr. Cooper noted that the Soviets have completed a number of experiments designed eventually to provide "over the horizon targeting" information directly from satellites to missiles launched from submarines and aircraft. Given the ability of both aircraft and missiles to achieve much greater accuracy by using satellite data for in-flight navigation, the military need for ASAT weapons to destroy such satellites will become stronger. See *The San Diego Union,* June 17, 1984, page C4.

[5] For the unclassified summary of the Fletcher panel report, see "The Strategic Defense Initiative: Defensive Technologies Study" (Washington, D.C.: Department of Defense, March 1984).

[6] James C. Fletcher, "The Strategic Defense Initiative," testimony before the Subcommittee on Research and Development, Committee on Armed Services, U.S. House of Representatives, March 1, 1984.

[7] Sidney D. Drell, Philip J. Farley, and David Holloway, *The Reagan Strategic Defense Initiative: A Technical, Political and Arms Control Assessment* (Stanford: Stanford University, 1984).

[8] See especially Ashton B. Carter, *Directed Energy Missile Defense in Space,* Office of Technology Assessment Background Paper (Washington, D.C.: U.S. Congress, 1984) and John Tirman, ed., *The Fallacy of Star Wars* (New York: Vintage Books, 1984).

[9] *The New York Times,* March 30, 1983, p. A14.

Alexander Flax

Ballistic Missile Defense: Concepts and History

IN THE HISTORY OF WARFARE there has been a continuing contest for supremacy between offensive and defensive weapons. Advances in technology or the military art have at various times made either the offense or the defense temporarily dominant, but sooner or later the pendulum has swung back the other way. It has thus become an axiom of warfare that for every weapon or tactic a counterweapon or countertactic can be found.

Had it not been for the invention of nuclear weapons, this dialectic of offense and defense might have been expected to continue indefinitely. The destructive power of nuclear weapons is so enormous, however, and their means of delivery so swift and effective, that much doctrine in the nuclear age has postulated a virtually permanent advantage for the offense. The advent of the ballistic missile as the principal means of delivery has reinforced this. A prevailing belief thus emerged in the mid-1960s that nationwide deployment of an anti-ballistic missile (ABM) system to protect the United States against heavy attack by Soviet ICBMs was not feasible.

Instead, a strategic policy evolved that came to be known as mutual assured destruction (MAD). It was based on the maintenance of strategic nuclear forces (ICBMs, SLBMs, and bombers) that could survive a first strike in sufficient numbers to pose an unacceptable threat to the adversary's population and industrial centers. In agreeing to the ABM treaty of 1972, the Soviet Union tacitly accepted the concept of offensive dominance, if not the entire concept of MAD.

Neither side has so far been entirely happy with the idea that its survival is dependent on retaliatory forces, that there is no such thing as an adequate defense. This concept sits especially hard with the Soviet Union, which is surrounded by potentially hostile neighbors and has suffered repeated devastating attacks throughout its history. It is not surprising, therefore, that a strong bias towards defense permeates Soviet military and political doctrine.

In the United States as well there has been growing discontent with MAD. One response has been the new emphasis placed on defensive technologies by the Reagan administration's Strategic Defense Initiative (SDI).

In examining the offense-defense relationship in the era of strategic nuclear weapons, this chapter seeks to illuminate the baseline from which the SDI must start.

U.S. BALLISTIC MISSILE DEFENSE PROGRAMS

At its inception, ballistic missile defense was regarded by many as a natural extension of the concepts concerning guided missile systems for air defense. The equipment would have to be more powerful and complex, but the basic architecture of defensive systems, the underlying military doctrine, and the degrees of effectiveness were assumed to be parallel.

At first, each military service sought to extend its own mission to include the BMD challenge. Different modes of BMD were explored by the Air Force, which used interceptor aircraft for area defense of the United States, and the Army, which had responsibility for protecting military and civilian targets with surface-to-air missiles (SAMs). It was quickly realized, however, that the distinction between area and terminal defense was less relevant in an age of nuclear-armed ballistic missiles, and the new BMD programs were assigned to the Army alone.

By early 1957, the Army was working on Nike Zeus, a system of radars and interceptor missiles for high-altitude interception of incoming ballistic missiles. By mid-1962, prototypes of the radars and missiles were ready for a test against a live missile, and in December 1962 the system successfully intercepted an Atlas D missile.

During this period, however, attention was also being given to the problems of countermeasures, i.e., those means an adversary could take to foil an ABM system. Indeed, the U.S. military was studying various types of countermeasures, including decoys (dummy warheads) and other "penetration aids," both to see what the Soviet Union might be capable of doing to breach our defenses and to ensure that U.S. ICBMs would be able to penetrate a possible Soviet ABM system. As a result of the R&D efforts given to penetration aids, the steadily increasing numbers of ballistic missiles in the Soviet inventory, and the rapidly evolving radar and computer technology, no persuasive case for deployment of an ABM system based on the Nike Zeus development could be made. There was also growing concern in the political sector over the effects that deployment of an ABM system might have in accelerating the arms race and potentially decreasing strategic stability in a crisis.

Research and development efforts for BMD turned next to seeking ways to counter decoys and other countermeasures. It had already become apparent that decoys could best be dealt with by intercepting the actual warheads at low altitude, since the atmosphere acts as a filter, slowing down or burning up less massive bodies, making it much easier to distinguish between decoys and actual warheads. Thus, a key component of an improved system would have to be a low-altitude interceptor capable of extremely high acceleration so it could be fired late in the reentry trajectory of the incoming warhead. Initial R&D on the needed components was well under way when, in 1963, the Army's BMD program was redirected toward a new system, Nike-X, that would incorporate just such a low-altitude intercept capability and use better radars.

The new high-acceleration missile, named Sprint, entered development in late 1963 and remained an essential component of subsequent systems. Equally important was the development of an entirely new set of phased-array radars to replace the original Nike Zeus set. The target track, missile track, and discrimination radars were all replaced by a single phased-array radar, the missile site radar (MSR). Phased-array radars, in addition to their ability to perform multiple functions, had the advantage of fixed installations that needed no mechanical scan and could be much more easily protected.

Sentinel and Safeguard

From 1963 to 1967, development work proceeded on Nike-X, and studies were made to evaluate the capability of the system. Among these were studies of terminal defense of Minuteman silos against a heavy Soviet attack and area defense against a light attack by a third country, specifically China. Nonetheless, the case for extensive deployment of the system was not persuasive. In the summer of 1967, though, it became apparent that something (perhaps the White House) had swung support towards deployment of a system designed primarily to protect U.S. cities against a light attack.

The system, which was called Sentinel, would incorporate an improved Nike Zeus missile, the Spartan, for high-altitude interception and the new Sprint missile for low-altitude interception. Both missiles carried nuclear warheads. The missile site radar developed for the Nike-X was to serve essentially the same functions as previously contemplated. A new phased-array radar, called the perimeter acquisition radar (PAR), was to perform long-range surveillance and also track incoming warheads to support Spartan interceptors.

As the Sentinel program proceeded toward deployment, public realization that the proposed sites, including some in urban areas, would have numerous missiles with nuclear warheads, stirred strong opposition. With the arrival of the Nixon administration, construction was suspended in early 1969 pending a presidential decision. On March 19, 1969, President Nixon announced that the Sentinel system deployment would be "substantially modified" and in its new form would be known as Safeguard. The Safeguard system was comprised of the same missile and radar components as Sentinel, but was to be deployed in different numbers and locations. The first priority for the deployment was the protection of U.S. nuclear forces against a Soviet attack; the second priority was the provision of a nationwide defense against a hypothetical Chinese attack—hypothetical, because China did not (and does not) have deployed ICBMs that could reach the United States. A subsidiary role was defense against accidental attack from any source. The Safeguard proposal rekindled political opposition to any ABM deployment and almost failed to receive Congressional approval.

Deployment proceeded on the initial phase, which provided for defense of the hardened Minuteman sites at Grand Forks Air Force Base in North Dakota and at Malmstrom Air Force Base in Montana. The signing of the ABM treaty in 1972, however, radically changed Safeguard deployment plans. The treaty, together with a subsequent protocol, permitted a total of only one hundred interceptors in only one of two types of deployment, for defense of a strategic missile site or for defense of the national capital. The U.S. quickly made the decision to defend only the Minuteman missile site at Grand Forks Air Force Base. This deployment became operational in mid-1975, but in the following year, on Congressional initiative, even this installation was phased out.

SOVIET BMD PROGRAMS

Justification for the pace of U.S. BMD activities has come as much from perceptions of Soviet defense activity as from what at first might seem a more appropriate motivation—the desire to counter Soviet offensive ballistic missile forces. The reasons for this are complex and include strategic policy, foreign policy, technical and economic feasibility, and domestic politics. Presidents Eisenhower, Kennedy, and Johnson showed great disinclination to have the U.S. take the lead in deploying ABM systems. The one disturbing factor was the threat of a massive deployment of an ABM system by the USSR. Indeed, by 1962, senior defense and political officials in the USSR had made public statements implying that effective ABM systems were being developed.

Origins of Soviet BMD

Following World War II, the Soviets, like the U.S., embarked on ambitious programs to deploy surface-to-air missiles, based in part on information and experimental hardware acquired from a defeated Germany. Their first operational anti-aircraft system, the SA-1 (Western designation), began to be deployed around Moscow in the mid-1950s. Like a similar U.S. system, the Nike Ajax, it was effective only against World War II–style bomber raids. This was followed very quickly by several new surface-to-air missile systems capable of interception at much higher altitudes and longer ranges than the SA-1.

Of these, the one that could be interpreted as having at least some ABM capability was the "Griffon" missile system that began to be deployed around Leningrad about 1960. This system, however, was not deployed at other locations. The early version of the more mobile SA-2 air defense system, which was being deployed in the same period, did not appear to most analysts to be a likely candidate for a major ABM role.

About 1963, another new missile called the SA-5 was being widely deployed in the USSR. Unlike the earlier missiles, it was estimated to have a long-range, high-altitude intercept capability. Features of the SA-5 suggested at least some ABM capability, although it clearly also had long-range capability against high-altitude supersonic aircraft of the U.S., such as the XB-70 that was in development at that same time. The XB-70, however, was never produced and deployed. The U.S., in fact, had publicly declared that its bomber penetration tactics would rely on very low-altitude flight so as to evade the coverage of radar and surface-to-air missiles. The SA-5 system appeared to have little low-altitude capability; the SA-2 and SA-3 missile systems, which were then entering the Soviet inventory in increasing numbers, seemed a much more effective response to low-altitude penetration of bombers.

To many U.S. analysts, the massive and expensive deployment of the SA-5 system for defense against high-altitude bombers made no sense. Therefore, it seemed prudent at first to assign at least a dual role to the SA-5 as both a high-altitude bomber defense and an ABM. Gradually, however, as more information on the system became available, the preponderance of expert opinion swung away from crediting the SA-5 system with significant ABM capabilities.

The Moscow ABM System

As perceptions of the SA-5 as an ABM system faded away, growing information on a new system being deployed in a ring around Moscow left no doubt that the Soviets were indeed deploying an ABM system; it is now known in the U.S. as the ABM-1 system. The interceptor missile, the Galosh, was first publicly displayed in 1964. As more information became available, it became obvious that Galosh, which is assumed to be nuclear-armed, was designed for long-range high-altitude interception.

Eventually, the ABM-1 system consisted of complexes at four sites around Moscow. Each site had sixteen launchers and two sets of missile-tracking and guidance radars, for a total of six radars per site. Initial tracking of incoming ICBMs, threat assessment, and target hand-over to missiles for defense was provided by a very large phased-array radar. Later, a second radar of the same type was added at another location.

In addition, the Soviets deployed eleven radars at six locations on the periphery of their territory. It remains unclear whether these radars contribute significantly to ABM performance, other than to give early warning. Like corresponding U.S. radars, these radars have a collateral tracking role for space satellites (a role permitted by the ABM treaty), and there is always some degree of ambiguity about the function of radars of this kind.

The ABM-1 deployment around Moscow is capable of defending a large area of the western Soviet Union. However, being a high-altitude intercept system, it is vulnerable to countermeasures such as chaff and lightweight decoys. Moreover, the frequencies of the system's radars are relatively low and subject to blackout from nuclear blasts, either offensive or defensive. In addition, the large radars are few in number and could be destroyed by a few missiles during a concentrated attack. Finally, since the ABM-1 system itself is limited by the ABM treaty to no more than one hundred launchers, it could be easily exhausted in a concentrated attack.

In an era when there are thousands of ICBM and SLBM warheads on each side, the only plausible rationale for the Moscow system seems to be as a marginal defense against a third country (Britain, France, China), as a precaution against accidental or unauthorized launches of a few missiles, and/or as an initial stage of a possible wider deployment, especially in the event of a breakdown in the ABM treaty.

ROLES FOR BMD

As weapons technology and doctrine have evolved, so too have the roles for BMD. In the early 1950s, the role of defense was to limit damage to the population, industry, and economy as well as government and military installations. This notion also fitted the prevailing concepts of air defense and civil defense, attitudes that were primarily a legacy of World War II. These concepts were carried over into the nuclear weapons era, at first without any real appreciation of the enormous difference between conventional weapon and nuclear weapon attacks.

Damage Limiting

Population defense is a special case of damage limiting. It makes little sense to protect a population without also providing the resources required for post-attack survival and recovery. And since large fractions of the population and the logistic support for the population are to be found in major metropolitan areas, it is these areas that would have to be given the greatest emphasis.

A definitive Defense Department study on damage limiting, including consideration of air defense, was carried out in 1964. Air Force Brig. Gen. Glenn Kent, then on the staff of the Directorate of Defense Research and Engineering, analyzed data from all Defense Department offices to arrive at a devastatingly negative appraisal of the cost-effectiveness of defense against attacks by nuclear-armed ballistic missiles. For Soviet missiles attacking U.S. targets in numbers ranging from a few hundred to a thousand, the cost-exchange ratio—that is, the cost to the defense of countering each additional offensive warhead—would be intolerably high, varying from three-to-one to ten-to-one. One of the key issues in the current debate on the merits of new types of ABM systems is whether new technologies have made it possible to obtain a reversal of this unfavorable ratio.

Defense of Military Command-and-Control Centers

One of the problems that has been most worrisome for military planners has been the threat of a successful attack on the national command centers and the communication systems connecting them to the strategic forces. Although the national command centers of both the U.S. and USSR are located in the national capitals, each country has developed alternate and mobile command centers and has arranged for devolution of command responsibilities. Nevertheless, there is concern that surprise attacks on the command centers could significantly reduce the effectiveness of a retaliatory attack.

Moreover, attacks on communications systems could involve the detonation of just a few nuclear weapons at high altitude. This would subject command and communication systems on the ground to electromagnetic pulses (EMP) of tens of thousands of volts per meter, large enough to burn out or damage electronic equipment.

Accordingly, any defensive system cannot rely only on active BMD defenses for individual critical centers, such as Washington, D.C., or the military command center at Cheyenne Mountain in Colorado. Protection against EMP, involving the hardening of electronic equipment and the use of alternate communication channels, as well as alternate and mobile command centers, will be just as essential, with or without active defense.

While it is possible that BMD systems could provide a subjective increase in confidence for defense, and that a parallel increase in uncertainty might help discourage an attacker from making a first strike, it can also be argued that such uncertainty could result in the offense enlarging its ballistic missile and other nuclear forces to assure achievement of its objectives. The issue then reverts to questions of cost-effectiveness and cost-exchange. In any case, arguments for the use of BMD for defense of command centers alone are hardly likely to justify the cost of an ABM system.

Light Defense Against Third Countries and Accidental Launches

While the utility of BMD against massive attacks by a heavily armed attacker is questionable, more persuasive cases have been made for BMD against missile attacks by third countries (i.e., other than the two major powers) and against accidental or unauthorized missile launches. As noted earlier, the need for light defense against a possible Chinese ballistic missile threat led to plans for the deployment of the Sentinel system, while Soviet perceptions of the limited threats from the British, French, and Chinese missile forces have probably played a part in their deployment of ABM-1.

The difficulty with light defense deployments by one of the major nuclear powers is that they might be used as a base for eventual heavier deployments. The thin BMD system might also be converted to achieve other objectives, such as greater protection of ICBMs. This can be accomplished in relatively short order with little hardware change, as exemplified by the metamorphosis of the Sentinel system into the Safeguard system.

Defense of Deterrent Forces

The Safeguard system, the only ABM system actually deployed by the U.S., had as its primary objective the defense of Minuteman ballistic missile silos. This was also a secondary objective of its predecessor, the Sentinel system, which used the same interceptor and radar components. But at the time of the decision to proceed with Sentinel, the Air Force objected to this approach on the grounds that the system's radars were too few and too soft. Moreover, because the system was not designed specifically for silo defense, it was a highly cost-ineffective way to "buy back" a threatened Minuteman silo; it would be cheaper to buy an additional Minuteman if that option were not foreclosed.

However, constructing super-hard silos makes active defense of missiles a much more tractable task. Very low altitude defense is also feasible, and this greatly alleviates the decoy problem, since the atmosphere can be relied on to sort out the decoys. Finally, relatively high rates of leakage can be tolerated, since with adequate silo separation each warhead can destroy only one silo. Thus, even if 20 or 30 percent of the incoming warheads penetrate, there may still be a formidable retaliatory force.

The key to the effectiveness of hardened silo defense is the possibility of low-altitude interception very close to the target silo. This should lead to smaller, lighter, and cheaper defensive missiles, engagement radars, and so forth, but such advantages have yet to be realized. Even less sophisticated systems for point defense of super-hardened silos have been suggested. Richard Garwin, for example, has proposed the "swarmjet" concept, which employs large numbers of short-range rockets to defend each silo. Such close-in defense systems are advantageous from an arms-control standpoint because they cannot be converted to protect large areas.

Using mobile missiles in combination with multiple silos or other protective shelters gives added leverage to systems for defending hardened silos. This concept was originally proposed by the Carter administration for deployment of the MX. An important aspect of this leverage is the fact that only those few silos occupied by missiles need be defended, while the attacker must target all the sites even though most are empty. The possibility of saturating defensive systems is thus greatly reduced.

Ballistic Missile Defense: Concepts and History 43

Tactical Ballistic Missile Defense

ABM systems that can intercept short- and medium-range ballistic missiles, such as those deployed in Europe, are not precluded by the ABM treaty. Almost any high-altitude SAM system, such as the U.S. Nike Hercules or the newer Patriot or Soviet SA-5 could, with relatively small modifications, be used to intercept short- to medium-range tactical ballistic missiles. Several demonstrations of this capability have taken place. The newest missile system to be credited with a tactical ABM capability is the Soviet SA-X-12, which is said to be nearing deployment.

Whether or not such systems will be effective against shorter-range ballistic missiles is another question. The shorter range and time of flight of these missiles, in conjunction with multiple warheads and countermeasures such as lower trajectories and radar jamming, could make the task of tactical defense as difficult as strategic BMD.

COUNTERMEASURES TO ABM SYSTEMS

The effectiveness of an ABM system is not simply a matter of whether the system can intercept and destroy ICBMs and their warheads. In the U.S., at least, a major consideration has been the projected effectiveness of the system in the face of plausible countermeasures. Most defense analysts agree that a countermeasure is plausible if it could be developed within the projected lifetime of the ABM system and could be deployed at a cost no greater than that of the defensive systems being countered.

Yet countermeasures, even if reasonably well-defined, will be viewed differently by attackers and defenders. The attacker may demand great certainty that the countermeasures will succeed (the offense conservative view), while the defender may demand great certainty that the defensive system will remain highly effective in spite of the countermeasures (the defense conservative view).

The gap between offense and defense conservative views of countermeasures leads to many points of contention, not the least of which is the effect of the uncertainty itself. Opponents of ABM argue that the opponent will adopt an offense conservative response and increase the size of his attacking force in an effort to make the introduction of any ABM capability counterproductive. Proponents, on the other hand, argue that the uncertainty brought about by introducing ABM can be a powerful contribution to deterrence.

Multiple Warheads

The best countermeasure against an ABM system is to increase the number of warheads carried by each missile. Increasing the number of warheads is highly effective against both soft-area targets (cities) and hard targets (missile silos), especially as improvements in accuracy have offset the lower yield of these warheads.

The first multiple warhead missile was the submarine-launched (SLBM) Polaris A-3, which carried three warheads. Later, multiple independently targetable reentry vehicles (MIRVs) were developed, with up to three on the Minuteman ICBM and up to fourteen on the Poseidon SLBM. Although the SALT agreements have placed limits on the number of warheads permitted, some Soviet ICBMs could carry as many as twenty to thirty warheads.

Decoys and Chaff

Relatively lightweight objects can be designed to appear to radar as additional warheads. Unsophisticated decoys (e.g., radar corner reflectors) may weigh only a few pounds, and a hundred or more of these can fit in the space required for a single warhead. As mentioned earlier, however, these lightweight decoys function well only in the early phases of detection and tracking and are not an important countermeasure against terminal ABM systems.

Heavier decoys, weighing perhaps up to 10 percent of a warhead's weight, can be much more effective against radars. With thrust augmentation, they can mask slowdown effects enough to degrade a substantial part of the high-altitude capability of an ABM system. These heavier decoys can also be very effective against mid-course ABM systems using radar or infrared sensors, even those that are relatively sophisticated.

Radar chaff, consisting of very thin wires deployed in long trails surrounding each warhead, is a relatively lightweight, simple countermeasure that is effective during mid-course and early reentry. Chaff is the cheapest and lightest approach, especially when a "bus" is used for MIRV deployment.

Ballistic Missile Defense: Concepts and History

Jammers

Jammers that generate radar noise to overwhelm meaningful signals can also be used against ABM radars. A large number of small jammers dispensed over an area covering the paths of incoming missiles can deny detection and tracking or generate false targets down to low altitude. Since jammers need only emit for the short period of attack, electronic components and power supplies can be very small and light.

Maneuvering Warheads

Most of the radar countermeasures described above (except for proliferation of warheads) function only against the high-altitude aspects of an ABM system. Low-altitude interception would not be affected by these countermeasures. One countermeasure that can be effective against low-altitude interception is a warhead that maneuvers during reentry so that prediction of intercept points and targets becomes difficult. Maneuvering warheads are especially effective against preferential defense tactics.

There are drawbacks to this countermeasure. Depending on the degree of maneuvering desired, this countermeasure may increase the weight of the warhead by anywhere from a few percent up to 50 percent. Maintaining hard-target kill accuracy also imposes design constraints on maneuvering and affects weight and cost. However, for ICBMs with large payloads such as the Soviet SS-18s, the tradeoffs may not be too severe, especially as improvements in accuracy permit reduction of warhead yield and size.

Infrared Countermeasures

The problems ABM radars have in seeing through chaff and in discriminating decoys from warheads led to investigation into the use of long-wave infrared sensors for these purposes. As a result, in the late 1960s and the 1970s, some interest developed in countermeasures against such infrared-based systems, and research and limited flight experiments were carried out.

In particular, lightweight balloons made of reflective and insulating surface materials can be designed to simulate warheads. Furthermore, using a technique known as anti-simulation, the warhead itself can be shrouded by an outer shell to make it appear to be a decoy.

Very large numbers of simple balloons can be deployed in combination with more sophisticated decoys to create large volumes of "traffic." Infrared chaff can also be dispersed.

Against these countermeasures, very high-altitude and mid-course ABM systems would have to employ a multitude of sensors to identify warheads on the basis of small or transient differences. Such capabilities go well beyond anything heretofore contemplated in ABM systems. This problem is being investigated by the Strategic Defense Initiative program, but no well-defined, workable, integrated system has yet been publicly described. Many analysts doubt that such a system can be devised.

Finally, infrared sensors can be blinded by nuclear explosions ("red-out") in the same way that radars are blinded ("blackout").

Radar Cross-section Reduction

In ABM systems, radar has been the primary sensor for early detection, target acquisition, and tracking of incoming missiles. Radar cross-section reduction has therefore been an important countermeasure, since it reduces the distance at which radar detection first takes place and decreases the time available for response.

Reduction of radar cross-section can be accomplished by covering the reentry vehicle with a shield that functions through mid-course and early reentry, or by reducing the cross-section of the reentry vehicle itself.

With powerful ABM acquisition radars, early detection of blunt objects with large radar cross-sections can occur a thousand miles or more from the point of interception. Even with designs of low radar cross-section, early detection by these powerful radars (e.g., those defined in the ABM treaty) takes place at distances of hundreds of miles. However, reducing the radar cross-section of missiles eliminates the ability of lesser radar systems (e.g., most SAM radars) to function as ABMs, and makes it easier to protect the missiles with chaff, decoys, and jammers.

SOVIET ABM TECHNOLOGY SINCE THE ABM TREATY

In the U.S., ABM research efforts after 1972 were placed under various informal, but nevertheless effective, Congressional enjoinders against proceeding too far toward engineering development. In contrast, Soviet R&D was perceived in the U.S. as being much more oriented to working toward engineering prototypes and final production models.

Soviet work included improvements in high-altitude interceptor missiles as well as the development of a new high-acceleration interceptor, broadly similar to the U.S. Sprint, for lower altitude kills. Extensive R&D on phased-array radar technology was conducted. Prototypes of transportable ABM radars (in contrast to mobile ones deployed on a vehicle—which are prohibited by the ABM treaty) were tested. The Soviet components under development were assessed by the U.S. military as providing not only the potential for upgrading the Moscow ABM system, but also for the rapid deployment of a nationwide system. The latter possibility is often described in the West as a potential "breakout" from the ABM treaty with or without the six-month notification required for formal withdrawal.

In addition to the extensive R&D on conventional ABM systems, the Soviets appear to have carried out large-scale and costly investigations into directed energy weapon technology, including both lasers and particle beams. However, with the limited information we have on the kind of overall system into which directed energy components might fit, it is difficult to predict the applications for which such devices are intended.

Modernization of the Moscow System

The Moscow ABM system has been in continuous operation ever since its initial deployment in the early 1970s. Several years ago, it became apparent that major modifications to the Moscow system were in the offing, and the Soviets subsequently announced that they would be dismantling elements of the existing system. It now appears that a two-layer system is being deployed, with high-acceleration missiles for low-altitude interception augmenting improved Galosh-type high-altitude interceptors at new sites. The new missiles are being placed in silos, thereby reducing their vulnerability, a partial remedy to one weakness of the original system. New engagement radars at the missile sites are also anticipated. It appears the number of high- and low-altitude interceptors on launchers will reach the total of one hundred permitted by the ABM treaty.

In parallel with these developments, a new large multi-face phased-array radar—commonly referred to as the Pushkino radar—has been under construction in the original Moscow ring. This is similar to, but larger than, the missile site radar once deployed by the U.S. as part of the Sentinel/Safeguard system. This radar may be used to perform a variety of functions for target acquisition and tracking, decoy discrimination, and battle management.

In recent years, the Soviets have been constructing a set of five large phased-array radars on the periphery of the country, and a sixth is under construction in the interior. Except for the latter, these have been construed as being peripheral early-warning radars permitted by the ABM treaty. However, inherent in modern radars of this type is a capability for serving ABM battle management roles if properly supported by (intrinsically unobservable) control, data processing, and communication functions. (It should be noted that the U.S. is, at the same time, deploying a new set of peripheral phased-array radars designated "Pave Paws" to improve warning against SLBM attacks.) Extensive coverage of the approaches to the western part of the Soviet Union, in an arc from the Kola Peninsula to the northwest around Siberia and down to the Caucasus, is provided by these radars in combination with the earlier peripheral radars. The Defense Department has expressed concern that this radar network could be used to support ABM deployments well outside the present Moscow ring, especially in connection with a possible rapid deployment of an expanded ABM system in the event of a breakout from the ABM treaty. However, since large radars similar to these are used for tracking satellites, there is always some uncertainty as to the primary purpose of such installations.

U.S. BMD RESEARCH SINCE THE ABM TREATY

Even though the U.S. decided not to maintain an ABM system at the Grand Forks ICBM missile field in North Dakota, as permitted by the ABM treaty, research on ABM technologies continued throughout the 1970s, concentrated particularly in two areas. The first was the development of more effective and less costly technology for defending hardened sites, such as ICBM silos or command-and-control centers. This program received a great deal of interest at the time that the Carter administration was considering a mobile basing system of

Ballistic Missile Defense: Concepts and History

multiple shelters for the MX missile. However, President Reagan's decision in 1983 to base the MX in existing Minuteman silos lessened interest in this type of mobile, hard-site defense.

The other area of R&D that has been pursued vigorously since 1972 is exemplified by the Homing Overlay Experiment. This was an effort to develop the technology for discriminating decoys at high altitudes using long-wave infrared sensors, either airborne or rocket-launched, and for developing non-nuclear-kill interceptors using long-wave infrared seekers. The culmination of this effort was the successful interception and destruction of a Minuteman-launched missile by an interceptor missile on June 10, 1984. This area of R&D is likely to be of increasing interest as a result of the SDI program.

"Unconventional" BMD: Early Concepts

Even in the 1950s, concepts different from those derived from the air defense tradition were being considered for defense against ballistic missiles. The Department of Defense was looking into so-called "exotic" technologies such as particle beams, and though these technologies were relatively immature at the time, research in such fields was being pursued for a variety of military applications, including BMD.

One program, mounted twenty-five years ago, was project BAMBI (ballistic missile boost intercept). This program foreshadowed the modern interest in using space-based weapons to intercept ICBMs in their initial boost phase, during which the rocket boosters emit infrared waves. The proposed system contemplated using rocket-propelled interceptor missiles on space platforms, in conjunction with sensors, to detect missile launch and to enable the interceptor missiles to home in on the ICBMs. Despite extensive studies, however, the Defense Department concluded in the 1960s that the state of technology did not permit a reliable and effective system. However, the High Frontier study of 1982 revived the BAMBI concept as the linchpin of a proposed new defense-oriented strategy.

By the 1970s, increased attention was also being given to the military applications of lasers and neutral particle beams for both tactical missions such as defense of naval forces, and for space-based BMD. During the Carter administration, there were debates within the Defense Department and Congress on the priority to be given BMD in the high energy laser program, and these debates spilled over into the early years of the first Reagan administration. Although these internal debates were effectively muted by President Reagan's 1983 Star Wars speech, they have not disappeared entirely.

The types of lasers being investigated by the SDI include not only advanced types of chemical lasers, but also free-electron and excimer lasers capable of generating shorter wave-lengths that lend themselves to smaller and simpler optical systems for producing high-intensity beams. Also under study in the Department of Energy laboratories are lasers powered (or "pumped") by nuclear weapon explosions that generate beams of X-rays. There is currently considerable difference of technical opinion on how well and how soon such devices might be incorporated into effective ABM weapons, especially when countermeasures are considered.

THE U.S. STRATEGIC DEFENSE INITIATIVE

President Reagan delivered his so-called "Star Wars" speech on March 23, 1983, calling for a national effort to move from a strategic deterrence policy based on offensive weapons alone to one more strongly based on defensive systems. Although the speech was general in nature, its direction was clear and it immediately evoked a controversy. The most compelling of the arguments against the president's proposal seemed to be over the question of whether such a system of the kind envisioned—even if it would "work" in a technical sense—could really be effective in enhancing national security, rather than simply causing a new and dangerous escalation of the strategic arms race.

The Defense Department, in response to the president's direction, initiated two independent but related studies; one on technology and systems, under Dr. James Fletcher, and the other on policy, under Dr. Fred Hoffman. The major recommendation of the Fletcher study was for a long-term R&D program on ballistic missile defense, only the first five years of which were described in any detail. No specific BMD systems were selected for ultimate deployment, but promising new technologies and system components were identified. Decisions about further R&D on system development for deployment would be made after the initial five years of effort. Although the tone of the recommendations was optimistic as to the likelihood of technical success, there were no firm predictions of when such success might be achieved for various system alternatives.

The Hoffman study addressed the policy implications of a U.S. ABM deployment in such areas as the U.S.-USSR strategic balance, the stability of deterrence, initiatives and responses of the USSR, and effects on relations with allies. The study concluded that deployment by the U.S. of BMD systems would be favorable in most respects, or at least preferable to other alternatives. The recommendations of the Hoffman study, in contrast to the Fletcher study, placed emphasis on the desirability of early deployment of BMD systems, even though their projected effectiveness would be considerably below that called for in the Fletcher study and implied in the president's speech.

The Department of Defense responded to these studies by establishing a new organization, the Strategic Defense Initiative (SDI), to carry out the expanded R&D program as recommended in the Fletcher study, although initially at a somewhat reduced level of funding. The debate on BMD will undoubtedly intensify as the program proceeds, and as funding requirements increase substantially in the next few years.

* * *

To some extent, the issues in the current debate over strategic defense do not differ from those of the ABM debates of the 1960s. As was the case then, there are important questions concerning technical feasibility and cost estimates of systems employing new technologies. These are important because one of the issues that must be decided is whether new technologies such as lasers and particle beams will make it possible to reverse dramatically the unfavorable cost-exchange ratios that have always burdened defensive systems. Closely related to this is the question of countermeasures and countertactics; of whether an opponent can overwhelm or negate a defensive system relatively easily and cheaply.

One of the new issues in the debate is that many of these new BMD systems would have to be deployed in space in order to be effective against Soviet missiles right after they are launched. However, space systems may be particularly vulnerable to countermeasures and direct attack, so their survival is a particularly critical issue.

Alexander Flax

The larger issues, however, remain primarily in the areas of policy and strategy. Can a BMD system change the reality of offense dominance brought about by the overwhelming destructive power of nuclear weapons? Will the uncertainty brought about by BMD efforts deter the offense, or will it simply lead to offsetting increases in offensive forces? On the other hand, can the U.S. run the risk of not proceeding with its BMD efforts, given that the Soviets may take the initiative and deploy them anyway? Finally, what is the role for arms control in avoiding a destabilizing BMD competition? In the coming months and years, these fundamental questions will need to be resolved if strategic defenses are to play an important part in promoting stability and deterring war.

REFERENCES

For more information on the technology and politics of ABM efforts in the 1960s and 1970s, see: Benson Adams, *Ballistic Missile Defense* (New York: American Elsevier, 1971); *ABM Research and Development at Bell Laboratories: Project History*, a report by Bell Laboratories, Whippany, N.J. (U.S. Army Ballistic Missile Defense Systems Command, 1975); Abram Chayes and Jerome B. Weisner, eds., *ABM: An Evaluation of the Decision to Deploy an Anti-ballistic Missile System* (London: Macdonald & Co., 1969); Fred Kaplan, *The Wizards of Armageddon* (New York: Simon and Schuster, 1983); Roger P. Labrie, ed., *SALT Handbook* (Washington, D.C.: American Enterprise Institute, 1979); Robert S. McNamara, *The Essence of Security: Reflections in Office* (New York: Harper & Row, 1968); and Gerard Smith, *Doubletalk* (Garden City: Doubleday, 1980).

Hans A. Bethe, Jeffrey Boutwell, Richard L. Garwin

BMD Technologies and Concepts in the 1980s

THE PUBLIC DEBATE CONCERNING strategic defenses against nuclear weapons, seemingly laid to rest by the 1972 ABM treaty, resurfaced in the early 1980s. This resurgence of interest was brought about, in part, by technological advances in a number of different fields: new possibilities for weapons systems employing laser and particle beams and kinetic energy projectiles, along with advances in guidance and sensing systems, optics, and computer processing seemed to make ballistic missile defense a more attractive and realistic proposition than it had been in the 1960s and 1970s.

This chapter will review how these new technologies might be incorporated into a multilayered BMD defense system of the sort being studied by the Strategic Defense Initiative. (For a more detailed assessment of these new technologies, see Appendix A.) Because the SDI is still in the research stage, we will not attempt to evaluate the ultimate effectiveness or final costs of such a system. Instead, we will focus on the potential strengths and vulnerabilities of the technologies that it would employ. We will be concentrating in this article on ballistic missile defense, although we emphasize that defense against other nuclear weapons delivery systems, such as manned bombers and cruise missiles, will also need to be addressed if the goal of a truly perfect defense is to be realized.

Before examining the BMD technologies and basing modes currently under consideration, it will be useful to review the flight stages of a ballistic missile in order to determine the tasks facing the various tiers of a multilayered BMD system.

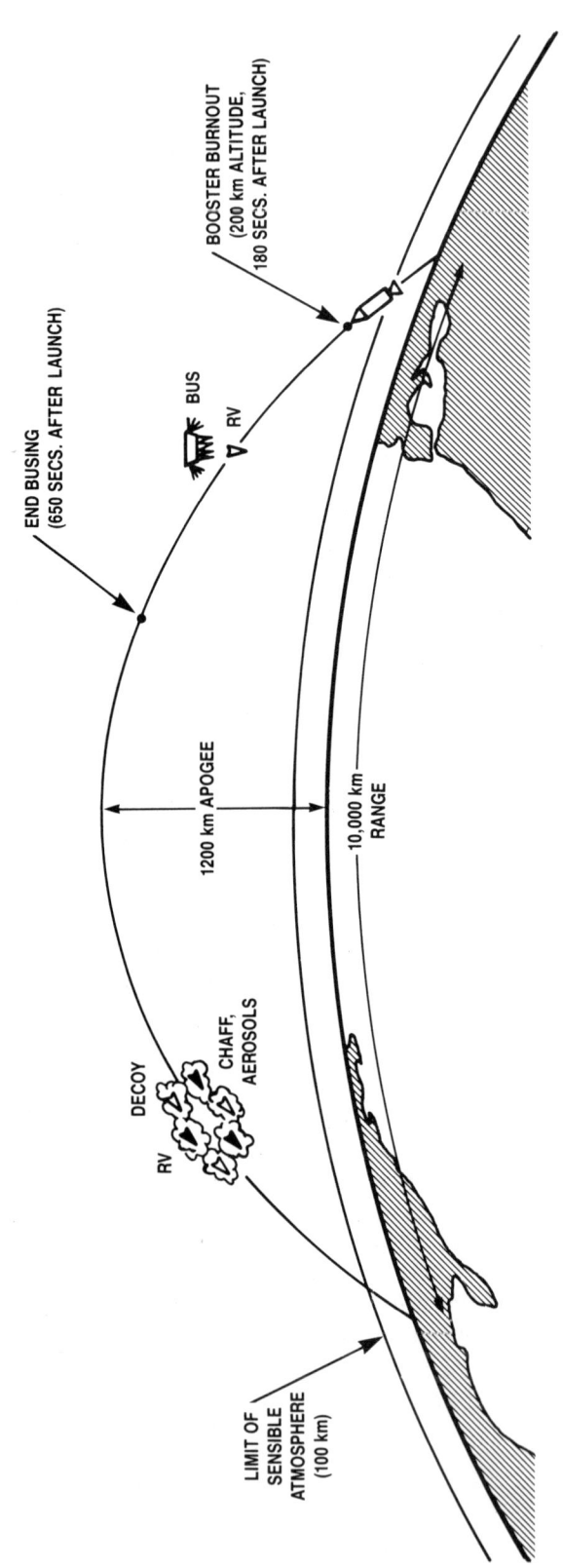

Fig. 1 Phases of Flight for an ICBM with Booster Characteristics Similar to the U.S. MX Missile

PHASES OF ICBM FLIGHT

The flight of a land-based intercontinental ballistic missile (ICBM) would follow a trajectory over the Arctic from one superpower to the other, last 25 to 30 minutes, and cover approximately 10,000 kilometers. When equipped with multiple warheads (MIRVs), the ICBM would have four phases of flight: a boost phase, during which large rocket boosters accelerate the missile to a velocity of seven kilometers per second; a post-boost phase, during which the missile deploys its warheads and decoys; a mid-course, where the warheads and decoys follow a ballistic trajectory through the vacuum of space; and a terminal phase, during which the warheads reenter the atmosphere (the decoys having been burned up, or slowed by friction) and descend toward their targets (see figure 1).

Unlike the ABM systems of the 1960s, which consisted only of terminal defenses, a multilayered BMD system would seek to attack enemy missiles during all four phases of flight. The ability to destroy ballistic missiles along the entire flight path, but especially during boost-phase before the missile has released its warheads, is what gives the multilayered BMD system its great conceptual attraction.

The same four phases of flight also characterize the flight of a MIRVed submarine-launched ballistic missile (SLBM), but SLBMs have particular characteristics that pose greater difficulties for the defense. Whereas the locations of ICBM silos are known with precision, submarines are moving and submerged. The uncertainty of where SLBMs would be launched forces BMD sensing systems to search a much larger area than it needs to for incoming ICBM warheads. Also, submarines can approach an opponent's coast and launch the SLBMs from within a few hundred to a few thousand kilometers, shortening the SLBM flight time to anywhere from seven to fifteen minutes. Finally, SLBMs can be given a flattened trajectory, making it more difficult for space-based BMD systems to target them.

CONCEPTS OF A MULTILAYERED BMD SYSTEM

Boost-phase intercept

The boost phase of currently deployed ballistic missiles lasts approximately three to five minutes for ICBMs and two to three minutes for SLBMs. For ICBMs, this means that the missiles will still be over the Soviet Union as they begin to release their warheads, so a BMD system would have to be deployed either totally or partially in outer space in order to have a clear line of fire.

Missiles in boost phase are particularly vulnerable to destruction by a BMD system for two reasons. First, the missile booster is a large and fragile target, easily damaged by directed energy weapons (lasers, particle beams) and kinetic energy projectiles. Second, the flame emitted by the booster rockets gives off intense infrared radiation, providing space-based defense satellites with a fairly accurate fix on the missile's location. This same infrared radiation is what provides certain U.S. and Soviet satellites with early warning of an enemy attack.

There is a clear and powerful advantage to destroying ICBMs in boost phase, as destroying a single missile will result in the destruction of all the warheads and decoys on that missile. Given that current ICBMs and SLBMs carry anywhere from three to fourteen warheads and might carry one hundred decoys, successful boost-phase intercept can give a BMD system important leverage over the offense by substantially reducing the number of warheads and decoys that will confront the mid-course and terminal defense systems.

On the other hand, the Soviet Union could take steps to reduce the vulnerability of its missiles during boost phase. One means would be to shorten the duration of boost phase so that the missile releases its warheads and decoys while still in the earth's atmosphere, where many BMD weapons (X-ray lasers, particle beams, and kinetic energy projectiles) either cannot penetrate, or are ineffective. The MX missile is reported to have a boost phase of 180 seconds, releasing its warheads just outside the atmosphere at an altitude of 200 kilometers. Much shorter boost phases (50 seconds and 90 kilometers) are judged feasible, with only limited adverse effects on missile weight and performance. Missile vulnerability during boost phase could be further reduced by shielding the ICBM against lasers and particle beams, rotating it so that these directed energy beams are not allowed to deposit their energy on one particular spot, or by greatly increasing the number of real or dummy ICBMs that would have to be targeted.

Post-boost phase

The post-boost phase of an ICBM or SLBM begins when the final rocket booster has separated and fallen away from what is known as

BMD Technologies and Concepts in the 1980s

the MIRV "bus," a small platform on top of the missile carrying the warheads and decoys. Depending on the number of warheads and decoys, this phase will last approximately two to five minutes, and can occur at altitudes of 100 to 1000 kilometers. During this period, small engine thrusters make slight corrections in the angle of the bus, allowing the warheads to be sent along predetermined trajectories toward their targets.

If a space-based BMD system attacked the MIRV bus during the post-boost period, it could hope to destroy a number of warheads and decoys along with the bus. On the other hand, the Soviet Union could shorten this phase through a technique known as "cluster release," in which all the warheads and decoys are ejected nearly simultaneously, with only a slight loss of warhead accuracy.

Mid-course phase

Once the warheads and decoys have been released, they travel through space for approximately 15 to 20 minutes (7 to 10 minutes for SLBMs) before reentering the atmosphere. There are now as many as fourteen (and, in the absence of arms control limitations, potentially twenty to thirty) warheads and one hundred or more decoys for every ballistic missile launched. The use of decoys to increase greatly the number of possible targets is especially effective, because in the vacuum of space the trajectories and velocities of warheads and decoys will be indistinguishable. These decoys could consist of suitably shaped pieces of metal or plastic, or aluminized balloons, which will appear to BMD sensing systems as actual warheads. In addition, the release of fragments of metal wire, called chaff, or clouds of aerosol could hide the actual warheads by overwhelming the BMD sensing systems.

A more fundamental problem with mid-course interception concerns how to locate precisely these warheads and decoys in the first place. Unlike missile boosters with their infrared signatures, warheads and decoys do not give themselves away to space-based satellite sensors. A BMD system must either search for the warheads and decoys, which emit modest amounts of infrared radiation and reflect the earth's faint infrared light, or it must illuminate these objects by reflecting radar waves or lasers off them.

One means being investigated by the SDI for reducing the problem of locating targets, and then distinguishing warheads from decoys, involves what is known as "birth-to-death tracking." A BMD sensor would lock on to a real warhead as it is being deployed from the MIRV bus and track it continuously through the mid-course and terminal phases of flight. The difficulty with birth-to-death tracking is that the release of decoys, chaff, and clouds of aerosol could interfere with the process. Moreover, warheads could be hidden inside decoys such as aluminized balloons, further complicating the problem of distinguishing actual warheads from harmless decoys. Finally, the initial identification of warheads as they are being released from the MIRV bus could be complicated by the use of a shield or the emission of smoke.

Terminal phase

The final portion of ballistic missile flight is the terminal phase, during which the warheads and decoys reenter the earth's atmosphere at an altitude of about 100 kilometers and approach their targets. Although the terminal phase is short, lasting only thirty to sixty seconds, there is no problem with target identification. Upon entering the atmosphere, air friction will burn up the lighter decoys (balloons) and slow the heavier ones, leaving only the warheads themselves traveling at a high velocity.

Advances made since the 1960s in radars and other sensors, computer processing, and the accuracy of BMD interceptors has made terminal defense a much more viable proposition than it was twenty years ago. In particular, striking improvements in radar and computer technology now make it possible to use non-nuclear interceptors instead of the nuclear warheads that were part of the Sentinel and Safeguard ABM systems, and which are still found in the Soviet Galosh ABM system around Moscow. An illustration of this is the Homing Overlay Experiment carried out by the U.S. Army in June 1984, in which a steel mesh projectile launched by a Minuteman booster destroyed a dummy warhead at an altitude of more than 150 kilometers over the Pacific Ocean. Other possibilities include the "swarmjet," a weapons system that would fire thousands of small darts at incoming warheads.

Yet warheads can be provided with devices to foil a terminal defense. One such method is known as "salvage fusing"—the warhead would detonate automatically when it sensed an approaching BMD interceptor such as a missile, or when it is the target of a

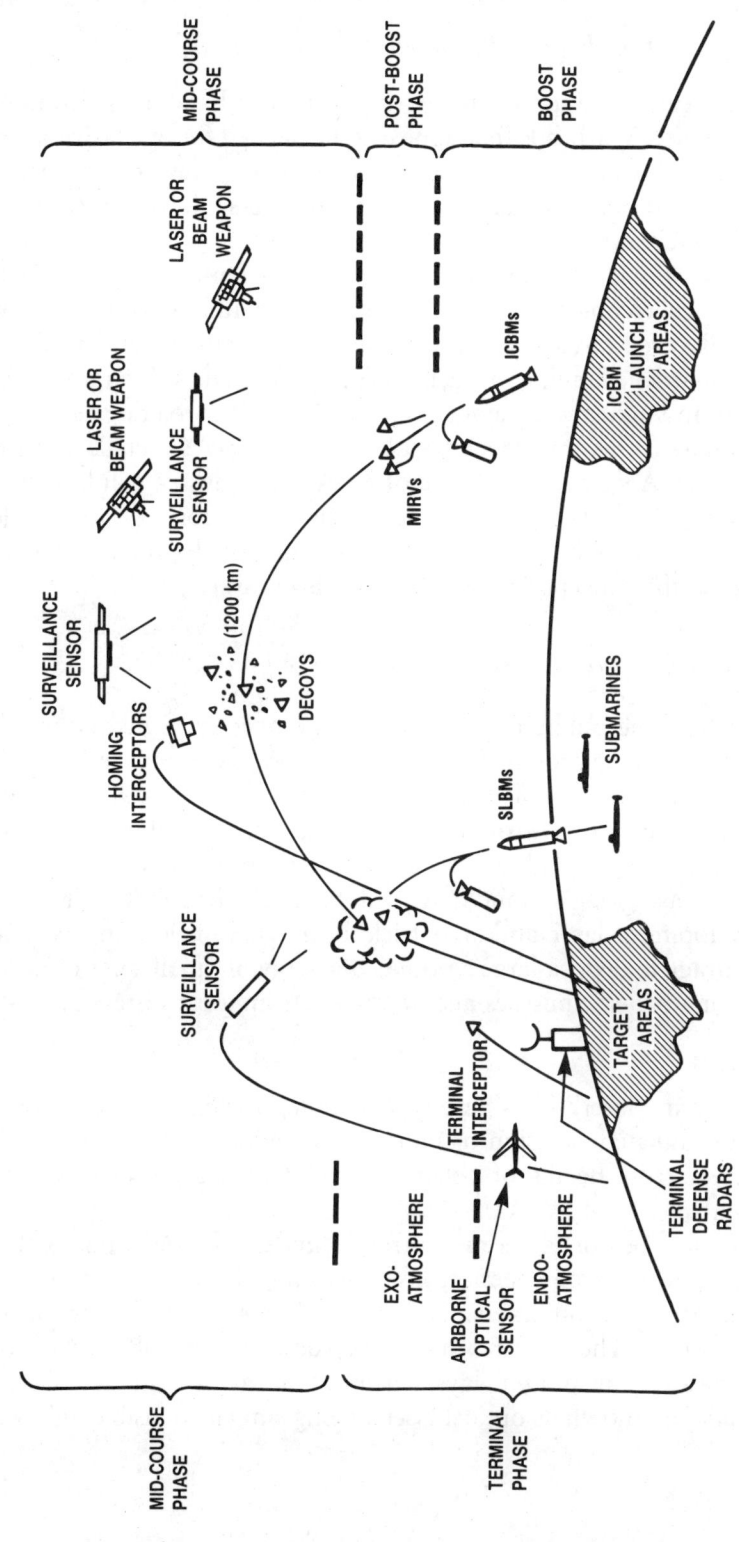

Fig. 2 Multilayered BMD System

laser beam. The use of maneuvering warheads (MaRVs) is another possibility. Warheads that appear to be headed for one target as they reenter the atmosphere could, using small winglets or specially configured noses, suddenly veer off at hypersonic velocities toward completely different targets.

The primary difficulties facing terminal defenses, however, are not so much technical as conceptual. To be a fully effective tier of a multilayered BMD system, terminal defense must be able to protect both hardened military targets (point defense) and soft targets such as cities and industry (area defense). A defender can tolerate destruction of many of his missile silos so long as a sufficient retaliatory force survives. A successful defense of cities and industry will have to be nearly perfect, however, since even the detonation of a few nuclear warheads high above cities will cause immense destruction. Figure 2 depicts the concept of a multilayered BMD system.

NEW BMD WEAPONS TECHNOLOGIES

Much of the public focus in the debate over strategic defenses has centered on the promise held out by new BMD weapons technologies, especially the so-called *directed energy weapons* (DEW).[1] These include various types of laser and particle beams, which travel at or near the speed of light (300,000 kilometers per second). Yet creating a promising DEW candidate will be a challenge: it will involve developing a laser and/or particle beam of sufficient intensity and accuracy to deposit large amounts of energy on small areas of rapidly moving ballistic missiles and warheads from great distances.

Lasers

The term "laser" is an acronym, standing for *l*ight *a*mplification by the *s*timulated *e*mission of *r*adiation. Essentially, lasers can provide very narrow beams of coherent light with sharply defined wavelengths.

Four types of lasers are currently under consideration for BMD missions: infrared chemical lasers, electrically driven ultraviolet lasers, free-electron lasers, and X-ray lasers produced by nuclear explosions. These laser beams could destroy a missile booster or a warhead in one of two ways. Chemical, ultraviolet, and free-electron lasers (all known as optical lasers, being similar to visible light rays) would use a concentrated beam of light to burn into the missile "skin." The X-ray laser works differently, penetrating only a very thin layer of the booster skin, but initiating a shock wave that could destroy the missile's internal components or knock it off course.

In the design of optical lasers, the problem consists in achieving suitably intense beams of parallel light and aiming these at targets thousands of kilometers away. A large and precisely machined mirror, perhaps ten meters across, or a system of mirrors, would produce a beam of concentrated light that would spread to only a meter diameter after having traveled thousands of kilometers through space. Once the target is reached, the laser beam must focus on the same spot of the rapidly moving object for the several seconds needed to heat it sufficiently and destroy it.

Unlike optical lasers, X-ray lasers are created from the explosion of a nuclear weapon.[2] The incredibly high temperatures produced in a fraction of a microsecond by the nuclear explosion bring about a "lasing" action in a number of very thin metallic wires that are embedded in a plastic around the nuclear device. This lasing process then produces beams of X-rays, just before the nuclear explosion itself destroys the satellite battle station.

Because of their short wavelength, X-rays differ from other lasers in three important respects. First, unlike optical lasers, powerful X-rays cannot be pointed by mirrors. Second, the X-rays generated from the wires cannot be as tightly focused as optical lasers. This means they can only be used from moderately short distances (a few thousand kilometers), and probably not from geosynchronous orbit (36,000 kilometers above the earth). Third, the soft X-rays do not penetrate the earth's atmosphere, and thus can be used only against missiles or warheads traveling through the vacuum of space.

On the other hand, a nuclear-powered X-ray laser device will be much lighter than optical laser battle stations, because the latter require large mirrors and more fuel. As will be seen below, this makes X-ray lasers a potential candidate for so-called "pop-up" systems, in which the X-ray laser would be lofted into space on top of a ballistic missile.

Particle Beams

Research is also being carried out on *particle beams* produced by particle accelerators (atom-smashers). These highly energetic beams are familiar research tools in physics laboratories. If the particles (hydrogen ions, for example) are accelerated to nearly the speed of light, they can penetrate deeply into material targets. Yet charged particle beams are not suitable for transmission over large distances because they are deflected by the earth's magnetic field. Scientists have instead proposed using neutral hydrogen beams, which might be obtained by accelerating negative hydrogen ions and stripping off the extra electron.

Particle beams have an advantage over lasers in that their deep penetration into the target could destroy the electronics and guidance systems of the MIRV bus. Another advantage is that the beam from the accelerator can be steered by electric and magnetic fields, without moving the accelerator itself; this is important if several targets are to be attacked in quick succession.

A main problem facing weapons designers is to produce particle beams of sufficient intensity and small divergence from accelerators that are lightweight enough to be lofted into space. In addition, particle beams have very limited ability to penetrate the earth's atmosphere. Collision with air molecules will strip the electron off the hydrogen atom, converting the neutral into a charged beam and causing it to fan out in the earth's magnetic field. This effectively destroys the lethality of the beam. Particle beams are therefore of only limited usefulness against missile boosters, especially if these are designed to release their warheads while still in the atmosphere.

Kinetic Energy Weapons

The final type of BMD weapon is the *kinetic energy* weapon. This is typically a rocket-propelled weapon that accelerates small projectiles. As with the F-15 anti-satellite weapon, these projectiles "home in" on the target using infrared sensors for guidance and tiny rocket thrusters to change direction.

These projectiles can destroy a booster or warhead simply by high velocity impact. The velocity of the missile and/or warhead actually contributes to its destruction; the projectile hurtles towards its target either head-on or at right angles, penetrating it to a depth many times its own length. The weapons themselves would be carried by a large satellite, and fired by either rockets or, possibly, electromagnetic railguns.

BMD Technologies and Concepts in the 1980s

Kinetic energy weapons have two drawbacks when it comes to boost or post-boost intercept. Because they are very much slower than directed energy weapons, they cannot travel very far during the first few minutes after enemy missiles have been launched. Moreover, if the target is still within the earth's atmosphere, friction caused by the projectile will heat the surrounding air and interfere with the infrared guidance system.

Accordingly, kinetic energy weapons are considered more appropriate for mid-course defense in the vacuum of space, where there is also more time to engage the target. Such weapons could be either ground-based or space-based. If space-based, they would have an advantage over directed energy weapons in not needing such fragile ancillary equipment as large space mirrors. Kinetic energy weapons are also very suitable for terminal defense, at least of hardened military targets, after the warheads have reentered the atmosphere.

DEPLOYING BMD SYSTEMS IN SPACE

Because of the curvature of the earth, and the fact that Soviet ICBMs will only be 100 to 400 kilometers above the USSR during boost phase, it is only from space that a BMD system will have a clear shot at missiles rising from Soviet silos. For this reason, BMD systems will either have to be on permanent station in space, on orbiting satellites, or be launched ("popped up") into space on notice of an attack.

In devising a constellation of orbiting space-based BMD platforms, however, a second physical constraint intrudes: only in geosynchronous orbit will a satellite remain over a given spot on earth during its entire orbit. Given the tremendous distance from a geosynchronous BMD platform to its target, it would be preferable to station the platforms just above the earth's atmosphere (at altitudes of 500 to 1000 kilometers). However, BMD platforms in low earth orbit will be circling the earth once every 100 minutes or so, and do not remain stationary over the same spot (e.g., an enemy's missile field) throughout their revolution. This means they would only be within range of enemy missile silos once or twice a day, necessitating an increase in the number of space platforms required for continuous coverage. This is known as the "absentee problem": of a certain number of BMD platforms orbiting the earth, only a fraction will be available at any given moment to attack ICBMs.

In designing a space-based BMD system, then, there are trade-offs that must be made between the placing of weapons in geosynchronous orbit (which demands effectiveness over tens of thousands of kilometers), and low-earth orbit (where dozens if not hundreds of platforms would have to be placed in order to ensure a sufficient number within effective range of Soviet silos and submarines).

There have been many preliminary proposals for the BMD systems mentioned above—among them, stationing infrared lasers in low-earth orbit; deploying ultraviolet (excimer) lasers in geosynchronous orbit; and basing the excimer laser itself on the ground and using a set of relay mirrors in both geosynchronous and low earth orbit to target the rocket booster.

The use of "pop-up" BMD systems, consisting of X-ray lasers launched into space quickly on top of ballistic missiles, is an alternative to maintaining BMD systems continuously in orbit. Here again, however, physical constraints intervene. First, because the earth is round, any pop-up system will have to be deployed fairly close to the Soviet Union in order to be effective for boost-phase intercept. The reason for this is that the popped-up system will have to climb very quickly to a point far above the atmosphere so as to have a clear line of fire against the enemy missile before that missile releases its warheads and decoys. Even if the X-ray laser could be launched nearly simultaneously with the firing of Soviet ICBMs, this BMD system would still have to be deployed within a few thousand kilometers of the Soviet silos. This means putting pop-up systems on submarines, and deploying the submarines in the Indian Ocean, the eastern Mediterranean, the Arctic Ocean, or the Sea of Okhotsk. Of course, deploying submarines so close to the Soviet Union could make them vulnerable to Soviet anti-submarine forces. Moreover, pop-up systems could still be countered by the development of fast-burn ICBM boosters, even if these release their warheads above the atmosphere. For instance, if the boost phase can be completed in 130 seconds, even a system located 4000 kilometers from the missile silo could not reach a high enough altitude in space to get a direct line of fire at the ICBM.

Command-and-Control of BMD Systems

One of the major tasks facing a multilayered BMD system will be the creation of appropriate command-and-control capabilities (see the essay in this volume by Charles Zraket). The range of tasks facing a BMD system will include acquisition and tracking of targets, discrimination of warheads from decoys, actual firing of the directed energy or kinetic energy weapons, and follow-up assessment of whether or not the target was destroyed. All these tasks must be coordinated by a reliable system of computers and high-speed data links. The issue of reliability is especially important given the issue of automated response, i.e., the fact that there may not be enough time for a president to give the command to destroy Soviet missiles during boost and post-boose phase.

Computers, which could be stationed both in space and on the ground, would have to be capable of analyzing and transmitting huge amounts of information reliably and securely. More important than the computer hardware itself will be the extremely sophisticated software needed to coordinate a BMD defensive network. Many experts believe that the task of developing this software, and ensuring its reliability so that the BMD system works perfectly the first time in conflict, will be even more difficult than building the actual BMD components.

Another important criterion for the success of a multi-layered BMD system is the ability of each layer to "hand over" those targets not destroyed to the next layer. Each successive layer must be "told" by the previous one, through use of a battle management system, which nuclear warheads have been destroyed and which have not. Given that lasers and particle beams do not always physically destroy their targets but sometimes merely damage internal components, the problem becomes more difficult. In short, the tiers of a BMD system will not act independently, but must be fully integrated by the command-and-control system to perform their intended functions.

COUNTERMEASURES

Resolving the related problems of offensive countermeasures and BMD system survivability will be even more important than demonstrating the actual technical feasibility of BMD weapons and other systems components, or ensuring the efficiency of the command-and-control system. One point to consider is that, while many of the technological and systems problems associated with new BMD concepts are far from being solved, a large number of different types of countermeasures are available now. Many of these countermeasures are far less technically sophisticated than the BMD systems they will try to overcome, and thus are less expensive as well. It may well be that, as new BMD systems are developed, methods of negating many of these countermeasures will be found. On the other hand, new countermeasures will almost surely be developed as well, leading to a continuing competition between offensive and defensive systems.

A second point to consider is that by their very nature BMD systems are reactive: they will be designed to work on short notice, at a time and place of the opponent's choosing, and against a wide range of possible countermeasures. This gives the offense an additional advantage, for it may choose which of the numerous available countermeasures it wishes to employ, while the defense will have to be prepared to deal with all of them. On the other hand, one of the conceptual attractions of a multilayered BMD system is that, in many cases, a countermeasure that is effective against one particular BMD tier will not work against another tier, thus increasing the number of countermeasures an attacker must have at his disposal to ensure the ability of his warheads to penetrate.

In considering the broad range of countermeasures, it is useful to group them into three categories of passive, active, and threatening.[3] Passive countermeasures are those that seek to aid the penetrating ability of incoming missiles and warheads without directly attacking the BMD systems. There are numerous types of passive countermeasures, most of which are designed to work against a particular tier of a BMD system. Examples include: fast-burn boosters and methods of shielding the booster (against boost-phase defenses); a cluster release of warheads and decoys from the MIRV bus (against post-boost defenses); the use of aluminized balloons, chaff, and clouds of aerosol (against mid-course defenses); and salvage fuzing and manueverable warheads (against terminal defenses).

Active countermeasures are those that attempt to interfere directly with the components of a BMD system or to destroy them altogether. In the 1960s, it was recognized that the large ground-based radars associated with terminal ABM systems were both vulnerable to direct attack and could be blinded by a few nuclear explosions very high in the atmosphere ("radar blackout"). These types of "defense suppression" tactics could pose an even greater threat to a space-based BMD system. Because the defense's sensors, space mirrors, command-and-control systems, and weapons platforms will be traveling in predictable orbits, such satellites, especially in low earth orbits, will be easier targets than the ballistic missiles and warheads they are designed to detect and destroy. In addition to "blinding" and "radar blackout," existing U.S. and Soviet ASATs, as well as small space mines, could be used to destroy BMD satellites. The Soviets could also use their Galosh nuclear BMD interceptors to punch holes in a space-based BMD system. Even the release of a swarm of small steel pellets in the path of an oncoming BMD satellite or unprotected space mirror would be sufficient.

Various methods are being investigated to enhance the survivability of satellites, including hardening them, giving them maneuvering ability for evasive action, and even deploying defensive satellites (DSATs) to actively protect them. While such measures are certainly feasible, they would impose extra costs and constraints on the defensive system. Hardening and maneuverability would add to the size and complexity of the BMD satellites, while deploying defensive satellites would be costly and increase the command-and-control problems of the overall system.

Other types of countermeasures consist of those that interfere directly with the complex operations of the BMD systems without necessarily damaging the system itself. For example, decoys traveling through space can be equipped with radar transmitters that send out false signals to confuse BMD sensor systems. In addition, both decoys and actual warheads can be given the ability to maneuver during mid-course in order to confuse passive tracking schemes that depend on these objects remaining on a ballistic trajectory. This type of maneuvering capability would, as we have seen, also complicate the task of terminal BMD systems.

The final category of threatening countermeasures is especially worrisome. Along with the active and passive means described above, an opponent could simply increase his offensive forces in an effort to overwhelm a multilayered BMD system. In essence, this is what happened in the early 1970s, when the U.S., followed by the Soviet Union, started adding multiple warheads (MIRVs) to their missiles so as to overwhelm the projected terminal ABM systems of the other side. Both sides could also build dummy silos and fake ICBMs that would look real to BMD sensor systems. Or, both countries could deploy additional cruise missiles and other types of nuclear delivery systems against which BMD systems would be ineffective.

In sum, a good deal of attention will have to be given to the interaction of new BMD systems and countermeasures as they evolve.[4] It will not be enough merely to demonstrate the technical feasibility of new BMD systems: we will need to ensure that such systems will be relatively invulnerable to direct attack or interference.

SDI EXPENDITURES AND COST-EFFECTIVENESS

Given that the SDI program is still in its early stages, it is difficult to estimate the final costs of deploying and maintaining a multilayered durable and effective BMD system. Tentative estimates that have been made range from $100 billion to more than $1000 billion (one trillion), but these must remain speculative until more is known of the configuration of the BMD system.

What can be said with some certainty is that deploying and operating strategic defenses against ballistic missiles, involving dozens if not hundreds of complex weapons platforms and sensing systems in space, extensive command-and-control networks, as well as numerous space vehicles to deploy and repair the system, will be a very expensive proposition. If comprehensive defenses against manned bombers and cruise missiles are deployed as well—an effort former Secretary of Defense James Schlesinger has estimated could cost $50 billion annually—then the overall cost could well be prohibitive.

In the near term, the costs of the SDI research program are projected to increase sharply.[5] For fiscal year 1986, the administration has requested $3.7 billion, a 160 percent increase over 1985. There are doubts, however, concerning the ability of the individual SDI research programs to make profitable uses of such increases. Schlesinger, for example, has testified before Congress that military R&D programs normally should not expand by more than 35 percent annually, and that SDI funding of $1.5 to $2 billion for 1986 would be "ample."[6]

If Congress does grant the administration's SDI funding requests, which is by no means certain, by fiscal year 1989 the SDI budget will be $7.5 billion, or 16 percent of total Defense Department R&D funding, a significant amount. If one adds in projected 1989 spending on other types of defense, such as anti-satellite weapons and air and civil defense, the total approaches $12.4 billion.[7]

For the period 1985–89, then, the administration has requested $26 billion for the SDI program, and upwards of $40 billion for strategic defenses in general. By 1989, this level of investment will have created a large military and industrial constituency with a vested interest in accelerating the program, even if many of the proposed technologies should fall far short of expectation. By the early 1990s, once the SDI has moved from the research stage to the development and testing of engineering prototypes, the annual budget for strategic defenses could well be in the tens of billions of dollars.

The question of whether such expenditures will ultimately produce worthwhile defenses cannot be answered now. Yet, as Lt. Gen. James Abrahamson, the SDI program director, has noted, it must be proved that a strategic defense system is both technically feasible and affordable, i.e., that it will cost the U.S. "substantially less" to build the system than it will for the Soviets to counter it.[8]

The notion of cost-effectiveness is based on the relationship of the cost to the offense of ensuring the ability of its warheads to reach their targets, compared with the cost of developing, deploying, and maintaining a defensive system. The standard method for analyzing cost-effectiveness uses a marginal cost-exchange ratio that asks, in effect: what marginal improvements will the defense have to make to offset corresponding improvements in the penetrability of offensive forces? (See the essay by George Rathjens and Jack Ruina in this collection.) If it can be shown that, in all likelihood, a defensive system will suffer a negative cost-exchange ratio (i.e., that it will be cheaper for the offense to upgrade its forces than it will for the defense to negate those improvements), then a strong argument can be made against spending the money to develop and deploy the defense in the first place.

For instance, Defense Department studies of marginal cost-exchange ratios done in the mid-1960s, before the actual deployment of U.S. ABM systems, demonstrated convincingly that Soviet offensive forces would enjoy anywhere from a three-to-one up to a ten-to-one cost-exchange advantage over a U.S. ABM defense. In other words, for every dollar spent by the Soviets to ensure the penetrability of its warheads, the U.S. would have to spend from three to ten dollars to offset that improvement. As Alexander Flax notes in his paper, this negative cost-exchange ratio for the defense figured importantly in arguments against the deployment of extensive ABM systems.

Whether or not the leverage provided to defensive systems by boost-phase intercept will be able to reverse these ratios will depend greatly on the feasibility of Soviet boost-phase countermeasures, and the ability of the Soviets to attack a space-based BMD system directly. Paul Nitze, senior arms-control advisor to the Reagan administration, has stressed that the two criteria of BMD cost-effectiveness and survivability are crucial; that "if the new technologies cannot meet these standards, we are not about to deploy them."[9]

A second major issue regarding SDI expenditures concerns possible opportunity costs strategic defenses could pose for other U.S. military capabilities. Even if BMD systems turn out to be marginally cost-effective vis-a-vis Soviet offensive forces, the expenditure of hundreds of billions of dollars on strategic defenses (if this figure is correct) could impose severe constraints on other U.S. military programs. Any benefits the SDI would bring to enhancing strategic nuclear deterrence could be offset if the program draws funds away from the continued modernization of U.S. conventional military forces and weakens conventional deterrence between the superpowers. While the issue of opportunity costs will not become serious until the SDI has moved from R&D to testing and procurement, which will probably occur in the early 1990s, some thought must be given now to how strategic defenses will mesh with the entire range of U.S. military capabilities, both nuclear and conventional.

In sum, despite remarkable advances since the 1960s in BMD related technologies, there are major uncertainties surrounding the ultimate feasibility of deploying and maintaining strategic defenses against ballistic missiles. Even if many of the remaining technical obstacles are overcome in the areas of weapons systems, optics, sensors, and guidance systems, there will still remain the issues of countermeasures, survivability, cost-effectiveness, and the integration of strategic defenses with overall U.S. military capabilities.

ENDNOTES

[1]See also Ashton B. Carter, *Directed Energy Missile Defense in Space* (Washington, D.C.: U.S. Congress, Office of Technology Assessment, April 1984); and Jeff Hecht, *Beam Weapons* (New York: Plenum Press, 1984).

BMD Technologies and Concepts in the 1980s

[2] It has been widely reported that, although the X-ray laser concept provided much of the impetus for President Reagan's interest in BMD and for his "Star Wars" speech, X-ray laser weapons have currently fallen out of favor because they entail the explosion of nuclear weapons; see, " 'Star Wars' Technology: It's More than a Fantasy," *New York Times,* March 5, 1985, p. 1.

[3] See especially the section on countermeasures in *Space-Based Missile Defense,* a report by the Union of Concerned Scientists, (Cambridge: Union of Concerned Scientists, March 1984), pp. 45–52.

[4] The Defense Nuclear Agency (DNA), for example, plans to spend $600 million over the next several years studying how the lethality of directed energy weapons and kinetic energy projectiles against rocket boosters and warheads could be offset by potential countermeasures the Soviets could employ by the year 2000. The DNA program, which will rely heavily on computer simulations, will look at six weapons candidates, including: ultraviolet lasers, excimer lasers, X-ray lasers, particle beams, kinetic energy projectiles, and microwaves. See Philip J. Klass, "Defense Nuclear Agency Directing Six-Year Lethality Assessment, *Aviation Week & Space Technology,* April 1, 1985, p. 65.

[5] One especially good analysis of short-term SDI funding is "Analysis of the Costs of the Administration's Strategic Defense Initiative 1985–1989," a staff working paper of the Congressional Budget Office (Washington, D.C.: U.S. Congress, May 1984).

[6] See "SDI Director Defends Rise in Research Funds," *Washington Post,* March 20, 1985, p. A12.

[7] See John E. Pike, "Paying for Star Wars—The Rest of the Picture," a staff research note of the Federation of American Scientists, Feb. 5, 1984.

[8] "Panel Told 'Star Wars' Needn't Breach Treaty," *Washington Post,* Feb. 28, 1985, p. A28.

[9] Paul Nitze, "On the Road to a More Stable Peace," speech delivered to the Philadelphia World Affairs Council, Feb. 20, 1985, p. 4.

Gerold Yonas

The Strategic Defense Initiative

IN HIS WIDELY PUBLICIZED SPEECH of March 23, 1983, President Reagan speculated on a new approach to national security. He did not put forth a concrete proposal, but instead posed a question: "Would it not be better to save lives than to avenge them?" He called upon the scientific and technical community "to give us the means of rendering these nuclear weapons impotent and obsolete." The only specific activity he proposed immediately was "a comprehensive and intensive effort to define a long-term research and development program to begin to achieve our ultimate goal of eliminating the threat posed by strategic nuclear missiles." The president also speculated that this R&D effort "could pave the way for arms-control measures to eliminate the weapons themselves." On April 18, 1983, he signed a national security directive mandating two intensive studies that would develop a plan for program implementation beginning in fiscal year 1985. The purpose of one study was to consider the policy implications of defense technologies; the other study was to deal with the technology itself.

The president clearly caught many people by surprise by calling for these studies in his address. It would have been less disturbing had the studies been carried out quietly *before* the president made his specific program request in public. Yet, unusual as this exercise of presidential leadership might have been, it may have been the most effective way of galvanizing a large and diverse group of experts to leave their jobs and invest several months of time and effort in a controversial project. While I cannot profess to know the president's precise motivations, I can cite several factors that influenced many of us to become involved in the technologies study. We were not only acutely aware of the restrictions existing on ballistic missile defenses (BMD) by the ABM treaty, but recalled the hope fostered by that treaty that our retaliatory forces would remain survivable. The United States stated in 1972 that the treaty would serve as the basis for "more complete limitations on strategic arms," and that "if an agreement providing for more complete strategic offensive arms limitations were not achieved within five years, U.S. supreme interests could be jeopardized. Should that occur, it would constitute a basis for withdrawal from the ABM treaty." The level of arms limitation that had been hoped for was not achieved. Instead, the Soviet Union's vigorous development and deployment of large ICBMs with multiple warheads now threatens our Minuteman missiles, with the result that our national interests are in greater jeopardy today than they were ten years ago.

In addition, the recent protracted search for a "survivable" basing mode for our MX missile had frustrated many of us who had worked in the strategic weapons arena. We began to feel that the continued buildup of Soviet forces was making the possibility of nuclear war all too real, and we believed that some solution would have to be found to the vulnerability of our land-based missiles. Nor can one work on weapons and be unaware, as well, of the strong, vocal, and broadly based freeze movement in this country that is calling for a halt to all further production and testing of nuclear weapons, or of the Catholic bishops' Pastoral Letter on War and Peace that expressed "profound skepticism about the moral acceptability of any use of nuclear weapons." Arms-control negotiations have been frustrated by the Soviet Union's resistance to significant reductions in nuclear weapons, and growing suspicion and mistrust on both sides had brought us to a new level of deterioration in Soviet-American relations. Indeed, the continued Soviet buildup of high-precision, short-reaction-time, offensive weapons had led to growing concern in the U.S. over the possibility of a Soviet preemptive strike and the effect this possibility would have on strategic stability. Many were beginning to question whether civilization could survive indefinitely in this condition.

Even before President Reagan's speech about the SDI, several public figures had been advocating revolutionary change in our approach to national defense. For several years, Senator Malcolm

Wallop (R-Wyoming) had vigorously supported a crash program to develop space-based lasers. His point of view had been expressed in publications, speeches, and proposed (but rejected) legislation dating back to 1979: "Technology now promises a considerable measure of safety from the threat of ballistic missiles."[1] His technologically optimistic point of view—undoubtedly known to the president—was exemplified by an open letter he wrote to the Senate following the president's speech, in which he offered to "integrate three mature technology programs into a space-based laser weapon demonstration as quickly as possible."[2]

Another voice was that of General Daniel Graham (retired U.S. Army), who presented the "High Frontier" proposal to deploy space-based rocket-launched projectiles to intercept boosters. This concept, he said, would "free America from the threat of Soviet ballistic missiles largely by using technology that was proven successful as far back as 1959."[3] Finally, Dr. Edward Teller, who had advised the president on scientific matters for a number of years, urged the development of nuclear-driven directed energy weapons that "by converting hydrogen bombs into hitherto unprecedented forms and then directing these in highly effective fashions against enemy targets would end the MAD [mutual assured destruction] era and commence a period of assured survival on terms favorable to the Western Alliance."[4]

It was also becoming increasingly obvious that the Soviet Union was investing far greater resources on strategic defense than was the United States. The Soviets were continuing tests of their operational anti-satellite system, making improvements in their conventional Moscow BMD system capability, constructing a new large radar in the eastern Soviet Union that, according to the U.S. government, is "almost certainly a violation of the ABM treaty," and making large investments in advanced long-term research (such as ground-based lasers) that could be applicable to BMD.[5] In addition, continued Soviet investments in anti-tactical ballistic missile development indicated the possibility of their establishing the type of infrastructure necessary for the rapid deployment of a nationwide defense system.[6]

While the Soviets were making substantial investments and clearly closing the technological lead that the United States had held ten years previously, the United States was continuing to pursue a low-level BMD research program, but without a well-defined underlying goal or strategy. Many of us in the weapons field believed that, because of continued starts and stops, changes in direction, disputes over accelerated or slow programs, and distribution of management among departments, military services, and agencies, the true potential of BMD had not been realized in spite of rather significant expenditures. We were aware that although technology had advanced considerably in the last ten years, our expenditures had not given us maximum returns on our investments.

Most of us who became involved in the technologies study did not feel that the president was urging us to pursue specific, short-term weapons development programs. Our understanding was rather that we were to initiate creative approaches that, although perhaps unavailable today or even in the near future, could someday provide entirely new and effective technological options that would "pave the way for arms control measures to eliminate the weapons themselves."

Many of us also came to recognize that we were faced with the need for more than just technological achievements. We realized that solving what may be the most serious problem facing us today would demand breakthroughs in policy and strategy, as well as in technology. When the president initiated the technologies study, there were many unresolved issues and many questions on program goals and approaches that were still unanswered. These questions were very much before the president, his advisers, and the study teams when they met in Washington on June 2, 1983, to begin their deliberations.

Two policy study teams and one technology study team were formed as a consequence of the president's national security study directive. The policy study group, which reported to Under Secretary of Defense for Policy Fred Ikle, was to examine "the impact of defensive technology on deterrence and arms control," and was composed of a government team chaired by Franklin Miller plus a team drawn from outside consultants and chaired by Fred Hoffman. (This group is often informally referred to as the Hoffman panel.) The Defensive Technologies Study Team (DTST), of which I was a member, reported to Under Secretary of Defense for Research and Engineering Richard DeLauer. (Our group is informally referred to as the Fletcher panel, after its chairman, James C. Fletcher.) These three teams, which met separately and reported to different Defense

Department officials, conducted their studies independently throughout the summer and were only afterwards integrated, by a senior interagency group. The DTST was divided into six main technical panels consisting of more than fifty full-time people drawn from industry, national laboratories, and universities.[7] Each panel held comprehensive reviews with hundreds of briefings, and after extensive deliberations prepared a report of its findings. Producing a consensus from a wide distribution of opinions required the individual members to examine and reevaluate their own views, as they worked alongside and compromised with others. An executive summary of the separate panel reports was also prepared that included the unified plans and budgets for two levels of research activity: one that would push ahead as fast as technological advances permitted, and the other that would proceed at a slower pace, set by limits on available funds. (As the executive summary is still classified, I will, in this article, be summarizing its contents for the general reader.)

The initial stage of the studies was the most difficult, involving as it did an attempt to clarify the goal that had been stated by the president, namely, to eliminate the threat of nuclear ballistic missiles. Should we pursue a nearly leakproof defense of our population, a less-demanding defense of our retaliatory forces, or some combination of increased defenses and arms control that would reduce the role of offensive forces? Should we emphasize readily available technologies as a hedge against Soviet abrogation of the ABM treaty, or should we scale down existing BMD programs and try for breakthroughs and a "grand solution"? Each choice had policy implications. But since the technology and policy team reports would be integrated only upon their completion, the task of technology evaluation and planning had to begin while the policy teams were still working to define the problem itself.

From the outset, the technologies study team emphasized a long-term program to research and develop a multi-tiered defense that would provide significant damage limitation. (The study team realized that a credible defense would have to have a low leakage of warheads, but no actual leakage number was defined.) It was also recognized that any U.S. R&D program to provide new options for defense would be met by a significant Soviet response, and the postulated countermeasures played a pivotal role in defining the R&D effort.

Since the technologies study team consisted of people active, or with past experience, in the BMD field, few members lacked strong convictions on the subject. Some were advocates and others detractors, but all shared a great deal of intellectual skepticism and a willingness to voice their opinions. The strong bastion of doubters resided in the countermeasures panel. They were duty-bound to suggest a myriad of possible methods to defeat any defense. The real issue was the relative costs of measures and countermeasures (the cost-exchange ratio), and here, more unknowns were uncovered than definitive answers. In the end, there was inadequate opportunity to complete a thorough measure/countermeasure/counter-countermeasure analysis, but the countermeasures panel was particularly successful in forcing the other panels to rethink and reshape their technology proposals.

The stronghold of "true believers" came from the established BMD community. They were convinced that rocket-propelled, homing, hit-to-kill vehicles could in a short time prove effective in hardpoint defense while requiring the lowest investment (see the essay in this collection by Maj. Gen. John Toomay). Such an approach, they felt, would also satisfy the need for a terminal phase of a multi-tiered defense, though it would have little effectiveness in area defense without the other tiers.

The bastion of technological optimists could be found in the directed energy (speed of light) weapons advocates, who had great faith in the ability of science and technology to make tremendous gains, given enough smart people, capable management, and sufficient funding. Because of the short time period in which missiles can be intercepted during boost-phase, speed-of-light weapons (or possibly very high velocity projectiles) were considered essential.

The approach adopted by the technologies team emphasized high payoff, long-range capabilities rather than systems that might be more easily realized but might, in turn, be easily countered by offensive technologies. We decided to sketch out a five-year research program that would thoroughly explore the limits of technology on both the offensive and defensive sides, in order to winnow out the least promising defense options and acquire sufficient knowledge so that a decision could be made in the early 1990s on whether to proceed with engineering prototypes.

The tasks faced by any strategic missile defense are formidable. To understand the defense concepts and technologies that are being investigated, it is useful to think of the flight of a ballistic missile as passing through four phases—boost, post-boost, mid-course, and terminal—each offering different opportunities and challenges to a defense system. A highly effective defense against a massive missile attack would require multiple tiers of defense, countering the offense at each phase until the fraction of warheads remaining is too small to be of military utility. (See figure on page 59.)

During the *boost phase* of a missile's flight, the advantage to the defense is that the flame from the booster's rocket motors provides a very bright marker that can be readily detected by space-based sensors and serve as an aim point for the defense's weapon. Moreover, a single successful intercept would destroy all of the warheads and decoys carried by the missile, yielding great leverage for the defense and reducing the task of subsequent layers of defenses. The challenge is that the boost-phase is a very brief time in which to intercept thousands of missiles—about 150–300 seconds for current ICBMs, 150–200 seconds for current SLBMs. Knowing this, the offense may time its attack to overwhelm the defense, or the offense may redesign and deploy entirely new missiles to reduce the duration of their boost-phase or harden them against damage by the defense.

During the *post-boost phase,* when perhaps ten or more warheads and hundreds of decoys are released from a maneuvering post-boost vehicle, or "bus," the defense may still have some advantage if it can intercept the bus before it releases all of its warheads and decoys. The challenge is that the bus' maneuvering engines are much smaller and more difficult to detect than booster motors, and the offense may redesign its post-boost vehicles to shorten the maneuvering phase or to mask the engines and the deployment of warheads and decoys from the defense.

During the *mid-course phase,* the advantages to the defense are the long time available to locate and destroy targets (roughly 10–15 minutes for ICBMs, 7–10 minutes for SLBMs), the opportunity to bring both space-based and ground-based interceptors to bear, and the chance to destroy warheads far away from soft targets such as cities. The challenge to the defense is the sheer number of warheads and decoys that may have been released—in the worst case, tens of thousands of warheads and hundreds of thousands of decoys. The offense can try to complicate the task by deploying not only clever decoys, but warheads inside of decoys, and so on. The defense must either discriminate warheads from decoys or make interceptions so cheap and easy that it can attack them all.

During the *terminal phase* between the warhead's reentry and its detonation over the target, the advantage to the defense is that the atmosphere will have sifted out most or all of the decoys, and the defense has the option of preferentially intercepting only those warheads that are still heading toward important targets on the ground. The challenge is that very little time is available, perhaps 30 to 40 seconds, to carry out the interception, and warheads that are "salvaged fused" (fused to detonate when intercepted) may still damage or destroy soft targets such as cities and would certainly complicate the problem of tracking those warheads not detonated.

The offense might try to defeat the defense in a variety of ways. Since there are many responses that can be postulated, and since an "unlimited" expenditure on offense can certainly defeat a "limited" defense, the real issue is to define the cost-exchange ratio between defenses and countermeasures, and to estimate how rapidly the adversary might deploy such countermeasures. These factors also have bearing on any arms-control agreements to limit offensive forces. To provide incentive for offense reduction agreements, the defensive technology response must enjoy a clear advantage in any cost-exchange analysis. To put it another way, if ten dollars of defense can be promptly countered by one ruble of offense, then we had better not invest our money. On the other hand, a favorable cost exchange would compel the Soviets to invest in defense and to seek offensive arms reduction in their own self-interest.

At each defensive tier, the defense must carry out four essential functions. One of these involves a combination of surveillance, acquisition, discrimination, and kill assessment. The second consists of pointing and tracking of defense weapons, the third, of interception and destruction, and the fourth, of battle management. All these functions must be integrated into an overall defense system. The concepts and technologies being investigated or identified by the Defensive Technologies Study Team as promising are summarized in table 1, opposite.

FUNCTIONS	BOOST	POST-BOOST	MID-COURSE	TERMINAL
DETECTION, ACQUISITION, DISCRIMINATION, KILL ASSESSMENT	SPACE-BASED INFRARED SENSORS	SPACE-BASED ULTRAVIOLET, VISUAL OR RADAR IMAGING SENSORS		AIRBORNE INFRARED OR RADAR SENSORS
POINTING, TRACKING	SPACE-BASED LASERS AND VISUAL IMAGING SENSORS			INFRARED TERMINAL HOMING
INTERCEPTION, DESTRUCTION	DIRECTED ENERGY WEAPONS, KINETIC ENERGY PROJECTILES			NON-NUCLEAR SURFACE-TO-AIR MISSILES
BATTLE MANAGEMENT	VERY HIGH SPEED INTEGRATED CIRCUIT HARDENED VS. RADIATION, NEW SOFTWARE DEVELOPMENT TOOLS			

Table 1 Elements of a Multilayered BMD System

Our present technological capabilities in each of the categories shown in table 1 are very far from what is needed for a robust defense against plausible Soviet threats and countermeasures in the future. Some of these shortcomings the study team regarded as critical; unless they could be overcome, we could not be certain that a robust defense could be achieved. We therefore proposed a long-term research program to determine whether the advances necessary in these critical technologies are feasible. These vital areas include the following:

SURVEILLANCE, ACQUISITION, DISCRIMINATION, AND KILL ASSESSMENT

At each phase in the missile's flight, the attacker may try to confuse and overwhelm the defense by increasing the number of warheads and by deploying decoys. Indeed, in setting out performance benchmarks for a defense system, we assumed that the Soviets might increase their offensive forces at a concerted rate so that their arsenal of strategic nuclear warheads might grow to two or three times its present size. The defense therefore must be able to detect and track tens of thousands of objects in flight, while rapidly and reliably distinguishing genuine targets from clever decoys.

Acquisition and discrimination during the boost phase will be greatly aided by the hot flame from the booster's engines; detecting colder warheads and decoys against the cold background of space and discriminating between them during post-boost and mid-course phases may be possible by using a variety of sensors—optical, infrared, and radar; space-based, airborne, and ground-based—and integrating the information derived from all of them. Decoys may perhaps be readily distinguished from warheads if we can observe the "birth" of decoys at the very instant they are deployed, keeping track thereafter of which objects are warheads to be intercepted, and which are debris or decoys to be ignored—so-called "birth-to-death tracking." All this will depend on important advances in sensor technology and the computer software necessary to process the volumes of data generated, as well as on careful studies of the characteristic signatures of warheads and decoys.

INTERCEPTION AND DESTRUCTION

The ability of any defense to respond effectively to an unconstrained Soviet threat is strongly dependent on the feasibility of boost-phase and post-boost interception. Two types of weapons are being considered for this task: directed energy weapons (speed-of-light weapons) and kinetic energy weapons.

Directed energy weapons could include space-based lasers, ground-based lasers, space-based particle beams, and X-ray lasers powered by nuclear explosives. The current goal is to construct laboratory-scale beam generators by the late 1980s or early 1990s and to demonstrate the feasibility of scaling them up to weapons-level performance by the early 1990s.

The critical parameter for beam weapons is beam brightness, the rate at which the weapon deposits energy on each square centimeter of the target. Brightness is an important determinant of the weapon's range and of how long it must dwell on each target; it determines the number of weapons needed in orbit to cover Soviet missile launching areas. Beams of great brightness may become available from short-wavelength lasers such as the electrically powered excimer and free-electron lasers, but both of these are in relatively early stages of development. The free-electron laser might be quite efficient and could be a good candidate for eventual deployment in space, given successful resolution of several physics and engineering issues. Technologically much simpler than the electrically driven lasers would be the invention of an entirely new, short-wavelength chemical laser exploiting recent chemical laser technology; an invention of this nature could revolutionize laser weapons. The search for just such innovative approaches was characteristic of the recommendations of the technologies study team.

Neutral particle beams would draw on a well-established technology base in the field of physics research accelerators. Such beam weapons would have the advantage of being difficult to shield against because of the penetrating power of atoms accelerated to nearly the speed of light. The sophisticated electronics of a missile or warhead might be disrupted or damaged by even low beam currents. The critical issue may not, in fact, be reaching the necessary level of brightness; instead, such problems as verifying kills, providing the power supply, and evaluating possible countermeasures may dominate the investigation of this promising approach. Neutral particle beams, for instance, become rapidly degraded and deflected once they enter even the thin atmosphere 100 kilometers above the earth, so development by the Soviets of a fast-burn booster that completes its boost phase within the atmosphere would defeat boost-phase particle-beam intercept. Replacing their arsenal of slow-burning liquid-fuel ICBMs with fast-burn boosters would be costly and time-consuming for the Soviets, and might well force them to reduce warheads. Nevertheless, the relative trade-offs in cost, effectiveness, and deployment time of entirely new Soviet boosters may be a critical issue for neutral particle beams.

Directed energy weapons powered by nuclear explosives are being investigated because they offer the prospect of supplying enormous quantities of energy in a short burst and at low cost. Concepts such as the X-ray laser are being examined to determine their feasibility. At the same time, the critical issues of how and where such weapons would be based are being studied, as well as the implications of possible Soviet use of such weapons to attack U.S. space-based defenses.

Kinetic energy weapons are the other devices being investigated for boost-phase intercept. These include so-called hypervelocity or electromagnetic launchers as well as more conventional chemical-fuel rockets, both of which would propel homing projectiles toward Soviet boosters at speeds of up to 10 kilometers per second. Based on orbiting platforms, such devices could also serve in mid-course interceptions and in defending space-based equipment from direct attack. The goal would be to design weapons that are not only effective but also very inexpensive so that if sensor technologies cannot reliably discriminate warheads from decoys, the defense could afford to engage them all.

There is little question about the lethality of, for instance, an 8-kilogram projectile hitting a target at 10 km/sec (the approximate force of 100 kilograms of high explosive). But the development of a low-mass guidance and control system for such a homing projectile—one that is able to survive the initial acceleration—as well as low-cost chemical rockets or a space-based power supply able to deliver hundreds of megawatts of electricity for electromagnetic guns, may be the most difficult tasks of all.

BATTLE MANAGEMENT

Integrating multiple layers of defensive weapons into a highly reliable and fault-tolerant system will require high-performance computers, computer software, and adaptable communications networks far beyond present capabilities. The system must keep track of tens of thousands of objects, from birth to death, and allocate defense weapons in the most efficient manner. A network of space-based computers, capable of performing millions of operations per second, surviving virtually maintenance-free for years in deep space, and able to adapt gracefully to failures within the network, is just one of the challenges facing a multi-tiered defense system.

The software requirements are awesome as well. Millions of lines of computer instructions must be written and then tested in an exhaustive way to ensure that they are error-free. Both tasks may themselves have to be performed by computer, requiring in turn the development of computer-controlled computer programming and debugging. The difficulties posed by this aspect of the defense program cannot be overemphasized, nor should the revolutionary nature of developments in this field be overlooked.

SURVIVABILITY

The strength of any defensive system rests on its ability to survive a direct attack, and to continue to function effectively even if degraded by attack. Space-based components that must orbit directly over the Soviet Union will face a host of possible threats, including direct-ascent ASAT weapons, ground- or space-based lasers, space mines, particle-beam weapons, and the effects of nuclear explosions. The tactics of survivability are familiar ones—hardening, active self-defense, concealment, proliferation, maneuvering—but applying them in a cost-effective way to future defenses will be a difficult challenge, one that may be pivotal to the outcome of the entire endeavor.

By no means does this short list of critical challenges exhaust the range of matters that must and will be addressed in the research and development program recommended by the study team. A whole set of support programs, for instance, must be pursued: determination of generic means for hardening boosters and the lethal effects of various weapons against those means (since we are not likely to get specific engineering design data on Soviet boosters); development of multi-megawatt space-based electrical power supplies, nuclear and non-nuclear, for weapons, sensors, computers, and so on; development of heavy-lift space launchers capable of carrying perhaps a hundred tons or more to orbit, rather than the space shuttle's current capability of twenty to thirty tons; and so on.

Those of us on the Defensive Technologies Study Team fully understood that present capabilities in all of the categories discussed above are very far from those needed for a robust defense against plausible responsive Soviet threats. As the research program evolves, however, new discoveries are likely to be made that may radically change the emphasis of the defense program. It is thus vital to stimulate, encourage, and fully investigate innovative approaches as early as possible in the program. These new ideas might arise from totally new sectors of the scientific and technical community, sectors that may not even be aware at the moment of the technical challenges we now face. To stimulate the broadest involvement of technical talent from government, industry, and universities, the Departments of Defense and Energy will provide funds to encourage the investigation of radically different concepts that offer enhanced capabilities, although bearing the admittedly higher risks associated with unexplored approaches. And since our discoveries might aid the offensive side as well, they could force us to eliminate particular defensive approaches along the way. Because we need to eliminate all unfeasible or impractical defenses as soon as possible in order to focus our efforts, it is essential to approach the overall program with a healthy skepticism, as well as with the creativity associated with the exploration of a new field.

The report of the Fletcher panel was submitted to the Department of Defense in October 1983. In March 1984, the secretary of defense announced the establishment of a new special program, the Strategic Defense Initiative, that was charged with developing and carrying out an expanded R&D program on strategic defense, incorporating numerous ongoing defense activities with new ones proposed in the study report. Lt. Gen. James Abrahamson was appointed director of the program.

The first phase of the SDI, from now until the early 1990s, will consist of research. During this period, decisions will be made on whether to begin engineering development of specific weapons. The second phase will focus on systems development, when prototypes of actual defense components will be designed, built, and tested (assuming a decision is made to go ahead with defenses). The third will be a transition phase, with incremental, sequential deployment of defenses, presumably by both the United States and the Soviet Union. At this stage, significant reductions in offensive missile forces would hopefully be negotiated. The final phase would be reached when defense deployments are complete, and offensive missile forces at their negotiated low point.[8]

The Defensive Technologies Study Team prepared a fairly detailed plan only for the first—research—phase. To permit a decision by the early 1990s on whether to advance to the next phase, the Reagan administration requested $1.99 billion in funding for the SDI in fiscal year 1985. $1.78 billion was to go to Defense Department programs, with $210 million earmarked for Department of Energy R&D on nuclear directed energy weapons. This was roughly 70 percent more than the 1984 request, although almost all the increase had been requested before the president's March 23 speech. (It is ironic that the amount actually appropriated for the SDI in fiscal year 1985 was $1.4 billion, an amount *less* than was initially requested for BMD programs before March 1983.) The total projected budget for the first five years of the SDI is $26 billion. The research phase could conceivably be completed as early as 1990, if one postulates a pace limited only by technological advances; or later if the program is slowed down because of fiscal restrictions. No attempt was made by our study to estimate the costs for the second phase of the program, but certainly that estimate would play an important role in the decision whether or not to proceed.

Assuming that the SDI is pursued, where might we hope to be in the early 1990s? The technologies study team was fundamentally optimistic in light of the spectacular advances that have taken place over the last two decades in the technologies of sensors, computing and software, and defensive weapons. We did not find any fundamental reason why such advances should not continue in many areas. In view of the richness of available options open to us, there was a general belief that even if certain approaches had to be eliminated, the overall defense requirements could be met.

As I discussed earlier, a defense system must not only work, it must maintain its effectiveness, at a favorable cost-exchange ratio, against an opponent who attempts to defeat it. Some critics have asserted a priori that the defense will always lose to the offense in such competition. We believe that mastering technical challenges such as mid-course discrimination, low-cost interceptors, and high-brightness beam weapons will enable us to negate simple countermeasures and force the Soviets to resort to very expensive offensive responses.

The policy studies done by the other study teams have provided additional insight into the possible outcome of the SDI program. They remind us that the decision to deploy widespread ballistic missile defenses does not rest with the United States alone. Given their history, doctrine, and military programs, the Soviet Union might be the first country to seize the defense initiative. If we were to lack adequate knowledge of advanced defensive technologies at the time, the situation we would face would be grave. On the other hand, the development of effective defensive measures on *both* sides might serve to reduce offensive investments and increase our leverage in seeking mutual offensive arms reductions. As George Keyworth, the president's science adviser, stated: "We see the investigation of strategic defense options as an absolutely vital catalyst to real arms control."[9]

We are therefore entering the research phase with a realization that such policy considerations will have to play an important role if this initiative is to succeed. The decision to implement effective defenses will depend not simply on technology, but will be the result of a complex combination of technology and policy. In this sense, the SDI cannot be likened to the Apollo program, the goal of which was well defined and not subject to change. Explaining this complex program and gaining support from the American people will present a significant challenge to its supporters. The key issue may be whether people will be willing for their government to embark on and commit itself to an admittedly ambitious program that cannot promise a guaranteed return on investment. In the final analysis, the decision may hinge on whether there is any alternative to the SDI as a means of assuring the nation's security.

By the time it had completed the reviews, planning, and report preparation, the Defensive Technologies Study Team still did not have a clear statement of its conclusion. The group therefore made a concerted effort to arrive at a consensus statement that could be presented to the president. It is appropriate here to quote the main conclusion that was finally approved by the DTST study:

By taking an optimistic view of newly emerging technologies, we concluded that a robust BMD system can be made to work eventually. The ultimate effectiveness, complexity, and degree of technical risk in this system will depend not only on technology itself, but also on the extent to which the Soviet Union either agrees to mutual defense arrangements and offense limitations, or embarks on new strategic directions in response to our initiative. The outcome of this initiation of an evolutionary shift in our strategic direction will hinge on as yet unresolved policy as well as technical issues.[10]

The paths our two nations will take in the pursuit of self-interest as well as common concerns will be influenced by technological changes, policy formulation, and strategic-arms negotiations. The consensus expressed in the DTST conclusions reflects a belief that even though no definitive prediction of the outcome can be made, the United States must begin a process that might offer a new direction and a new hope. Short-term, narrowly defined program elements were generally deemphasized by the DTST in favor of long-term issues that were poorly understood, but which we felt offered a greater prospect of achieving an effective defense. Our recommenda-

older ages, whether measured in terms of survival rates or average years lived, have been markedly greater in this period, although the relative declines in age-specific death rates have been somewhat smaller. In the 1900–1950 period, mortality indicators were consistent in showing lesser gains at the older ages.

Even though life expectancy at birth has been steadily increasing, the human life span may be fixed at about 100 to 105 years. The curve of survivors, based on annual death rates, has become increasingly rectangular in shape.[8] When overall mortality was relatively high, death rates were much higher at the younger ages than now, and the curve of survivors sloped downward at roughly a 45-degree angle, as in the 1900–1902 curve. As death rates have fallen at the young and middle ages, the survival curve has become increasingly level over most of the age span and has fallen more and more sharply at the higher ages, as in the 1983 curve. At its theoretical limit, the curve would assume a 90-degree angle, with virtually every member of the cohort surviving to age one hundred and then dying within the short time span suggested by the above age range. Fries and Crapo have added the notion that the period of chronic morbidity in later life is also being compressed as life expectancy and life span merge.[9]

We can measure the progress toward this theoretical limit, i.e., the complete "squaring" of the survival curve, as follows: In 1900–1902, when life expectancy was forty-nine years, it fell short of its potential "maximum" of about one hundred years by fifty-one years. By 1983, the number (and percent) of years lost had been cut in half to twenty-six (i.e., 100 minus 74). For persons who reached age sixty-five, the corresponding figures are twenty-three years (i.e., 100 minus 77) for 1900–1902, and eighteen years (i.e., 100 minus 82) for 1983.

Both the rectangularization of the survival curve and the associated hypothesis on the compression of the period of morbidity have been questioned.[10] Complete rectangularization of the survival curve cannot be expected for many decades at best, since it would require much progress in the treatment of chronic illness. In the meantime, the human life span may be slowly rising and, according to Walford and others, there is a reasonable possibility of extending it in the next few decades by fifteen to thirty years.[11] The implications for our society of a life expectancy near one hundred, and a life span of 115 to 130 years, have yet to be thoroughly explored.

make use of our technical talents in innovative ways. We must not make decisions on the basis of preconceived notions or, still worse, prejudice, but must do so on the basis of knowledge of the facts and an understanding of their broad implications. The purpose of the Strategic Defense Initiative is to provide the hard data as well as the in-depth insight to allow truly informed decisions.

ENDNOTES

[1] M.A. Wallop, "Opportunities and Imperatives of Ballistic Missile Defense," *Strategic Review*, Fall 1979, pp. 13–21.
[2] M.A. Wallop, "Letter to the Senate," July 8, 1983, unpublished.
[3] Daniel O. Graham, *The Non-Nuclear Defense of Cities: The High Frontier Space-Based Defense Against ICBM Attack* (Cambridge, Mass.: Abt Books, 1983).
[4] Edward Teller, private letter to the president, July 23, 1983.
[5] U.S. Department of Defense, *Soviet Military Power* 1984, pp. 34 and 36.
[6] Sayre Stevens, "The Soviet BMD Program" in A. Carter and D. Schwartz, eds., *Ballistic Missile Defense* (Washington, D.C.: Brookings Institution, 1984), p. 215.
[7] Two other contributors to this volume were involved in the study: Maj. Gen. John Toomay and Alexander Flax were deputies to James Fletcher (Alexander Flax was also chairman of the Soviet Countermeasures and Tactics Panel), and I was chairman of the Directed Energy Weapons Panel. The other panels were Battle Management and Data Processing; Surveillance, Acquisition, and Tracking; Weapons—Interceptors, Guns, and Other; and Systems Concepts.
[8] Lt. Gen. James Abrahamson, "The President's Strategic Defense Initiative," testimony before the Subcommittee on Defense Appropriations of the Senate Appropriations Committee, May 15, 1984.
[9] George A. Keyworth, "The President's Strategic Defense Initiative," testimony before the Committee on Foreign Relations, U.S. Senate, April 25, 1984.
[10] James C. Fletcher, "The Strategic Defense Initiative", testimony before the Subcommittee on R&D of the Committee on Armed Services, House of Representatives, March 1, 1984.

Donald L. Hafner

Assessing the President's Vision: the Fletcher, Miller, and Hoffman Panels

I T WAS A VERY APPEALING VISION that President Reagan invited the public to share in his March 23 speech—"a vision of the future which offers hope," a world free of nuclear dread, where nuclear weapons will be rendered "impotent and obsolete."

The president chose a surprising way to present his vision. In the closing moments of a televised speech arguing for a larger military budget, Reagan abruptly changed topics and announced what was no less than a revolution in American strategic doctrine. At the heart of his vision was a repudiation of deterrence—the bedrock of American strategic nuclear policy for three decades. "I've become more and more deeply convinced," the president remarked,

> that the human spirit must be capable of rising above dealing with other nations and human beings by threatening their existence.... If the Soviet Union will join with us in our effort to achieve major arms reductions, we will have succeeded in stabilizing the nuclear balance. Nevertheless, it will still be necessary to rely on the spector of retaliation, on mutual threat. And that's a sad commentary on the human condition. Wouldn't it be better to save lives than to avenge them?

This was not just presidential whimsy. Three years before, when Reagan was stumping in the presidential primaries, he had spoken with the same profound concern and disappointment that our security rested on mutual threats, and with the same diffuse confidence that American technological strength offered an escape.[1] Nor did it seem to be just political posturing, for domestic or foreign consumption. Reagan has shown rare animation and commitment whenever he discusses his proposal—the marks of a true believer.

It was perhaps this enthusiasm that prompted the president to rush ahead and present his vision to the nation before its implications had been fully considered. He had drafted the closing paragraphs of the speech himself, only five days before it was televised, and they were seen by only a handful of advisers.[2] It was not until ten weeks after the speech that he appointed three panels of experts to assess the feasibility of achieving the goal he had already set out. By October, the three panels had submitted their reports. By January 1984, the administration was fashioning a concrete defense program. In March, Secretary of Defense Weinberger released declassified summaries of two of these panel studies, along with his own overall report. He concluded that "a robust BMD system can be made to work eventually," and sketched out a five-year research program dubbed the Strategic Defense Initiative (SDI).[3]

None of these reports is the stuff of popular reading. The prose is strictly utilitarian, the jargon impenetrable; the general reader would need a glossary to make sense of the glossary. Even a casual reader would notice right away, however, that much of the prose in Weinberger's report was copied directly from the panel summaries, fostering an impression of broad agreement between the president's vision, the advisers' evaluations, and the Defense Department's SDI program.

That impression would be wrong, for Reagan's advisory panels in fact offered contradictory advice. The points of contradiction are vital, for they go to two assumptions at the heart of the president's vision: that an escape from deterrence through retaliation is possible; and that the path of escape is technological, not political. In fundamental ways, the panels' reports are less an endorsement of President Reagan's vision than a reflection of a basic skepticism about its core assumptions.

THE FLETCHER REPORT

Devising a strategic defense so nearly leakproof that it makes *all* nuclear weapons "impotent and obsolete" means confronting not one challenge, but several—specifically, land- and submarine-launched ballistic missiles, bombers, and cruise missiles. And if parts of our defenses are based in outer space, defending them against anti-satellite (ASAT) weapons poses an additional challenge.

President Reagan did not address all these challenges in his March 23 speech. The only weapons he mentioned specifically were strategic ballistic missiles; the only program he proposed was research and development on ballistic missile defenses (BMD). Days later, Secretary of Defense Weinberger asserted that the president's vision encompassed all nuclear weapons. Yet in June 1983, when the three study teams were appointed to assess the technological prospects and security implications of strategic defenses, their mandate focused only on ballistic missile defenses. When the SDI was announced in March 1984, it too addressed only ballistic missile defenses.[4]

Even working within this narrowed focus, the advisory panels came back with strikingly different evaluations of the path that ought to be followed. The Defensive Technologies Study Team (DTST; known as the Fletcher panel, after its chairman, James C. Fletcher) opened the body of its summary report with a rhetorical question: "What has happened to justify another evaluation of ballistic missile defense as a basis for a major change in strategy?" Perhaps the bluntest answer would have been that the president had already announced his commitment to comprehensive defense, and the panel had been told that, whatever its findings, it was not to embarrass the president. The answer provided in the panel report instead said that new technologies now offer the prospect of reaching out and intercepting ballistic missiles almost from the moment they are launched. To give at least some shape to the president's general goal of comprehensive missile defense as a focus for research, the Fletcher panel sketched out a hypothetical, multi-tiered defense system based on these prospective technologies, with each tier engaging missile warheads at a different phase of flight.

What security burden could such a system bear? Could it really liberate us from deterrence, as President Reagan hopes? On this, the report was politely evasive. "Meaningful levels of defense" might be feasible in the 1990s, it said, if we deployed terminal and mid-course layers of defense. Constructing an "effective defense," however, would be "strongly dependent" upon engaging thousands of Soviet missiles during their boost phase, before they could release thousands of warheads and perhaps hundreds of thousands of decoys. Boost-phase intercept requires stationing part of all of our missile defenses in outer space, where they can get a clear shot at Soviet missiles in the first minutes of flight. For this task to be within reach, the panel report continued, a list of "critical technologies" would have to be tackled, "technologies whose feasibility would determine whether an effective defense is indeed possible."

Three things are notable about this list. One is that deficiencies are to be found in every defensive tier. Even in mid-course and terminal BMD technologies, on which the United States has already lavished decades of effort and billions of dollars, current performance was judged by the panel as inadequate for a comprehensive missile defense. A second notable point is the panel's estimate that R&D programs lasting ten to twenty years might be necessary before we are able to solve critical technological problems and begin deployments. Such time scales are very different from our past experience. In 1941, two years after crucial scientific work by Szilard and Fermi suggested an atomic bomb might be feasible, the Manhattan Project began its task with promise from the project advocates that a workable nuclear weapon could be ready within two years. It took four. In May 1961, President Kennedy committed the United States to landing men on the moon within nine years. It took eight. Ten-year system acquisition cycles for advanced weapons are not uncommon now, and the Defense Department has experience with them. A ten- to twenty-year development cycle for exotic SDI technologies, however, would be quite a novel case, and the Fletcher panel openly acknowledged the great uncertainties involved.[5]

Most notable was the panel's silence on how effective it expected this "effective defense" to be. In touting the virtues of a multi-tiered defense, the report noted that if each layer in a three-tiered system allowed even 10 percent of its targets to leak through, the overall leakage rate for the whole defense system would be only 0.1 percent. But read carefully in context, this 99.9 percent effectiveness figure proves to be only an illustration, not a prediction. It has been reported that the panel's classified report rejects its own illustration, and asserts instead that "it is not technically credible to provide a ballistic missile defense that is 99.9 percent leakproof."[6] In practice, whether each layer could handle the leakage from the previous tier would depend very much on the devices deployed and on how self-sufficient the layers were. If several layers relied on common components (for instance, common sensor satellites, a central com-

Assessing the President's Vision

puter, or identical software programs), then they would not be genuinely independent tiers. A catastrophic failure of one tier or component might so overwhelm the next tiers that they too would fail just as catastrophically. The extent to which multiple tiers could raise the nation's confidence in defense performance, then, would depend upon technology and design choices that the Fletcher panel said could not be made for years or even decades.

While the Fletcher panel's summary report does not dwell on such problems, neither does it ignore or deny them. In its own muted fashion, the report is a rather candid document from experts who made every effort not to embarrass the president, and yet who knew from their own technical experience that they could not promise that his ultimate goal was within reach. The summary report hints of the conflicting pressures that existed on and within the panel. On the one hand, the tone of the closing paragraphs in the declassified version is upbeat, reportedly more upbeat than the panel's twelve supporting volumes of classified technical studies might warrant: "The members of the Defensive Technologies Study Team finished their work with a sense of optimism. The technological challenges of a strategic defense initiative are great but not insurmountable." On the other hand, when he presented the report to the Senate, panel chairman James Fletcher stated not that the panel *reached* an optimistic conclusion, but rather that "by *taking* an optimistic view ... we concluded that a robust BMD system can be made to work eventually"[7] (emphasis added). Moreover, the report speaks with cold candor of technological challenges that would give any reader pause. For instance:

Developing [computer] hardware will not be as difficult as developing appropriate software. Very large (order of 10 million lines of code) software that operates reliably, safely, and predictably will have to be deployed. *It must be maintenance-free for 10 years* ....[8] (emphasis added)

Even some of the panel's optimism is startling:

Analyses ... suggest that a properly constructed shield [to protect space-based components] could provide effective armor against small kinetic energy weapons and most directed-energy threats. ... Such a shield could reach 100 or more metric tons ... For the quantities of material required, two other sources are feasible. Material from the lunar surface or from nearby asteroids can be brought to the vicinity of the Earth ....[9]

Throughout, the report reiterates in a soft but firm voice that the technical feasibility of even one aspect of the president's vision—defense against ballistic missiles—is far from certain.

THE MILLER REPORT

The Fletcher panel was only one of three groups advising the president. The other two panels examined the policy import of new defense technologies for U.S. strategic objectives, and the effect this defense might have on our relationships with allies and the Soviets. Unfortunately, the report of one of these policy groups, an interagency group known as the Miller panel (after its chairman, Franklin C. Miller, director of Strategic Forces Policy in the Pentagon), has not been publicly released. Potentially, it could tell us the most about the future of the SDI, for even when done after a policy is announced, more as an analytic prop than a foundation for the policy, such interagency studies are revealing in the way they crystalize the distribution of opinion and influence within the bureaucracy. They have a subtle influence on public policy by indicating to all agencies where the meridian of opinion lies and by establishing a vocabulary of discourse. Time and again, as decisions must be made, the bureaucracy uses such studies as its compass and plagiarizes their phrases. The Fletcher and Hoffman panels, after all, were disbanded at the end of the summer, 1983; the members of the Miller panel returned to their desks within the bureaucracy, to begin implementing the president's strategic defense program.

THE HOFFMAN REPORT

The other policy group, the Future Security Strategy Study team (known as the Hoffman panel, after its chairman, Fred S. Hoffman), did issue a declassified summary of its study.

There are striking contrasts between the Hoffman and Fletcher reports. The first is simply the discrepancy in length. The Hoffman panel's unclassified summary is barely twelve pages long, almost identical, we are told, to the full classified version. If this indeed represents all that the panel felt need be said about the policy impact of President Reagan's proposal, it is a remarkably terse statement.

What is more striking about the Hoffman panel report is that after an exceedingly brief, even perfunctory, endorsement of Reagan's long-term goal of a nearly perfect defense, the report turns to quite a different concern: buttressing retaliatory deterrence by deploying "partial [missile defense] systems, or systems with more modest technical goals" as soon as this is feasible. Such an approach, the panel members argued, would offer "a hedge against the possibility that nearly leakproof defenses may take a very long time, or may prove to be unattainable in a practical sense against a Soviet effort to counter the defense."[10]

What the Hoffman panel proposed were progressive stages of defense deployments, starting most immediately with anti-tactical ballistic missile (ATBM) defenses for Europe, based upon available technologies for terminal defense. The next stage of defense buildup would come as soon as more advanced terminal and mid-course technologies were available; these defenses could be deployed to protect critical military targets in the United States. The third stage would be reached with the availability of exotic boost-phase defenses; these would be deployed to exert "leverage" on the Soviets, "even if they prove unable to meet fully sophisticated Soviet responses," that is, even if they fell short of the president's goal.

It is difficult to interpret the Hoffman report as anything but a skeptical dismissal of President Reagan's proposed comprehensive defense for the American people. The panel had little or nothing to say about what strategy and security might look like in a world of nearly perfect defenses. Their silence on this point was perhaps unavoidable, given their views of the Soviet Union's strategic goals and tactics. They described the Soviets' primary strategic objective as "domination of the Eurasian periphery"; the "preferred mode in exploiting their military power is to apply it to deter, influence, coerce—in short, to control—other states." Following this view to its logical extension, the panel concluded that the Soviets would make vigorous efforts to defeat any U.S. defenses.

But what of the new BMD technologies touted by the Fletcher study? The Hoffman panel could envision a high level of *technical* performance for a multi-tiered defense. After all, it noted, even if each layer intercepted no more than half its targets, a four-tiered defense could still be 94 percent effective overall. The problem was, "such a leakage rate is ... sufficient to create catastrophic damage in an attack of, say, 5000 reentry vehicles (RVs) aimed at cities. It would mean 300 RVs arriving at targets—sufficient to destroy a very large part of our urban structure and population." (The Fletcher panel, of course, had assumed the future Soviet threat might reach 30,000 warheads, which could mean 300 RVs arriving at targets even with 99 percent defense effectiveness.) Clearly, one could be optimistic about exotic BMD technologies, yet pessimistic about a strategic revolution. As if anticipating that the United States would therefore never reach the ultimate goal of nearly perfect defense, the Hoffman panel's report confined itself to being principally a brief for stabilizing retaliatory deterrence—rather different in scope from the president's call for a security posture that "did not rest upon the threat of instant U.S. retaliation."

The Hoffman panel's proposals for partial defenses, deployed sooner rather than later, are considerably less ambitious than those of the Fletcher group, still less ambitious than the SDI. And yet at the same time, the panel served to broaden the rationale for the SDI by widening the issues beyond simply "Star Wars" technologies, beyond simply "nearly perfect" defenses. Many of the defense tasks proposed in its report do not depend on exotic space weapons, and its arguments for partial defenses are familiar from the ABM debates of the 1960s and 1970s. The debate has been resurrected, however, in a rather different context. Voices of alarm over the future stability of nuclear deterrence have been heard in the United States for more than a decade now, from both the Right and Left; the possibilities of exotic space-based technologies for defense have sparked public imagination; and President Reagan has recast public expectations with his own ambitious vision. In this current context, partial defenses of the sort proposed by the Hoffman panel may well find new life as an achievable compromise, technically and politically, between deploying no defenses and embracing fully Reagan's ambitious and uncertain vision.

DETERRENCE AND THE PRESIDENT'S VISION

Defenses designed to buttress deterrence by protecting military command posts and retaliatory forces certainly sounds like a retreat from the president's goal of living "secure in the knowledge that . . . security did not rest upon the threat of instant U.S. retaliation to deter

Assessing the President's Vision 99

a Soviet attack." But are they? The question really cannot be answered without a careful review of what security and strategy would look like in a world of perfect defenses—a review the Hoffman panel was supposed to present but did not.

In principle, deterrence and defense are opposite. Deterrence works to dissuade the Soviets from attacking us or our allies by threatening punishment in retaliation; it works on the Soviets' intention and will. Defense, on the other hand, works by thwarting a Soviet attack when it occurs; it focuses on the Soviets' ability to do us harm, rather than on their intentions.

If either deterrence or defense worked perfectly, would we need the other? A perfect defense would seem to make deterrence irrelevant, make us indifferent to the plans and actions of the Soviets; it would simply shield us from attack. Of course, if the Soviets were convinced they could not penetrate our defenses, they would presumably change their minds and not bother to attack. In this sense, a perfect defense would also "deter." Nonetheless, perfect defense is *not* just another variety of deterrence. President Reagan formulated it correctly: defense saves by protecting, deterrence saves by threatening vengeance. In a world of perfect defenses, then, the United States would be liberating itself, as the president said, from the specter of retaliation, from mutual threat.

Or would it?

Even if a defense worked perfectly and stopped all incoming warheads, invoking the defense against an attack would use up valuable equipment and would leave us vulnerable to other attackers until our defenses were replenished. So we would still have an interest in dissuading others—by threat of retaliation—from taking "free shots" at us or from coercing us while our defenses were being restored. In addition, there might be occasions when we would wish to project force, rather than simply to defend ourselves. Offensive nuclear weapons are rather blunt instruments for diplomacy or military action, but they might prove necessary in protecting an ally from nuclear coercion in a distant place to which our own defenses did not extend perfectly.

Even in a world of perfect defenses, it appears, we could not give up offensive nuclear forces or retaliatory deterrence. And the further our defenses fell from perfection, the more our security would depend on offensive forces, either for deterrence or for damage limitation through a preemptive attack on Soviet forces.

Yet another reason to keep offensive forces stems from the crucial role of boost-phase interception in any comprehensive defense system. Boost-phase intercept systems can work only if the defense weapons and their sensors are based in space or popped up into space at the moment of attack. Any space-based missile defense technology, however, will be very capable of destroying other objects in space—including other space-based defenses—long before it is ready for use as an effective BMD system. The side willing to shoot first with its BMD-turned-ASAT, and to follow up with an offensive strike, would have a great advantage—unless the other side maintained formidable retaliatory forces.

One way the United States could try to liberate itself from dependence on offensive nuclear forces would be to maintain a constant technological lead over the Soviet Union in offensive missile and BMD-turned-ASAT technologies. A diffuse confidence in the perpetual superiority of American technology is an important part of President Reagan's defense vision. Yet from a practical point of view, nothing so complex as a multi-tiered BMD network could be continually modernized. The initial investment in equipment would be too high to be scrapped wholesale with each real or feared breakthrough in Soviet technology. Modernization would have to proceed in cycles: the latest U.S. defense technology would be deployed; as it aged and Soviet technology advanced, incremental modifications would be made in our defenses to sustain a desired effectiveness level; when modifications could no longer keep pace, a major new U.S. system would have to be deployed. During these modernization cycles, the technological advantage would seesaw back and forth between the United States and the Soviet Union. The United States would have to fall back on threats of retaliation during those periods when confidence in our defenses was undercut.

In short, deterrence based on retaliatory weapons is a difficult condition to escape in a nuclear world. Even if the United States and the Soviet Union had comparable and near-perfect defenses and no other security interests beyond protecting themselves from each other, they would still have an incentive to keep offensive nuclear forces. If our defenses were less-than-perfect, giving up retaliatory forces would be foolhardy.[11]

ARMS CONTROL AND THE SDI

What remains, then, of the president's vision?

A world in which deterrence still mattered might nevertheless become "defense-dominant," that is, a world in which (within limits) it was easier to be a defender than an attacker. We might arrive at this point by two paths. One would be through technological innovations that would make defenses cheaper than *both* offensive forces arrayed against them *and* weapons designed to attack the defense directly. Making comprehensive defenses cheaper than anti-defense weapons will be difficult, however, because of the paradox noted above: every development in boost-phase defense technology can be used by the opponent as an excellent *anti*-defense weapon.

The other path would be through arms control—mutual formal or informal agreement between the superpowers to limit offensive forces and anti-defense weapons, but not defenses. This brings us back to the Fletcher and Hoffman reports. Following its instructions, the Fletcher panel narrowed its focus to technical issues and returned with a vision of technological innovations that might give the defense significant leverage over the offense, "leading to the final, low-leakage system." And yet, recognizing that what our own technology gives us, Soviet technology can take away, the panel concluded that maintaining the defense's dominance over the offense could not be accomplished by technology alone. It would require arms control.

> The ultimate effectiveness, complexity, and degree of technical risk in this system will depend not only on the technology itself, but also on the extent to which the Soviet Union either agrees to mutual defense arrangements and offensive limitations, or embarks on new strategic directions in response to our initiative.[12]

Advice like this had been heard before. One month after the first atomic bomb test in 1945, scientists with the Manhattan Project (Oppenheimer, Fermi, Compton, and Lawrence) offered a similar bit of counsel to Secretary of War Henry Stimson:

> We believe that the safety of this nation—as opposed to its ability to inflict damage on an enemy power—cannot lie wholly or even primarily in its scientific and technical prowess. It can be based only on making future wars impossible. It is our unanimous . . . recommendation to you that . . . all steps be taken, all necessary international arrangements be made, to this one end.

The Fletcher panel did not probe Soviet motives, and so did not explain why the Soviets might be willing to resurrect arms control years after limits defined in the ABM treaty had been abolished to make way for missile defenses.

The Hoffman panel, on the other hand, squarely addressed Soviet motives but came away with little optimism for arms control, unless a wholesale transformation of Soviet character occurred:

> Current Soviet policy on arms agreements is dominated by the Soviet Union's attempt to derive unilateral advantage from arms negotiations and agreements. . . . There is no evidence that Soviet emphasis on competitive advantage over mutual benefit will change in the near future, unless a fundamental change occurs in the Soviet Union's underlying foreign policy objectives.
>
> If the new defense technologies offer sufficient leverage against the offense . . . the Soviets may accept a reduction in their long-range offensive threat against the West, which might be reflected in arms agreements. . . . Their current program emphases suggest that they would be more likely to respond with a continuing buildup in their long-range offensive forces.

In this panel's view, if anything could induce the Soviets to negotiate seriously (presumably, even after the demise of the ABM treaty), it would be a combination of Western toughness and technological dominance that threatened to thwart Soviet objectives. "In that event, it might also be possible to reach agreements restricting offensive forces so as to permit defensive systems to diminish the nuclear threat."

Where does this leave us? Members of the Fletcher panel argued that the United States could not assert and preserve defense dominance without arms control. The Hoffman panel doubted Soviet openness to arms control, unless the U.S. could assert technological dominance. Yet the defense secretary's report, released at the same time and ostensibly summarizing the advisers' findings, asserted that "defense against ballistic missiles offers . . . new opportunities and scope for arms control." This, of course, turned the Fletcher panel's conclusion on its head and transformed the Hoffman panel's pessimism into optimism, without providing any rebuttal of the panels' judgments.

The discrepancy in opinions between the Fletcher and Hoffman panels regarding arms control is not surprising, since members of the

Defense Department supervising the panels were reluctant to have them come in contact or coordinate their findings. Having two panels work separately on the same problem is an excellent way to get independent assessments. But here the panels were being asked to work in isolation on parts of a puzzle that were supposed to fit together. Precisely how the "policy" panels were to assess the political implications of strategic defenses, ignorant of the technologies and performance levels being recommended by the Fletcher panel—or how the DTST was to sketch out technology paths and requirements, ignorant of "policy" constraints—is not clear. In the end, staff members in Under Secretary of Defense Fred Ikle's office (which supervised the Hoffman panel and, through Franklin Miller, the Miller panel) objected to the Fletcher panel's remarks on arms control, on the grounds that this was a "policy" matter and not a technology issue. The deadline for submitting the reports arrived before the disagreement between the staffs of Ikle and DeLauer, and the discrepancy between the reports, were resolved.

CONCLUSION

The conflicting advice of the Hoffman and Fletcher panels, and their scaled-back visions of what the Strategic Defense Initiative might achieve, cut to the very heart of President Reagan's vision. "Diminishing the nuclear threat"—in the Hoffman panel's words—is a rather different vision from rendering nuclear weapons "impotent and obsolete." Even granting that the very purpose of the SDI is to narrow the technical uncertainties about what can and cannot be achieved, there seems little doubt that both the SDI and the public debate over strategic defenses would have greater clarity if the fundamental issues raised by the Fletcher and Hoffman reports had been confronted, debated, and then integrated into our strategic defense policy. Surely, it matters to the public whether the risks and sacrifices entailed in pursuit of the president's vision are likely to yield less-than-perfect rather than nearly perfect defenses, defenses of retaliatory forces rather than cities, a world free of mutual nuclear threats rather than one of buttressed deterrence. Yet long after the panels finished their work, these issues were still not placed in sharp focus. For instance, a White House pamphlet issued under President Reagan's signature in January 1985 seemed to speak squarely on the matter of whether we are pursuing a true defense—living "free in the knowledge that ... security did not rest upon the threat of instant U.S. retaliation"—or merely buttressed deterrence:

> U.S. policy has always been one of deterring aggression and will remain so even if a decision is made in the future to deploy defensive systems. The purpose of the SDI is to strengthen deterrence.

But the pamphlet also reiterated the president's own phrase, "What if a free people could live secure in the knowledge ..." as a banner across page one, and promptly blurred the meaning of "deter" on the same page by asserting "a very real possibility that future Presidents will be able to deter war by means other than threatening devastation to any aggressor—and by a means which threatens no one."[13]

In a sense, the failure of the panels to bring sharper focus to the debate rests with the nature of the panels themselves. From the beginning, they were contending with a president who had already committed himself publicly to a goal of nearly perfect defenses. Although members of the panels clearly believed that more modest expectations were warranted, they had no reason on that point alone to confront Reagan and force him from his commitment. The panel members were in fact quite content with more modest goals than "perfect defense," and thought that even modest goals justified a vigorous BMD research program. Moreover, either by design or good fortune, the defense system and research program sketched out by the Fletcher panel also seemed to discourage debate and hard decisions rather than to provoke them. Almost everyone's favorite BMD technology or strategic purpose could find a place within the multi-tiered defense system.

If any of the panels were to provoke debate or dissuade the president from his commitment, logically it would be the Hoffman panel that would do it. Certainly the Hoffman group had a more coherent argument for its less-ambitious recommendations than did the Fletcher panel. But an argument is not an inducement, and in the way of inducements to confront choices or to change policies the Hoffman panel offered Reagan little or nothing. The Hoffman panel's strident image of unrelenting Soviet competition was not necessary for a president already persuaded that the Soviets should be dealt with firmly; neither was it welcome to one also in a mood to offer Americans hope of a brighter future. And telling the president the journey to a brighter future should begin with a request to the Europeans that they accept American ATBM missiles, was certainly no inducement.

From the president's standpoint, the hardest choice the Hoffman panel posed to him was whether to set aside the exotic technologies and his long-term vision of population defense—since escaping deterrence was doubtful—in favor of more prosaic technologies and near-term defense of retaliatory forces. But President Reagan was not about to make such a choice. The prevailing view in the White House was that the United States had been down that path already in the 1960s, and had stumbled on two problems: inadequate BMD technology, and domestic opposition to having BMD sites in the nation's neighborhoods. While new technologies might perhaps overcome previous deficiencies, the recent political obstacles encountered in finding a home for the MX missile showed that siting problems were, if anything, greater in the 1980s than in the 1960s. Moreover, since the U.S. could defend the bulk of its retaliatory forces only by scrapping or renegotiating current ABM treaty limits, and thus liberating the Soviets as well from ABM restraints, this made little sense if the only defenses at hand were prosaic, ground-based technologies that the Soviets, unfettered by domestic opposition, might exploit more readily than the United States. Space-based defenses could circumvent such problems, play to the American strong suit in exotic technology, and avert unpalatable choices between defense of retaliatory forces and defense of population. So long as the Hoffman panel was willing to say polite things on behalf of space-based technologies and the president's long-term vision, Reagan was under no compulsion to make hard choices. His vision could remain intact.

ENDNOTES

[1] In interviews with a *Los Angeles Times* reporter in the spring of 1980, candidate Reagan remarked: "I think the thing that struck me here was the irony that here, with this great technology of ours, we can do all of this yet we cannot stop any of the weapons that are coming at us. I don't think there has been a time in history when there wasn't a defense against some kind of thrust, even back in the old-fashioned days when we had coast artillery that would stop invading ships if they came. . . . I do think that it is time to turn the expertise that we have in that field—I'm not one—but to turn it loose on what do we need in the line of defense

against their weaponry and defend our population, because we can't be sitting here." Quoted in Robert Scheer, *With Enough Shovels* (New York: Vintage Books, 1983), pp. 233–234.

[2] Robert Scheer, "Teller's Obsession Became Reality in 'Star Wars' Plan," *The Los Angeles Times,* July 10, 1983, section VI, pp. 6–9.

[3] Texts of the unclassified summary reports from the Defensive Technologies and Future Security Strategy study teams and from Secretary Weinberger can be found in "Strategic Defense and Anti-Satellite Weapons," U.S. Senate Hearings before the Committee on Foreign Relations, 98th Cong., 2nd sess., April 25, 1984, pp. 94–175. Page references to the reports below are to the publicly released versions, with corresponding page numbers in the Senate Hearings noted in brackets [ ].

The Defensive Technologies panel had some sixty members. Formally, it reported to Richard DeLauer, under secretary of defense for research and engineering; informally, the panel chairman, James Fletcher, discussed the group's progress directly with Secretary of Defense Weinberger. The Future Security panel had twelve members, plus a senior policy review group of an additional nine members, and it submitted its findings to Franklin C. Miller, director of strategic forces policy in the office of Fred Ikle, under secretary of defense for policy. Both panels were drawn from the national weapons-research laboratories, defense-related private industries and research firms, universities, and the military services. The panels' members and their affiliations are listed in the reports.

Although an unclassified version of the Miller panel's report has not been released, brief testimony by Franklin C. Miller on the panel's work may be found in the Defense Department Authorization for Appropriations FY1985, U.S. House of Representatives, Hearings before the Committee on Armed Services, 98th Cong., 2nd sess., part 6: "The Strategic Defense Initiative," March 8, 1984, pp. 2; 948–59.

[4] Studies of defenses against bombers and cruise missiles were also begun at this time within the Defense Department. But they were not part of, nor integrated into the conclusions of the Fletcher or Hoffman panels, nor are they part of the Strategic Defense Initiative.

[5] The SDI may also prove novel in its final cost. The Manhattan Project cost $2 billion between 1941–1945 (about $13 billion in 1984 dollars), roughly 0.2 percent of total GNP of those four years. The Apollo project cost $25 billion between 1961–69 (about $57 billion in 1984 dollars), roughly 0.4 percent of total GNP of those years.

The nation's GNP in 1984 was $3664 billion. Assuming that a real growth rate (after inflation) of 3 percent is achieved for the next 20 years, total cumulative GNP over those years would be roughly $98,500 billion. So, to equal the same portion of total GNP as the Manhattan Project, a comprehensive defense system would have to cost no more than $200 billion; to equal Apollo, no more than $400 billion. In contrast, a Defense Department study submitted to Congress in 1982 estimated that the cost of a space-based laser BMD system capable only of "damage denial" (that is, less-than-perfect defense of the population) would be $500 billion. See Senate Foreign Relations Committee hearings, "Strategic Defense and Anti-Satellite Weapons," op. cit., p. 67.

[6] The Fletcher report, p. 10 [p. 151], and "U.S. Strategic Defense Options: Study Urges Exploiting of Technologies," *Aviation Week & Space Technology,* Oct. 24, 1983.

Assessing the President's Vision   107

[7] See testimony by James C. Fletcher during the Hearings before the Committee on Armed Services, op. cit. (footnote 3), p. 2919. Fletcher was quoting from the classified version of the panel's summary report, and this passage was a summary judgment very carefully crafted so that all members of the panel could endorse it.

[8] The Fletcher report, p. 19 [p. 160].

[9] Quoted in *Aviation Week & Space Technology,* Oct. 17, 1983, p. 19, from the classified version of the Fletcher report.

[10] The unclassified version of Hoffman report may be found in the Senate Foreign Relations Committee Hearings, "Strategic Defense and Anti-Satellite Weapons," op. cit., pp. 125–140.

[11] Within days after the declassified versions of the Fletcher and Hoffman reports were released, Defense Department officials confirmed that they did not foresee a future in which the U.S. would be able to give up its offensive nuclear forces, even when defenses were deployed. According to Under Secretary of Defense Fred Ikle, the only way to escape our need for offensive nuclear weapons would be through a combination of defenses plus arms-control agreements. See the testimony of Ikle, Under Secretary of Defense Richard DeLauer, and director of Defense Advanced Research Projects Agency (DARPA), Robert Cooper, Hearings before the Senate Armed Services Committee, cited in footnote 7, p. 2924.

[12] James C. Fletcher, testimony, cited in footnote 7 above.

[13] "The President's Strategic Defense Initiative," pamphlet released by the White House, Jan. 1985, pp. 3 and 1.

Charles A. Zraket

Strategic Defense: A Systems Perspective

IN DISCUSSIONS ABOUT PRESIDENT REAGAN'S Strategic Defense Initiative, the popular tendency is to speak as if inventing new technical components, such as directed energy weapons, is the same thing as creating a ballistic missile defense system, and, furthermore, as if devising a missile defense system is the same thing as providing strategic security to the nation and our allies. The task is more complicated than that. The new defense *weapons* that have been receiving so much attention are to be but one part of a complex strategic *system;* planning wisely for strategic defense, then, requires taking a systems perspective.

Let us use a simple analogy: we can compare a systems perspective with the approach taken by a good architect, who begins by assessing the purpose a new building is to serve and then asks, "What kinds of materials and components, assembled in what sequence and manner, would best serve this purpose, given the building's environment?" In a similar way, the first step in planning a ballistic missile defense should be to assess our strategic purposes, including the role we expect nuclear deterrence to play in our security.

In assessing where and how missile defenses would fit in this overall posture, we should begin with the premise that a perfect defense against all nuclear threats is feasible neither technically nor economically, at least not in the foreseeable future. On this, there now seems to be general agreement throughout the defense community.[1] Consequently, since our security will continue to rest upon a combination of offensive and defensive forces, before we are in a position to decide which kinds of ballistic missile defenses are desirable, we need to consider a variety of combinations and evaluate their impact on nuclear deterrence, the stability of international relationships, and our security priorities. Performance requirements must be defined broadly; obviously, the system must be able to intercept missiles, but it must also effectively perform such functions as command-and-control, self-defense, and operational maintenance and testing.

Once we have a system design in mind, we must investigate the best way of deploying such a defensive system, remembering that no complex system can be deployed quickly. We must therefore devise a pace and sequence of construction that does not leave us vulnerable during deployment. This brings us back to the problem of determining the proper mix of defensive and offensive forces, given limited resources.

The multilayered BMD system described by the administration's Defensive Technologies Study Team (the Fletcher panel) offers an example of the complexities we face. The figure on page 59 shows in schematic form the various layers of the system: the boost and post-boost phase where multiple warheads are released by the attacking ICBMs or SLBMs, the mid-course phase where large numbers of decoys may travel with the warheads, and the terminal phase where the warheads are a minute or less from their targets. The hypothetical BMD system that opposes the attack could contain many tens or even hundreds of terminal phase elements, tens to hundreds of warning and tracking satellites, and several hundreds to thousands of weapons-carrying satellites. The BMD system must have a self-defense capability, using both its own sensors and weapons and separate defense satellites for special threats. Ground-based elements of the system could include both the terminal defense elements shown and surveillance probes and weapons units that could be launched into space after warning of attack to make intercepts during the late mid-course phase.

Figure 1, opposite, shows the relationships among the various elements of such a BMD system and their connections to defense higher authorities, allied forces, and offensive forces. The shadow indentations in the diagram illustrate either backup elements or multiple space-based and ground-based elements. Each layer would

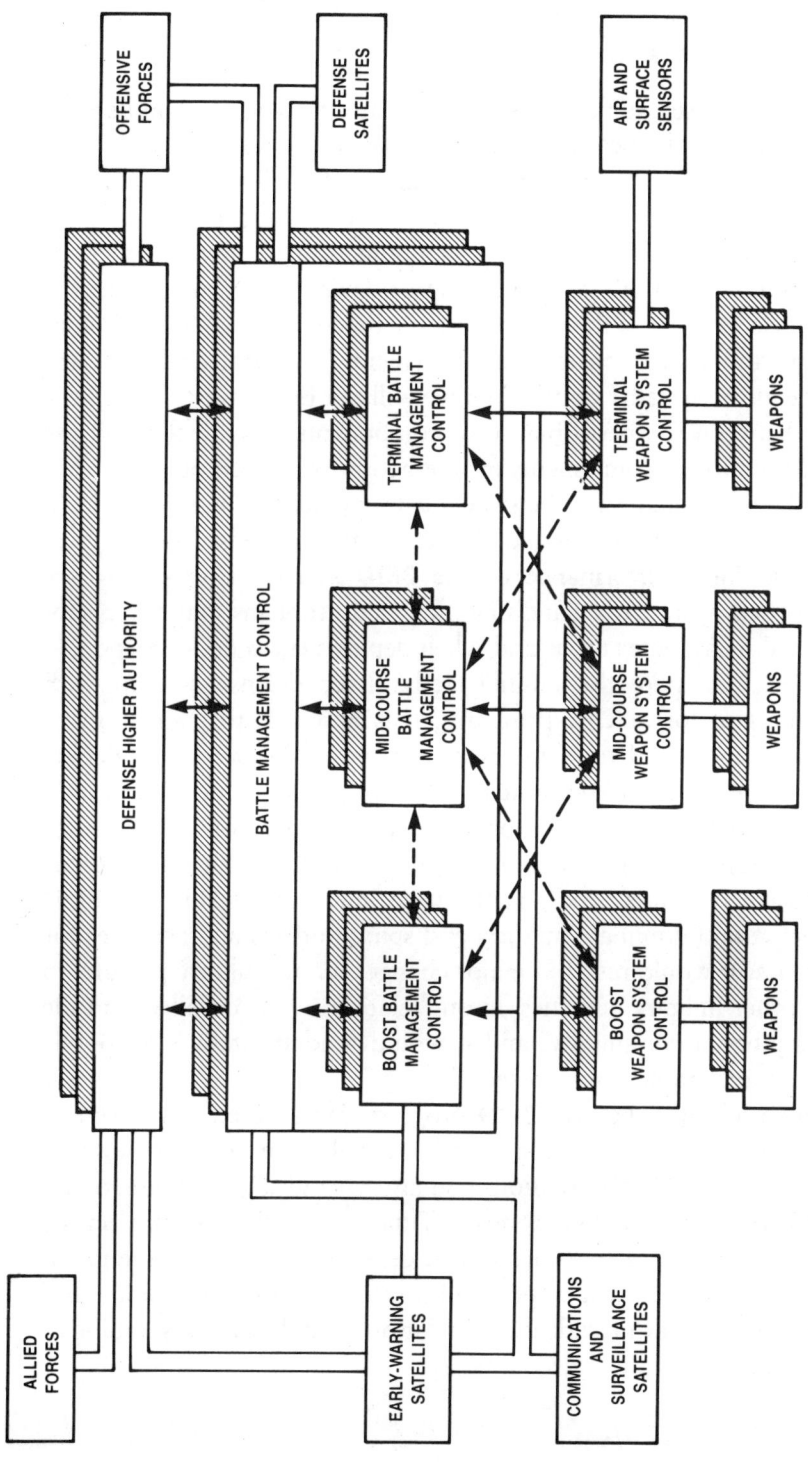

Fig. 1 Command-and-control of Multilayered BMD System

ideally be as self-sufficient as possible, and the layers would be integrated by multiple communications networks.

At the moment, the Strategic Defense Initiative is only at the research stage. The next hurdle will come at the component stage. Since the actual performance of components can differ by orders of magnitude from the ground-testing of devices, the challenge facing scientists is to close the inevitable gaps between design and development, and development and deployment. The final challenge will come at the systems stage, when we will not only have to bridge the gap between the performance of components individually and the performance of components in interaction with each other, but we will have to make sure the system can operate in a hostile environment of countermeasures and physical attack. These issues are critical for the final performance of the BMD system. Unless they are addressed at an early date, unanticipated problems may force substantial delays later on, if and when deployment is decided upon.

If we compare the multilayered missile defense systems with previous major defense programs, it is apparent that an unprecedented amount of time and resources would be needed to deploy a comprehensive BMD system. The effort would be significantly greater, for example, than our programs in the 1950s and 1960s to develop and deploy comprehensive air defenses, or our strategic offensive systems of the 1950s, 1960s, and 1970s. This new BMD effort would dominate military and space budgets and activities for decades. It would result in major changes in our military posture, in our other military activities, in our relations with our allies and the international community, and it would fundamentally alter public perceptions about national security.

Such a comprehensive BMD program can succeed in the United States only with broad public support. Public support, however, is contingent on finding solutions that are generally agreed upon, or are based on workable compromises. Timely research programs on the system capability aspects of strategic defense are therefore urgent, not only because they will be needed for informed decision-making, but also because they provide the foundation for public understanding of and support for these decisions.

THE STRATEGIC PURPOSES OF DEFENSE

We have noted that a systems perspective should begin with a definition of the purposes that strategic forces must serve. One difficulty in defining these purposes stems from the current controversy over strategic doctrine and the proper role of strategic defenses—particularly BMD—within that doctrine. For example, the report of the president's policy advisers (the Hoffman report) envisioned a mix of strategic defensive and offensive forces and emphasized the role of BMD in strengthening nuclear deterrence. The implication of the report is that nuclear deterrence based on the threat of retaliation or countervailing action will remain an essential element of our strategic doctrine, regardless of the offensive/defensive mix in the force structure.

A lively point of debate today is the extent to which deterrence depends on the capacity of our strategic forces and command system to survive a massive nuclear attack. There seems to be general agreement that improving the survivability of our offensive system is both desirable and feasible. But there is great debate over the desired character and size of our offensive forces, the extent to which protracted endurance is needed and feasible, and the variety of strike options to be kept open. This debate is complicated by disagreements over the contribution that arms control might make in bolstering deterrence.

The proposed deployment of a substantial BMD capability raises these issues again, but in the context of a much more complex strategic posture involving a mix of offensive and defensive systems. A BMD system could help offensive forces survive. However, the BMD system itself must be able to survive and work at least as long as the things it is designed to protect. If the Soviets increase and improve their offensive forces and deploy their own BMD and ASAT systems—and that is highly probable—this will further complicate the situation.

Uncertainty has its advantages, however: it would greatly complicate matters for a Soviet attacker, and some proponents of BMD (the Hoffman panel, for instance) believe that adding uncertainty is one way BMD can strengthen deterrence. But uncertainty also impedes our ability to determine whether deterrence based on various offense/defense mixes is stronger and more stable than deterrence based only on offensive forces. We must consider that the ability to survive and to fight, implied by BMD, might well lower the nuclear threshold rather than strengthen deterrence.[2] For instance, in a complex world of BMD deployments on both sides, how confident will we be of our own capability to resist an attack and retaliate, and how might this weaken deterrence and our sense of security? Do we have some idea of what our operational doctrine would be in this more complex world, if deterrence should fail as a result of either a calculated or an irrational attack? Which elements of domestic and allied strength are to be given the highest priority if not all can be perfectly defended, and how cost-effective would less-than-perfect defenses be in these cases? Would BMD, especially if limited to defense of our retaliatory forces, reduce the utility of first-strike attacks against us and raise uncertainties enough to deter them; or would it have the opposite effect, raising the scale of possible limited nuclear attacks?

The acquisition and integration of new strategic defenses and offensive forces should, of course, be guided by some coherent strategy, which means we must try to grapple with these uncertainties. This will require the use of gaming, dynamic engagement models, or other methods to analyze the potential war outcomes when both sides deploy varying mixes of strategic offensive and defensive forces. Static cost-exchange ratios[3] cannot be used to assess the cost-effectiveness of BMD systems unless the value and the number of the targets to be defended are considered, and the design and performance of the systems being analyzed are well known over a wide range of contingencies. Performances, however, will never be knowable in advance, in part because these systems cannot be tested realistically (we will return to this point later).

In sum, the range of uncertainties introduced by BMD may in fact result in *incalculable outcomes,* producing a different kind of deterrence for which our ideas and analyses thus far have not prepared us. Whatever its character, to be credible this deterrence will require a mix of both offensive and defensive forces with much more complex, diverse, survivable, enduring, and high-performance wartime capabilities in a nuclear environment than anything we now have in our strategic forces.

Strategic Defense: A Systems Perspective 115

DEFENSE PRIORITIES

Because ballistic missiles are not the only nuclear threat we face, BMD systems alone will be insufficient for the full spectrum of strategic defense objectives. Unless and until defenses can be constructed to cope reliably with other threats—from aircraft and cruise missiles, for example—any strategic defense posture will be seriously vulnerable.

The current emphasis being given to BMD seems to imply that an attack by ICBMs is probably the most stressful defense situation we are likely to face. The reasoning appears to be that if this threat can be mastered through a multilayered defense, then meeting all other military threats will involve lesser challenges and smaller efforts. But this judgment is unproved, and perhaps unprovable. We must bear in mind that the future is certain to bring major technological advances in aircraft, cruise missiles, and space platforms used for communications, surveillance, and anti-satellite missions. On-station or alert times of such new offensive forces—both nuclear and conventional—may well be high, with proliferation greatly broadened and maneuverability and other performance measures greatly enhanced; their ability to avoid detection by electronic means such as radar may be greatly increased. All this is certain to heighten our concerns in other security areas such as the readiness and modernization of our own forces, the adequacy of our air defenses worldwide (including our naval task forces), the balance of conventional forces within NATO, the adequacy of our anti-submarine warfare forces, and so on.

Many new BMD technologies may themselves prove useful in certain of these other security areas, and this may permit us to avoid some hard choices and trade-offs. Conversely, the same BMD technologies, deployed by an adversary, could pose severe new threats in these other areas. For example, the use of lasers on the conventional battlefield or as anti-satellite weapons would create wholly new problems. Comprehensive planning that takes seriously these wider impacts of ballistic missile defense is needed to accompany the BMD research efforts that will provide empirical data for such assessments.

OFFENSE/DEFENSE INTERACTIONS

The operational relation between offenses and defenses is something a systems perspective must also consider. It is particularly important in our planning to consider to what extent the operational performance of our BMD system in the midst of a nuclear attack will dictate in what way and how soon we plan to use our offensive forces.

In the strategic nuclear situation, the responses of defense and offense are tightly interrelated. This is illustrated by the following example: assume that both the United States and the Soviet Union have each deployed BMD systems whose performance in actual war situations is unknown. If our BMD turns out to be highly effective under attack, we might want to respond only lightly with our offensive forces. If our BMD is highly ineffective, however, delays in our response may well jeopardize the very forces we need to retaliate. And if the attacker has a BMD system that proves effective against our response, our resulting vulnerability may be even greater. Of course, we cannot know precisely how effective our defense is until an attack is underway, so we may need to formulate and execute our response at the instant of attack. Yet consistent with current U.S. policy, we must assume that a fully automated, offensive response would never be part of our operational plans.

From a conceptual point of view, this situation may appear to differ little from the one we currently have that relies on offensive forces alone. Yet, the absence of sure knowledge about the performance capability of *both* sides' BMD systems greatly complicates our task of developing appropriate offensive forces and responses to an attack. For example, right now we cannot know for certain how many of our ICBMs and SLBMs might survive a Soviet attack, but we do know that those that do survive will, with high probability, reach their targets; in a BMD world, even this would become uncertain, increasing the difficulty of our operational planning. At some point in the course of battle, our own forces may need to pass through our own defensive screens, and untested "rules of engagement" for our defenses that allow our own forces through, but not those of the attacker, will add another uncertainty.

DEPLOYING OUR DEFENSES

As we noted above, a complex defense network cannot be deployed quickly. The planning of any ballistic missile defense must consider the best methods of developing, deploying, and testing such a system in stages, through a transition period that may last for decades.

As the Soviet threat and the international political environment change, our defense systems may be perpetually in transition. Since a poor defense might actually undermine our security rather than minimally enhance it, we must be sure at every point in the transition that each defense component we have deployed functions well enough to promote stability in crises and military postures.

To achieve this end, we may need to take oblique paths towards our ideal system; in fact, planning the actual transition may be even more challenging than devising a detailed ideal system. For instance, once a multilayered BMD system is fully deployed, it is possible for the performance of each layer to be fairly modest yet the whole quite effective: four layers, each intercepting only 50 percent of its targets, could be more than 90 percent effective overall. Such modest performance levels for each layer are likely to be within easier technological reach and more cost-effective. But if the system as a whole becomes adequately effective only when *fully* deployed, should we wait and deploy the layers all at once, or deploy one at a time? If one at a time, will such low performance levels be acceptable and stable during the transition, or must the initial layers be made more effective (and more costly) than will eventually be needed? What changes in our strategic and operational doctrines will have to accompany the changing levels of effectiveness, and how must these be coordinated with changes in the configuration of our offensive forces? Where and when are defenses against other nuclear threats (air defenses, for instance) to be deployed so that we can always be assured of balanced protection and stability? Assuming the Soviets will try to shape their offenses to exploit transitional weaknesses in our defenses, how adaptable can or must our ideal system be during the transition period, to avert possible vulnerabilities?

The nature of nuclear deterrence and the stability of international political relationships are bound to change during the time that strategic defenses are being created and deployed. What is therefore needed is a far better, more rigorous understanding of the relationships between continually changing military postures and stability.

COMMAND-AND-CONTROL OF STRATEGIC DEFENSES

An effective strategic weapons system is one that constantly monitors and evaluates its own preparedness throughout periods of peace, so that it can respond effectively in the event of war.[4] The complexity of this task for a multilayered BMD system was stated by the Fletcher panel: the system must be prepared to "bring to bear several hundred space, airborne and surface assets to counter several tens of thousands of offensive threat objects in a period of about thirty minutes following the start of an offensive nuclear attack." Moreover, it must do this in an environment in which anti-satellite weapons (such as laser and beam weapons), electronic jamming, space mines, and nuclear weapons (fired from surface-based missiles, for example) could be attacking the defense itself. Note that in a multilayered system, the initial defense against Soviet ICBMs would occur over *their* territory, where Soviet ground-based countermeasures could be concentrated.

To be ready for battle, the defense system must have fulfilled three conditions. It must have performed reliability testing and self-repair on perhaps hundreds of remote, unattended, and long-enduring space-based weapons stations; it must have established and enforced minimum keep-out zones in space for potentially lethal objects or countermeasures; and it must have tested in as realistic a manner as possible all of the system's battle functions, including coordination with offensive forces and national command authorities, and its ability to sustain effectiveness when portions of the system are damaged or destroyed by hostile action. It must also assure weapons safety, arming, and release. If the whole system is deployed in stages over a period of decades, the command-and-control procedures must constantly be adapted and tested throughout the transition, to assure effectiveness and security. Yet command-and-control has never been practiced on such a scale, and has never had to contend with such complexity.

The challenges described above can be illustrated by a more detailed look at several crucial tasks the system must perform.

BIRTH-TO-DEATH TRACKING

Accurate discrimination and tracking of genuine warheads and rapid kill assessment are crucial to the efficient use of weapons by any defense. But engineering data on the targets' characteristics may be uncertain or scarce; active countermeasures (such as electronic jamming and laser blinding of sensors) must be anticipated; targets and

Strategic Defense: A Systems Perspective 119

decoys may not move on predictable ballistic trajectories; and two or more objects may be in the same discrimination cell of a surveillance sensor.

To cope with all this, the defense system will have to gather data from a great number of sensors and integrate it into a coherent picture. Surveillance sensors will require power levels and source/receiving structures at least one order of magnitude in performance and size beyond today's technological capability. Also, the speed, reliability, and intelligence that the data-processing components will require far exceeds the ability of today's computer hardware and software. This brings us to a second challenge.

SOFTWARE DESIGN AND DEVELOPMENT

The computer software of a complex defense system must be error-free and must operate reliably twenty-four hours a day in peace and war. Software throughout the system would also have to be designed, implemented, and tested years before operation. We would need new design and testing techniques to provide for correct operation under all feasible contingencies and at all component and system levels, as well as new techniques to modify and retest software remotely. The task of replacing or modifying software installed in computers on board satellites would require secure communication links possessing near absolute accuracy.

The tasks of managing the operational complexity of the BMD system and avoiding unforeseeable breakdowns would be handled by the computer software distributed throughout all the elements of the system. Software could thus become the limiting factor in the performance and reliability of BMD. Little is currently known about the potential size and complexity of the computer program codes that would make up the software for a multilayered BMD system. The Fletcher panel has estimated that the command-and-control software would be about 10 million lines of interrelated operational code, but this estimate was made at a time when a definitive system design did not exist.[5]

The nation's experience in developing and deploying software for large-scale, real-time control systems is limited. The SAGE air defense system begun in the 1950s, the National Air Space System for air traffic control built by the FAA in the 1960s and 1970s, and the

Safeguard anti-ballistic missile (ABM) system built by the Army in the late 1960s and early 1970s are notable examples.[6] Each of these systems had about 500,000 interrelated operational lines of software code in their final computer programs. This does not include the support software that was needed for assembling, testing, and verifying the operational code. The support software, which was not as interrelated and complex as the operational code, comprised at least as many lines of code and sometimes much more.

Designing, developing, implementing, and testing each of these operational software systems took over one thousand man-years of effort by professional staff over periods of up to a decade. Almost twice as much effort was required to plan, design, test, and verify the design, and then specify for the software the detailed operational performance of each overall mission system (air defense, air traffic control, ABM) as to produce the operational software code itself. This is typical of the development of large-scale computer-based systems.[7] And this does not even take into account the design and development of hardware.

If we make a linear extrapolation from our past experience, the overall effort needed to plan, design, and specify the performance of a comprehensive BMD system, and then to develop, implement, test, and verify the operational software code to meet these specifications would be several tens of thousands of man-years.[8] Extrapolation may be misleading, however, as the air defense, air traffic control, and ABM systems developed in the 1960s and 1970s tracked only a few hundred objects and made only tens of intercepts at a time. Our postulated BMD system would have to track several tens of thousands of objects and direct hundreds to thousands of intercepts at a time. Furthermore, it would have to carry out these tasks while many complex, software-dependent interactions were taking place among the various layers of the defense system and with offensive and allied forces. It is possible that a team of a few thousand professionals working for up to twenty years might be needed to accomplish such a feat. This would be an undertaking unprecedented in the history of software development and implementation.

Techniques for developing and producing software that we can now project for the 1990s include automatic programming and artificial-intelligence support programs—computer programs that write and test computer programs. While these techniques may ameliorate the task of writing software code, they will not simplify the tasks of designing the overall software program itself and confirming its reliability. If the interrelated operational lines of code needed for a BMD system could be reduced to less than 2 million or so, rather than the 10 to 20 million now estimated, we may with some confidence estimate that a reliable program of some 2 million lines of code might be produced in a reasonable time. Simplification could be achieved in a number of ways: by using common software packages in defense components such as satellites; by using approximate rather than optimal algorithms and rules for weapons allocation and intercept control; and by preassigning defense elements to known threat corridors. If these measures prove feasible and relatively efficient, then the complexity of the BMD system could be reduced significantly. Given the current uncertainties about the design of a BMD system, however, the feasibility of such a simplification effort is highly uncertain now.

Thus, our ability to build thoroughly reliable and high-performance command-and-control software could be the greatest challenge to achieving an effective BMD.

DEFENSE SYSTEM ACTIVATION AND SUSPENSION

Judicious control of weapons involves striking a balance between readiness for use and assurance that use occurs only when and as intended. Offensive nuclear weapons, of course, pose unique and serious questions in this regard. In the case of the non-nuclear defensive weapons as envisioned for the SDI, different questions arise. Space-based, unattended weapons would have to fire early enough to be effective during the boost phase of attacking missiles, a period lasting no more than tens or hundreds of seconds, lest the mid-course and terminal layers of the defensive system become saturated. Given the short response times available and the difficulty of reaching decisions within these times, we must confront the question of "automated response."

The need for a high degree of automation in the weapons release process of a boost-phase-intercept system was noted by the Fletcher panel in its studies. Many people within the defense community have considered such procedures and have concluded that fully automated responses by machines, with programs designed long before operation, would be acts of folly. In the case of a defense system made up of non-nuclear weapons, there is one mitigating factor—the consequences of a false start need not be catastrophic. However, given their experience with computers, the public and the politicians have shown little trust in assigning machines such decision-making roles. Expeditious responses in a combined defense/offense system would thus require what we have today in our nuclear weapons control procedures: a combination of human judgment, delegated responsibility and defined procedures, rapid execution capabilities, and rules of engagement suited to particular levels of crisis or threat and the amount of strategic warning-time available.

Another problem involving the control of weapons concerns the termination or suspension of a conflict. We would always hope that if deterrence fails, any conflict, especially one involving strategic nuclear forces, will be stopped as soon as possible. But a pause in battle may represent a true termination, or only a temporary suspension of hostilities. In either case, our ability to assess rapidly our defense system's status and performance, and to poise it to resume operations may be crucial to preserving the pause in hostilities. Thus, it would be necessary for our system design to include rapid stop/restart capabilities in both our defensive and offensive forces. The system-design implications of incorporating stop/restart features deserve study and research.

SYSTEM MAINTENANCE, TESTING, AND EXERCISING

Assuring that a highly automated BMD system works as intended, and maintaining it in peak condition for years, would constitute a new and unprecedented challenge. Innovative forms of acceptance testing and of routine exercising and operational testing would have to be devised for the overall system in order to assure thorough reliability and effectiveness. System-operational testing would be exceptionally difficult, given that many parts of the overall system could not be tested together operationally except under the hostile environmental conditions of an actual war. Since we would have to rely on simulation of war conditions to evaluate BMD system performance, and because such exercises can never be thoroughly realistic, the operational performance of the system under major attacks of varying kinds would be highly uncertain.

Strategic Defense: A Systems Perspective

Maintenance of the operational system would also be difficult. Defense satellites, for instance, would be large, numerous, complex, expensive, and in some cases located as far away as tens of thousands of kilometers. Manned or robot repair and replacement of parts would be necessary. Systems research is essential to define maintenance and testing requirements and concepts, especially during the long periods of transition to full deployment, and to make tradeoffs among these requirements.

Testing and exercising of a BMD system would very probably require major growth in our current test range and space launch facilities. Given public resistance over the last two decades to the Safeguard and MX deployment schemes, the question of where to site such extensive facilities would need to be addressed at an early stage of the system design.

Constructing and maintaining a large space-based defense system would also require significant investment in space transport. These requirements would vary, of course, depending on whether the Soviets complicated our construction tasks by attempting to prevent or to delay our deployments. If we were to use space shuttles of roughly current capacity, the deployment of a comprehensive system over the period of a few decades would take at least several hundreds of flights, and probably more than a thousand. This would require an expanded ground-based infrastructure for launching and recovering shuttles, and a replenishment and repair infrastructure to replace and fix satellites. Because of the long lead times involved, a plan for space transport, replenishment, and repair should be part of a comprehensive BMD deployment plan.

Finally, of course, those elements of our defense system that are undergoing exercises or routine maintenance will not be combat-ready. A larger defense force with duplication of vital elements may therefore be needed to assure that some indispensible combination of elements is always operational.

CONCLUSION

While the political and scientific challenges posed by a comprehensive strategic defense system have been widely discussed since President Reagan's March 1983 speech, a third sphere—system capabilities—

has received less attention and yet involves challenges that are at least as great.

The sheer technical and operational complexity and scale of a multilayered BMD system, compounded with the changing threat and countermeasures it would give rise to, are daunting. The feasibility of a survivable and reliable system, with perfect or near-perfect performance, is far from certain. Research and experimentation in new systems architectures and subsystem designs are needed that can meet more limited objectives. The addition of BMD to our strategic posture would initially, and perhaps indefinitely, require a mix of both offensive and defensive forces each having more complex, diverse, survivable, sustained, and high-performance wartime capabilities than we now have. It is not yet clear whether such a mix of forces can provide stronger nuclear deterrence than the deterrent capabilities of an offense-only system.

The operational interactions of strategic offense and defense are complex and not yet well-enough understood for us to assess the cost-effectiveness of proposed BMD systems. Using new methodologies based on gaming and interactive simulation, we should investigate a number of mixes of offense and defense to assess which combinations might strengthen deterrence of nuclear attack while also minimizing generation of crises. The transition period to a well-developed, two-sided offense/defense situation would require decades, and uncertainties about effectiveness could continue even longer. Research is needed to understand better the dynamics involved in the relationship between stability and changes in military postures over time.

It is questionable whether multilayered BMD is the top priority defense problem facing the United States. The modernization of strategic offensive forces, or the readiness of all forces, for example, may be more important. Yet the development of BMD technologies and systems will have a significant effect on the availability of technologies and systems to these other national security areas. It might be possible, for example, to use laser weapons in air defense and conventional warfare or to employ space-based sensors capable of superior surveillance of space for warning and characterization of attacks. We will need to reappraise *all* our national security requirements continually, as we learn what is possible and feasible in BMD.

Strategic Defense: A Systems Perspective 125

Assuring the performance of individual elements of a multilayered BMD system, and marshaling the effort and resources needed to develop, deploy, test, verify, and maintain an effective overall system will be an enormously complex and difficult task. Building such a system would dominate the defense and space activities of the United States for decades and would require extraordinary organizational and funding support. Given the potential costs and duration of the program, achieving and maintaining public support will be essential.

ENDNOTES

[1] Senate Committee on Foreign Relations, *Strategic Defense and Anti-Satellite Weapons,* 98th Cong., April 25, 1984. This document contains the testimony of various administration witnesses and experts on the Strategic Defense Initiative. See also the Defensive Technologies Study Team (James C. Fletcher, et al.), *Report of the Study on Eliminating the Threat Posed by Nuclear Ballistic Missiles,* unclassified summary, Pentagon, March 1984; the Future Security Strategic Study (Fred S. Hoffman, et al.), *Defense Against Ballistic Missiles: An Assessment of Technologies and Policy Implications,* unclassified summary, U.S. Department of Defense, April 1984; and Ashton B. Carter, *Directed Energy Missile Defense in Space,* background paper, Office of Technology Assessment, U.S. Congress, April 1984. These three reports are reprinted in the *Senate Hearings.*

[2] Richard Nixon makes this point with respect to BMD: "Such systems would be destabilizing if they provided a shield so that you could use the sword," (*Los Angeles Times,* July 1, 1984). So did President Reagan in his March 23 speech: "I clearly recognize that defensive systems have limitations and raise certain problems and ambiguities. If paired with offensive systems, they can be viewed as fostering an aggressive policy; no one wants that."

[3] The cost-exchange ratio is the cost to the United States of preventing the expected damage if the Soviets send one more attacking warhead, against the marginal cost to the Soviets of sending that additional warhead. Some defense analysts believe that there is little prospect of achieving a cost-effective BMD in the foreseeable future unless the offensive threat is tightly constrained technically and greatly reduced numerically through offensive arms-control agreements. See, for instance, Donald Hafner's discussion of the Fletcher panel in this volume.

[4] Charles A. Zraket, "Strategic Command, Control, Communications, and Intelligence," *Science,* June 22, 1984.

[5] Brockway McMillan, et al., *Battle Management, Communications, and Data Processing* (vol. V of the Fletcher panel report, op.cit.). A line of computer code is a line of instructions, written in a form or "code" the computer can understand, telling it what the next step is it must perform.

[6] See Charles A. Zraket, et al., "SAGE—A Data-Processing System for Air Defense" and Herbert D. Benington, "Production of Large Computer Programs," in a special issue of the *Annals of the History of Computing,* Oct. 1983; Gustav E. Lundquist, *The United States National Airspace System, NAS Stage A,* proceedings of the International Symposium on Traffic Control, Versailles, France, June

1–5, 1970; Subcommittees of the Committee on Appropriations of the U.S. House of Representatives, *Hearings on the Safeguard Anti-Ballistic Missile System,* part IV, Department of Defense Appropriations for 1971.

[7] F. P. Brooks, *The Mythical Man-Month* (Reading, Mass.: Addison-Wesley, 1982), chap. 8.

[8] See Herbert Lin, *Considerations in the Development of Software in BMD,* working paper of the Center for International Studies, MIT, Dec. 1984; and Barry W. Boehm, *Software Engineering Economics* (Englewood Cliffs, N.J.: Prentice-Hall, 1981), chaps. 5–9, describes the COCOMO model for estimating the amount of effort needed to produce software systems.

Paul Stares

U.S. and Soviet Military Space Programs: A Comparative Assessment

SINCE THE LAUNCH OF SPUTNIK IN 1957, the United States and the Soviet Union have steadily expanded their military use of space to a point where they now rely heavily on satellites to enhance their national security. While space is often referred to as the "new high ground" or a "new dimension of the arms race," there is nothing novel about its use for military purposes. Military satellites are used for reconnaissance, early warning, communication, navigation, weather forecasting, and geodetic data gathering. The current development of anti-satellite (ASAT) weapons and the growing interest in space-based ballistic missile defense (BMD) systems, however, clearly indicate a new trend in the militarization of space. Although it has been an integral part of the arms race between the superpowers for more than twenty-five years, space has generally been considered a "sanctuary" from the deployment and use of weapons.[1]

While the United States and the Soviet Union both use space for broadly similar purposes, there are significant differences in the manner in which they utilize their satellites, and the value they choose to place on them. These differences will become especially important when we come to examine the implications of a conflict in space, implications that will force the superpowers to seek new precautions as threats to their satellites increase.

EVOLUTION OF THE U.S.-SOVIET MILITARY EXPLOITATION OF SPACE

After a short period of uncertainty following Sputnik, when many feared and predicted that space would become an arena of superpower military competition, the United States and the Soviet Union settled down to a relatively stable relationship in their military exploitation of space. Compared with other areas of the arms race, and in sharp contrast to their rivalry in space exploration generally, the policies of both superpowers toward the military use of space has been remarkably free from competition. With few exceptions, both countries have developed space systems in accordance with their own terrestrial security requirements, rather than in response to the activities of the other. Moreover, the usual attributes of a military competition—the presence of weapons systems—has so far been absent from space. As military satellites are essentially supportive of or ancillary to activities on earth, space has remained, in effect, an adjunct to the arms race.

While military satellites are often regarded as "passive" because they are not lethal devices, they can enhance the lethality of other weapons and the effectiveness of military forces. This inherent duality, which has been aptly described as Janus-like, is apparent in the current applications of military satellites. Photographic and electronic reconnaissance satellites, for example, provide indispensable national technical means of verification for arms control and early warning of an attack. In this respect, they are considered benign and stabilizing. Yet they can also pinpoint military targets for attack during a conflict and provide valuable information for electronic countermeasures. Ocean reconnaissance satellites, on the other hand, do not have such redeeming virtues; their only function is to detect, track, and target naval vessels.

The U.S. nuclear explosion detection satellites were also initially developed to support arms control—specifically, to monitor compliance with the Limited Test Ban Treaty of 1963. While the latest system designed to detect nuclear bursts will continue to perform this role, its primary mission now is to facilitate post-attack assessment and re-targeting after a nuclear exchange. In other words, it has become an aid to warfighting.

Even communication, navigation, and weather satellites, which appear to be harmless, can be considered indirectly threatening to an adversary. Communication satellites may be able to increase a country's control over geographically dispersed forces and provide a valuable link for crisis management purposes, but in the same way they can also improve the overall performance of military forces. Space communication links, for example, provide the United States with an almost instantaneous method of collecting, collating, and disseminating information on the whereabouts of Soviet nuclear submarines, a capability that is obviously important for anti-submarine warfare (ASW) operations. Submarines, in turn, use navigation satellites to update their inertial guidance systems, which improves the accuracy of their ballistic missiles. These same satellites will increasingly be able to guide a range of strategic and tactical weapons to their targets with near-perfect precision. Meteorological satellites can be used to improve the effectiveness of military strikes by providing timely information about the weather over target areas. Even geodetic satellites, which map the earth's shifting gravitational fields, can provide valuable targeting data for strategic missiles. The only satellites that are unequivocally pacific are the early-warning systems that detect the launch of ICBMs and SLBMs. If ballistic missile defenses become a reality, however, these satellites will be used to acquire and track targets for interception and to assess the success of an engagement.

The dual nature of military satellites has become more apparent in recent years as the range of applications has expanded. For the first ten to fifteen years of the space age, satellites primarily benefited strategic forces, but since the mid-1970s, space systems have increasingly provided general-purpose forces with battlefield surveillance, communication facilities, and targeting information. The Soviet Union has as a rule lagged behind the United States in the development of military space technology, despite its early lead with Sputnik. Table 1 on the following page gives the dates for the introduction of various categories of U.S. and Soviet satellites and summarizes their current orbital deployments.

Despite a common appreciation of the benefits of military satellites, the U.S. and the Soviet Union have not always shared the same attitudes or policies towards the exploitation of space. Since the Second World War, four phases can be discerned.[2]

TYPE OF SATELLITE	TYPICAL DEPLOYMENTS		DATE OF FIRST DEPLOYMENT	
	U.S.	SOVIET	U.S.	SOVIET
PHOTOGRAPHIC RECONNAISSANCE	2–3 LOW ORBIT	1–5 LOW ORBIT	1959	1962
ELECTRONIC INTELLIGENCE (ELINT)	1–2 GEOSYNCH	6 LOW ORBIT	1962	1967
OCEAN SURVEILLANCE	6 LOW ORBIT	2–4 LOW ORBIT	1976	1967
EARLY-WARNING	3 GEOSYNCH	9 MOLNIYA	1960	1971
NUCLEAR DETECTION	6 SEMISYNCH[a] 2 SUPERSYNCH	??	1963	??
COMMUNICATIONS	11 GEOSYNCH 2 MOLNIYA	27 LOW ORBIT 12 MOLNIYA 12 GEOSYNCH	1959	1964
NAVIGATION	5 LOW ORBIT 6 SEMISYNCH[b]	10 LOW ORBIT 9–12 SEMISYNCH	1959	1967
METEOROLOGY	2 LOW ORBIT	3 LOW ORBIT	1963	1963

[a] consists of sensors on Navstar navigation satellites [b] the Navstar navigation system will contain 18 satellites when completed

Table 1 Categories of U.S. and Soviet military satellites and year of first deployments; chart includes satellites dedicated to military use (some civilian satellites may also be used for military purposes). For definitions of different orbits, see pp. 148, 150.

EARLY VISIONS AND PREPARATIONS: 1945–1957

The potential military benefits of outer space were recognized as early as 1946 in the United States, but the dramatic contraction in the U.S. military budget after World War II discouraged experimentation in new ideas and research programs. Early satellite and ballistic missile development proposals, especially, fell victim to military conservatism induced, in part, by fiscal constraint. Without large boosters, there could be no space program.

By the early 1950s, though, attitudes began to change as the cold war intensified. The need for a more reliable source of strategic intelligence provided the raison d'être for reconnaissance satellites, and the demand for additional strategic delivery systems sparked interest in the ballistic missile program. Neither of these, however, was pursued very urgently. It is also ironic that the U.S. commitment in 1955 to launch a "scientific" satellite as a contribution to the International Geophysical Year (IGY) also hindered the progress of the U.S. space effort. Eisenhower's decision to forbid the use of military boosters for the IGY program led to the disastrous choice of the Vanguard, a small and relatively undeveloped launch vehicle. Had the Eisenhower administration chosen the more advanced Army Orbiter project, with its Redstone (later Jupiter) rocket, the U.S. would have been the first country to place a satellite into orbit.

In contrast, ballistic missile development and the space program were receiving the highest political support in the Soviet Union. Not only was this compatible with the revolutionary scientific ethic, but strategic missiles offered a way to improve "the correlation of forces" with the West and an opportunity to reduce the size of Soviet conventional forces.[3] Furthermore, Soviet Premier Nikita Khrushchev saw clearly the political and propaganda benefits of being the first country to launch an artificial satellite. While the U.S. program was proceeding slowly, handicapped by a self-imposed separation of military and civilian space research, the Soviets stole the lead in developing the world's first intercontinental ballistic missile—the SS-6—and the world's first artificial satellite, Sputnik.

EXPANSION AND UNCERTAINTY: 1958–1967

The immediate crisis of confidence caused by the Soviet coup de main in space led to a rapid expansion of the U.S. space program. Sputnik also generated considerable public anxiety that the Soviet Union might use space to threaten the United States. Believing that space represented a new dimension of warfare and that the emerging Soviet threat would need to be countered militarily, the armed services and the aerospace companies actively canvassed for a variety of space weapons, including anti-satellite devices, orbital bombardment systems, and space-based ballistic missile interceptors.

Preliminary research was carried out on many such proposals (including the world's first ASAT demonstration in 1959) as an insurance against hostile Soviet activities in space, but Eisenhower and, later, Kennedy rejected further development on the grounds that it would spur competition with the Soviet Union, and ultimately destroy the "peaceful" image the U.S. was trying to foster for its space program. Although both presidents genuinely wanted to prevent an arms race in outer space, their policy stemmed principally from a desire to legitimize U.S. satellite reconnaissance and reduce the likelihood of Soviet countermeasures.

This need to protect reconnaissance satellites became even more imperative when satellites became the primary source of strategic intelligence following the demise of U-2 overflights in May 1960. Despite some veiled threats, however, the Soviets never attempted to rebuff U.S. reconnaissance satellites in the same way they had the U-2 flights. Instead, in 1962 the Soviet Union introduced a resolution in the United Nations to outlaw "espionage" from space. The United States countered that satellite reconnaissance was a peaceful and legitimate use of space, thereby implicitly distinguishing it from non-peaceful and illegitimate activities. A vigorous U.S. space weapons program would only have contradicted and jeopardized this strategy.

Aside from this primary motive, space weapons offered few if any attractions to the United States at this time. Until an unequivocal Soviet threat developed in space, anti-satellite weapons had little use. As table 1 indicates, the Soviet military space program was still in its infancy. Orbital bombardment systems also offered no meaningful advantages over ballistic missiles as strategic delivery vehicles; if anything they were inferior. Where a military rationale could be demonstrated, such as with space-based BMD systems, the United States considered further development prohibitively expensive and technically infeasible. The only exception to this stance came in 1963 when President Kennedy, as a precaution against the possible deployment of Soviet orbital bombs, authorized the development of two ground-based anti-satellite systems, one using the Nike Zeus ABM missile and the other the Thor IRBM. No more than a handful of these missiles were ever deployed. Although the Thor system remained operational until 1975, its fixed launch site and nuclear warhead severely limited the circumstances in which it could be used.

The Soviet Union dropped its opposition to satellite reconnaissance in the fall of 1963. At the same time, it also agreed to a U.N. resolution banning the deployment in space of weapons of mass destruction. This resolution subsequently became the basis for the Outer Space Treaty, signed in 1967. Although there were some remaining doubts, particularly when the Soviet Union began testing a fractional orbital bombardment system (FOBS) in 1966, it appeared to many that the Soviets had accepted—if only tacitly—the U.S. conception of space as a sanctuary from certain military activities.

This was partly true. Soviet diplomatic efforts to ban satellite reconnaissance had failed to gain wide support in the U.N., and the use of force against U.S. satellites would not have been easy. More important, continued opposition to reconnaissance from space would have been counterproductive to the Soviet Union's own satellite reconnaissance program, which began to return photos in 1962. The overriding requirement of responding to the Kennedy administration's offensive strategic buildup no doubt made further confrontation and a potential military competition in space especially undesirable for the Soviets. Nonetheless, they still considered it prudent to take out some technological insurance, and, in the process, gain the ability to counter U.S. space systems in the event of war. For all these reasons, the Soviet Union in the early 1960s embarked on a satellite interceptor program.

CONSOLIDATION AND STABILITY: 1968–1975

It was testament to the stability of the apparent modus vivendi in space that even when the Soviet Union began testing its satellite interceptor in 1968 the United States seemed remarkably unalarmed—something that would have been unthinkable a few years earlier. The little concern that did exist soon dissipated when the Soviets ceased their ASAT tests at the end of 1971 and conducted no further tests for the next four-and-a-half years. The United States was also encouraged by the progress of detente, which produced a number of agreements reinforcing the tacit ground rules for the military exploitation of space. The SALT I interim agreement, signed in 1972, contained important clauses prohibiting interference with national technical means of verification, including satellites. The ABM treaty signed at the same time also banned the testing and deployment of BMD systems in space.

After the rapid expansion of the early 1960s, the U.S. military space program settled into routine operation. The sophistication of satellites improved steadily, but the general level of space activity dropped off from its frenetic pace at the beginning of the decade. Even the space budget began to decline in real terms. These trends were partly due to greater efficiencies in the space program, and partly to the military's shift in attention away from space to the war in Southeast Asia.

The Soviet space program, on the other hand, underwent rapid expansion in the late 1960s, deploying new types of satellites with greater frequency. By the early 1970s, Soviet space activity had surpassed that of the United States in both number of launches and total payloads orbited. Soviet use of reconnaissance satellites during the major international conflicts of this period indicated clearly that space systems were becoming an integral part of Soviet crisis monitoring and, ultimately, Soviet war planning.

THE EMERGING ARMS RACE IN SPACE: 1976 TO PRESENT

By 1976, the increasing military use of space on the part of the Soviets had become a source of concern to the United States. This concern in many respects mirrored the U.S. reaction to the Soviet strategic buildup that also occurred during this period. The resumption of Soviet satellite interceptor tests in 1976 was the primary catalyst for the United States' reconsideration of the usefulness of anti-satellite weapons. While the new Soviet tests were not markedly different from the earlier series (and if anything had an inferior performance), the tenor of U.S.-Soviet relations had changed significantly. The Soviet attainment of strategic parity and its perceived adventurism in the Third World made the United States especially sensitive to any new Soviet military activities. As a result, President Gerald Ford in one of the last acts of his administration authorized the development of a new U.S. anti-satellite system. While the ostensible rationale was to counter the indirect threat from Soviet military satellites—especially ocean reconnaissance satellites—the real reason appears to have been an unwillingness to accept any imbalance in U.S.-Soviet ASAT capabilities.

President Jimmy Carter continued with the ASAT R&D program to gain bargaining leverage in ASAT arms-control negotiations and to provide insurance in case such negotiations failed. Three rounds of ASAT negotiations were held in 1978–79, but no agreement was reached. Although some progress was made, differences over the scope of an agreement, the diversions of the SALT II debate, and the suspension of arms-control negotiations following the Soviet invasion of Afghanistan thwarted any chance of success.

For nearly four years, the Reagan administration opposed resumption of ASAT negotiations, despite the Soviet Union's announcement of a unilateral testing moratorium and its proposal of two space weapons treaties to the United Nations. By the summer of 1984, however, growing congressional pressure and concern over the possible electoral repercussions resulting from an alarming deterioration in U.S.-Soviet relations forced the Reagan administration to reconsider its stance on the ASAT issue. A Soviet offer to begin discussions in September 1984 came to naught when the sides could not agree on the scope of the talks nor on an ASAT test moratorium. The Soviets wanted to include all space weapons and to impose an immediate test ban—an obvious effort to restrict the U.S. ASAT program and the BMD research provided by the Strategic Defense Initiative. The United States insisted on at least testing its new ASAT system—the air-launched miniature vehicle—before considering any ASAT limitations, and it insisted the talks should cover strategic and theater nuclear forces as well. Following the 1984 presidential election, a U.S. proposal for "umbrella" arms-control discussions, to include space weapons, was accepted by the Soviets. While some type of agreement relating to space weapons cannot be discounted, the

A COMPARATIVE ASSESSMENT OF U.S. AND SOVIET MILITARY SPACE PROGRAMS

Since outer space offers the same inherent opportunities to both the United States and the Soviet Union, it is not surprising that they should eventually find themselves using satellites for broadly similar purposes. There are real distinctions, however, between the two countries' programs. The most noticeable difference is the rate at which each country launches payloads into orbit. Throughout the 1960s, the United States led the Soviet Union both in total space launches and total payloads. During the 1970s, however, the number of U.S. space launches progressively declined, while that of the Soviets increased. In 1984, the Soviet Union conducted ninety-seven launches, of which 80 percent were identified as military related; the United States had twenty-three launches, of which 39 percent were for military purposes. These statistics are often cited to highlight the dynamic nature of the Soviet space program and indicate implicitly that the United States has fallen behind in the exploitation of space. Actually, a high rate of launches is more indicative of a lower capability. As the U.S. Department of Defense states, the fact that the "Soviets routinely conduct about four to five times as many space launches per year as the United States . . . is necessitated primarily by the shorter system lifetimes and poorer reliability of most Soviet satellites."[4] For example, Soviet reconnaissance satellites, which account for nearly 60 percent of all its military space launches, stay in orbit on average only about three weeks.[5] Most of these satellites are deliberately returned from orbit to recover the exposed film. In contrast, the two main U.S. systems, the KH-9 (Big Bird) and the KH-11, last approximately thirty-five weeks and one hundred weeks respectively. The KH-9 returns its film by recoverable capsules, and the KH-11 transmits images electronically. Although these two recovery methods were mastered by the United States in the 1960s, the Soviets began to use recoverable capsules only during the late 1970s and are only now developing an electronic transmission system. The additional argument that the Soviets' higher launch rate better enables them to replace satellites that are lost during hostilities may be true, but this smacks of making a virtue out of necessity.

Another reason the Soviets have a higher launch rate is because it generally takes a greater number of Soviet satellites to provide the same level of service as their U.S. counterparts. For example, the Soviets maintain three distinct constellations of satellites for the purposes of space communication, two for tactical and one for strategic communication. The tactical communication satellites are small and simple, operating at low altitudes. They are what is known as "store dump" satellites—they receive information as they pass over one part of the globe, transmitting it later at an appropriate time and place; they are not used for urgent communications. For these low-altitude satellites to provide continuous coverage of the world, the Soviets must cluster them into two constellations, with three satellites in one constellation and twenty-four in the other. In addition, the Molniya strategic communication system contains another twelve satellites, in a highly elliptical orbit, that provide continuous coverage of the northernmost latitudes of the Soviet Union where many of the country's key military installations are located. Because of the curvature of the earth, geostationary communication satellites placed 36,000 kilometers above the equator would not provide satisfactory coverage of these areas. More recently, the Soviets have begun using geostationary satellites for communication, some of which is almost certainly military related. In contrast, except for two Satellite Data System (SDS) satellites used for polar communications in an orbit similar to the Molniyas, U.S. tactical and strategic military communications are handled by approximately eleven high-capacity, long-lasting geostationary satellites. The United States also uses the geostationary orbit for early warning and electronic intelligence-gathering satellites.

Besides these differences in the duration and distribution of U.S. and Soviet satellites, there are other practices that distinguish the two space programs. U.S. satellites more often perform dual functions: the Navstar GPS (global positioning system) navigation satellites and the Defense Support Program (DSP) early-warning satellites also carry nuclear-explosion detectors, other satellites carry special strategic communication transponders, and the Defense Department's meteorological satellites carry a variety of sensors with different applications.⁶ Such multipurpose satellites are not common in the Soviet space program.

The Soviet Union, on the other hand, has placed greater emphasis on manned military space operations. The manned Salyut space station has been used for reconnaissance purposes and possibly for other military-related experiments. Currently, the only manned U.S. spacecraft is the space shuttle, which stays in space for relatively short periods to ferry satellites into orbit. The U.S. Air Force will begin shuttle operations from its Vandenberg launch complex in late 1985 or 1986. The Soviet Union is currently working on its own shuttle, but a full-scale vehicle has yet to be tested.

RELATIVE DEPENDENCE ON MILITARY SPACE SYSTEMS

Assessing the relative importance of military space systems to the two superpowers is considerably more difficult than comparing their operational styles. It has often been said that the United States is more dependent than the Soviet Union on the services of military satellites. Statements about the relative dependency on satellites, however, must be interpreted with some care. To begin with, there is a vital difference between dependence based on choice, and dependence based on necessity. While satellites may be relied on to support certain missions, there may be other ways to achieve this. Also, in situations where satellite service is unique, it may not be critical or always important. The amount of dependence therefore varies with the different types of satellites, with their peacetime and wartime roles, and with the circumstances in which they are used. While it is true that accounting for all these variables requires a more complicated assessment, it would be undeniably more meaningful than merely "static" comparisons. This can be illustrated by examining four categories of satellites: reconnaissance, communication, early warning, and navigation.

Reconnaissance Satellites

Given the closed nature of Soviet society, the United States is more dependent on reconnaissance satellites for its strategic intelligence gathering and for verification of arms-control agreements. In the United States, no other method of verification can rival in importance the use of photographic and electronic reconnaissance satellites. We must be careful not to exaggerate, however, this difference between the superpowers. The Soviets, too, rely on satellites for monitoring activities in China, for strategic targeting, and for corroborating information gained from other intelligence sources. The Soviets also depend on satellites for crisis monitoring, as indicated by changes in the orbits of Soviet reconnaissance satellites and their higher launch/recovery rate during major international conflicts. The United States, on the other hand, can use high-altitude reconnaissance aircraft deployed at numerous bases around the world. The Soviets also have long-range reconnaissance aircraft, but they do not have as many bases worldwide and therefore cannot achieve as much coverage.

Levels of dependence on reconnaissance satellites during wartime are more difficult to assess. It is likely that satellites will play a vital early-warning role before hostilities, but subsequent use will depend on the geographic location of the conflict and the type of warfare. For example, the initial dependence on satellites during a conventional war in Europe would likely be low because ground-based in-theater reconnaissance systems, such as aircraft and, in the future, remotely piloted vehicles, would be available. Given the expected high attrition of these systems, however, satellites may become the primary source of information in a prolonged conventional war, provided that the equipment necessary to receive and use the data survives. In conflicts occurring outside Europe or the immediate periphery of the superpowers, conventional reconnaissance systems may be less available, creating greater dependence on satellites. Since the United States—with its global military presence and large number of overseas bases—is likely to have more alternatives available to it than the Soviet Union, it will probably have a lower overall dependence on satellites.

In the event of a nuclear war, the relative importance of reconnaissance satellites will depend heavily on the role that each country envisages for its space-based systems, and on the length and intensity of hostilities. The development of satellites designed specifically for post-attack assessment reflects a desire on the part of the U.S. to ensure flexible and discrete nuclear attack options to meet a variety of contingencies. In contrast, as one analyst has pointed out, Soviet doctrine favors massive initial attacks and evinces little concern for post-attack assessment and follow-up attacks.[7] After a prolonged nuclear exchange, satellites arguably cease to have much relevance, especially as the ground-based centers that receive and use their data will almost certainly be destroyed.

Communications Satellites

A statistic often cited to illustrate U.S. dependence on space systems is that more than 70 percent of all U.S. military communications are transmitted via satellite. Again, such a figure is meaningful only when put into proper perspective. In peacetime, a good deal of routine message traffic passes through military communication satellites simply because they offer an inexpensive way to transmit information. In wartime, when message traffic would be pared down to a vital minimum, satellites may be useful, but not necessarily critical; there are other ways of communicating information, including a variety of dedicated military radio systems and civilian channels.[8] Here again, geographical factors will determine the alternatives available to both sides with Soviet dependence on satellite communication likely to be higher than that of the United States in geographically remote areas.

Considerable redundancy also exists among methods of communicating with strategic nuclear forces. The Soviet Union has invested massively in redundant command, control, and communication facilities—including an extensive network of buried cables and radio relay systems—many of which appear to be designed to function during and after a nuclear war.[9] The United States, similarly, has a wide range of strategic systems to ensure unbroken communication at least in the initial stages of a nuclear war.[10] On the face of it, therefore, dependence on space systems for strategic communication appears to be low for both countries. This may change, however, as both sides begin to rely more heavily on highly dispersed and mobile strategic forces.

Early-Warning Satellites

It is difficult to predict which superpower would benefit more from the extra warning time provided by early-warning satellites. The low peacetime alert rates of the Soviet strategic forces suggest that they would gain more. Yet early-warning satellites can provide only tactical warning, so the benefits may be marginal, at best, for strategic forces not already on alert. While the Soviet bomber force would need early warning to get off the ground before an attack, the greater role bombers play in U.S. strategic forces suggests that early warning might be comparatively more important to the United States. The flight times of forward-deployed SLBMs and INF missiles are so short, however, that even the extra warning time provided by satellites may not be sufficient for bombers to clear their bases before attacking warheads arrive.

Navigation Satellites

Both the United States and the Soviet Union are deploying remarkably similar satellite navigation systems (Navstar GPS and GLONASS, respectively). This appears to indicate that the two nations have an equal interest in, and place equal importance on, navigation systems. But apart from its ability to pinpoint nuclear explosions, Navstar will almost certainly have a wider range of applications at both the strategic and tactical levels, making it the more valuable of the two (it has already demonstrated its ability to guide "dumb" weapons with the same accuracy as inherently "smart" precision-guided munitions).[11]

In conclusion, while space is clearly important to both the U.S. and the Soviet Union, the relative dependency on satellites varies between different categories of satellites and in different contingencies.

FUTURE APPLICATIONS

The importance of space shows every indication of growing in the years ahead as the military applications of satellites continue to expand. At the tactical level, the use of satellites to detect aircraft and direct air defenses—a role very similar to that performed by AWACS—is currently being considered. Local commanders will also be able to use such satellites for deep strike operations. With the added precision of navigation satellites, long-range military targets can be attacked with high confidence of success. High-accuracy theater and strategic missiles will also increase the potential for preemptive counterforce attacks. Perhaps the most portentous development will be the use of satellites for strategic anti-submarine warfare operations. Current efforts to develop space-based sensors that would make the ocean depths transparent will undermine the last truly invulnerable segment of the strategic triad. The impact of such systems, however, might be mitigated by the deployment of ballistic missile defenses. Regardless of the final configuration of a BMD system—whether for limited point defense or comprehensive population protection—space systems will be used for the early detection, tracking, and discrimination of attacking warheads. More ambitious systems would make outer space the first line of defense, through orbiting or "pop-up" battle stations.

Although the trend towards relying heavily on space systems to support a wide range of military missions will continue in the foreseeable future, the wisdom of pursuing this course must be evaluated in light of their growing vulnerability to a determined adversary's countermeasures.

VULNERABILITY AND PROTECTION OF SPACE SYSTEMS

The expanding role of military satellites has naturally increased their attractiveness as military targets. Weapon systems designed specifically to disable satellites, however, represent just one approach among several for thwarting a satellite's usefulness. Since every satellite is part of a complex web of communication links and ground facilities, it is vulnerable on a number of fronts—a fact well appreciated by the designers of ASAT systems. For example, the mission payload (such as sensors or communication transponders), power source (solar cells or a nuclear generator), communications package, and control system, which are all essential for the satellite's continuing operation, are vulnerable to specific ASAT efforts. The satellite itself can be threatened by homing vehicles, fragmentation warheads, directed energy weapons, and the effects of a nuclear blast.

The radio communication links to and from a satellite, which are required for routine station-keeping and the transmission of mission-related information, are also vulnerable to the effects of nuclear explosions, electronic jamming, "spoofing" (the giving of false commands and information), and "takeover" (unauthorized use by an adversary). Finally, the ground sites of a space system, which include launch sites, space surveillance networks, command-and-control centers, and receiver stations, are vulnerable to direct military attack and terrorist activities.

Given the variety of methods for disabling or interfering with satellites, a distinction is often made between *dedicated* ASAT systems (i.e., weapons designed solely or primarily for this purpose, such as the current U.S. and Soviet ASAT systems) and *residual* systems (i.e., weapons that have a dual or potential ASAT capability such as long-range missiles, electronic jammers, and ABMs).

Although the threats to space systems are therefore numerous, there are many ways that a country can prolong the survival of its satellites. The first method at its disposal is deterrence of a potential aggressor. As in all relationships based on deterrence, it is necessary that the side wishing to protect its space systems has a credible retaliatory capability, and that the attacker has a meaningful stake in the uninterrupted use of his own satellite services. If the latter is uncertain, the threat of retaliation could instead be made against whatever *non*-space systems are most highly valued by the attacker.

A second method is defense of the satellite and support systems by active and passive means. Active measures are designed either to destroy a threatening ASAT system—for example, with shoot-back lasers or interceptors—or to deflect, deceive, or evade an attack through the use of jammers, flares, decoys, and emergency maneuvering systems. Passive defensive measures—armoring the satellite against nuclear effects and directed energy weapons, adding antijamming devices and sensors to warn of an ASAT attack, and increasing the autonomy of satellites to avoid dependence on instructions from ground stations—can also be employed.

The survivability of a space system can be further enhanced by the use of reconstitution or redundancy of individual satellites. This involves either replacing disabled satellites with spares already prepositioned in space, or developing an emergency launch-on-demand capability. Communication links can be maintained by internetting procedures, while the vulnerability of ground terminals can be reduced by mobile systems. The ability to switch to non-space alternatives without creating a meaningful drop in performance would place a country in an extremely advantageous position. (See table 2, next page, for illustration of how survivability countermeasures offset threats to a space system.)

SPACE SYSTEM SEGMENT	THREAT	PROTECTIVE COUNTERMEASURES
ORBITAL SEGMENT	NUCLEAR: ICBMs, SLBMs, ABMs	ELECTROMAGNETIC PULSE (EMP) SHIELDING, NEUTRON HARDENING, IN-ORBIT SPARES, GROUND BACKUPS
	NON-NUCLEAR: FRAGMENTATION CHARGES, IMPACT WEAPONS; DIRECTED ENERGY: LASERS, ETC.	WARNING SENSORS AND MANEUVERABILITY, FLARES, JAMMERS, IN-ORBIT SPARES, GROUND BACKUPS, REFLECTIVE COATINGS/SHIELDING
LINK SEGMENT	JAMMING	ELECTRONIC COUNTERMEASURES (ECM), FREQUENCY HOPPING, ANTENNA NULLING, GREATER SATELLITE AUTONOMY
	TAKEOVER	AUTHENTICATION (ENTRY CODES, ETC.)
	SPOOFING	ENCRYPTION
GROUND SEGMENT	MILITARY OR TERRORIST ATTACKS ON: LAUNCH FACILITIES	REDUNDANT FACILITIES, SECURITY PRECAUTIONS, HARDENING, REPLACEMENT LAUNCHES FROM SLBMs OR ICBM SILOS
	SPACE SURVEILLANCE RADARS AND TELESCOPES	REDUNDANT FACILITIES, INTERNETTING
	COMMAND-AND-CONTROL FACILITIES	MOBILE SYSTEMS, SATELLITE AUTONOMY

Table 2 Threats to Space Systems and Protective Countermeasures

Although a wide range of survivability options exists, the efforts devoted to satellite survivability must be kept in proportion to the nature and probability of the threat, and the value of the satellite. While it is obviously desirable to protect a satellite against a crippling "cheap shot," ensuring survivability becomes progressively difficult and expensive as the ASAT threat grows and becomes more sophisticated. Beyond a certain threshold, it would probably be simpler and cheaper to build low-cost reconstitutable satellites or use non-space alternatives, rather than attempt to protect each and every existing satellite against the whole host of ASAT techniques.

* * *

The militarization of space has resulted in what many regard as two incompatible trends: the first is the steady expansion of the role of military satellites; the second is the inevitable response of anti-satellite weapons. This raises a dilemma. To continue to rely on increasingly vulnerable space systems is to court disaster. Yet, to restrict the development of ASAT weapons is to make outer space safe for potentially threatening satellites.

In a sense, though, the alternatives are not actually this stark. It is true that if the number of dedicated ASAT weapons were controlled, the existence of residual ASAT capabilities would still pose a threat to satellites. Nonetheless, we face a fundamental choice: we can enjoy the benefits—and tolerate the attendant hazards—of allowing satellites to operate in a relative sanctuary; or we can opt for a situation in which the full potential of space will never be exploited because we are reluctant to expose our satellites to the threat of unrestrained ASATs. Such a choice can only be made after careful weighing of the benefits and hazards satellites pose to our country. This choice is fast becoming moot, however, as the development of space weapons gains momentum, and reduces the likelihood of meaningful arms control.

ENDNOTES

[1] The term "sanctuary" needs some qualification: both superpowers have had the capability, from the beginning of the space age, to disrupt and disable satellites with electronic jamming or by detonating nuclear warheads nearby.

[2] Unless otherwise stated, information for the following section is taken from Paul Stares, *The Militarization of Space: U.S. Policy 1945–1984* (Ithaca, N.Y.: Cornell University Press, 1985).

[3] See Robert P. Berman and John C. Baker, *Soviet Strategic Forces: Requirements and Responses* (Washington, D.C.: The Brookings Institution, 1982), p. 46–47.

[4] U.S. Department of Defense, *Soviet Military Power*, 3rd ed. (Washington, D.C.: Government Printing Office, 1984), p. 46.

[5] See Nicholas Johnson, *The Soviet Year in Space: 1983* (Colorado Springs: Teledyne Brown Engineering Corp., 1984), p. 10.

[6] Thomas Karas, *The New High Ground: Strategies and Weapons of Space Age War* (New York: Simon and Schuster, 1983), pp. 137, 145.

[7] See Stephen M. Meyer, "Soviet Military Programmes and the 'New High Ground,' " *Survival*, Sept.–Oct. 1983, p. 207.

[8] For a comprehensive discussion of the variety of communication systems used by the superpowers, see William M. Arkin and Richard Fieldhouse, "Nuclear Weapon Command, Control and Communications," in *SIPRI Year-Book 1984* (London: Taylor and Francis, 1984), pp. 455–516.

[9] See Colin S. Gray, *American Military Space Policy: Information Systems, Weapon Systems and Arms Control* (Cambridge, Mass.: Abt Books, 1982), p. 5.

[10] Charles A. Zraket, "Strategic Command, Control, Communications, and Intelligence," *Science*, June 22, 1984, pp. 1306–1311.

[11] See U.S. Congress, Senate Subcommittee of the Committee on Armed Services, *Department of Defense Authorization for Appropriations for Fiscal Year 1981*, 96th Cong., 2nd sess., part 5, 1980, p. 2674.

Kurt Gottfried and Richard Ned Lebow

Anti-Satellite Weapons: Weighing the Risks

THE MISSION OF ANTI-SATELLITE WEAPONS (ASATs) is to destroy vital components of an adversary's network for intelligence gathering and for the command and control of his own forces. This raises a fundamental question: does the military usefulness of ASATs overshadow their capacity for exacerbating crises and conflicts? Any device that is able to destroy satellites has this dual nature; its targets' roles change dramatically as the world moves from peacetime, through crisis, to war. Satellites that can transmit images of military units electronically (in "real time") could be invaluable during negotiations to resolve a crisis and in building confidence that promises are being kept. Yet if diplomacy should fail, that same data could be used to optimize attack. A similar ambiguity characterizes satellites that give early warning of missile launches: in peacetime they provide reassurance, but in war they could be used to help target a retaliatory strike.

ASATs play a correspondingly complex role. On the one hand, they enhance a nation's military capability and thereby its ability to deter attack; on the other hand, the very fear of an ASAT attack on vital satellites could contribute to the escalation of a crisis or low-level conflict. An evaluation of ASATs must consider how they will affect this delicate balance between the somewhat contradictory goals of fighting a war, and preventing one.

This article will seek to strike that balance by examining the possible military and political roles of ASATs in a variety of crises and conflicts. To do this, we shall have to distinguish between the near and long term, since ASAT technology is still in its infancy. Assessing future confrontations involving systems that do not yet exist is a speculative venture. Nevertheless, we still conclude that all the technological trends indicate that ASATs possess a considerably greater capacity for transforming a crisis into a war, and for enlarging wars, than they do for assisting in military missions or enhancing deterrence. Given this conclusion, we shall argue that our security would best be served by a comprehensive ASAT test-ban treaty, complemented by feasible satellite protective measures.

SATELLITES AND ASATS

ASATs do not currently pose a threat to all satellites, though by the next decade they might be able to. At the moment, the vulnerability of satellites is in large measure determined by their orbit. These orbits can be divided into four distinct categories, as shown in table 1, opposite.

Low orbits. Satellites in orbits below 5000 kilometers are the only ones vulnerable to the current generation of ASATs. Both superpowers use such orbits for photo-reconnaissance, electronic intelligence (ELINT), meteorology, geodesy, and manned spacecraft. The Soviets also use these vulnerable orbits for some communications satellites and for their radar ocean surveillance satellites.

Highly elliptical, or Molniya orbits. Satellites in Molniya orbits are used extensively by the Soviets for strategic early warning and for military and civil communication. Because these satellites remain in sight of ground stations at northern latitudes for about eight of their twelve-hour orbital periods, they would be potentially vulnerable to the forthcoming U.S. ASAT, if the United States should stage its attacks in the Southern Hemisphere where Molniya satellites dip to their lowest altitudes. To counter this threat, the Soviets could conceivably use geosynchronous satellites instead, and relay signals through semi-synchronous satellites to reach ground stations at high latitudes.

Semi-synchronous orbits. Satellites in these orbits, at altitudes of about 20,000 kilometers, are beyond the range of ASATs expected to

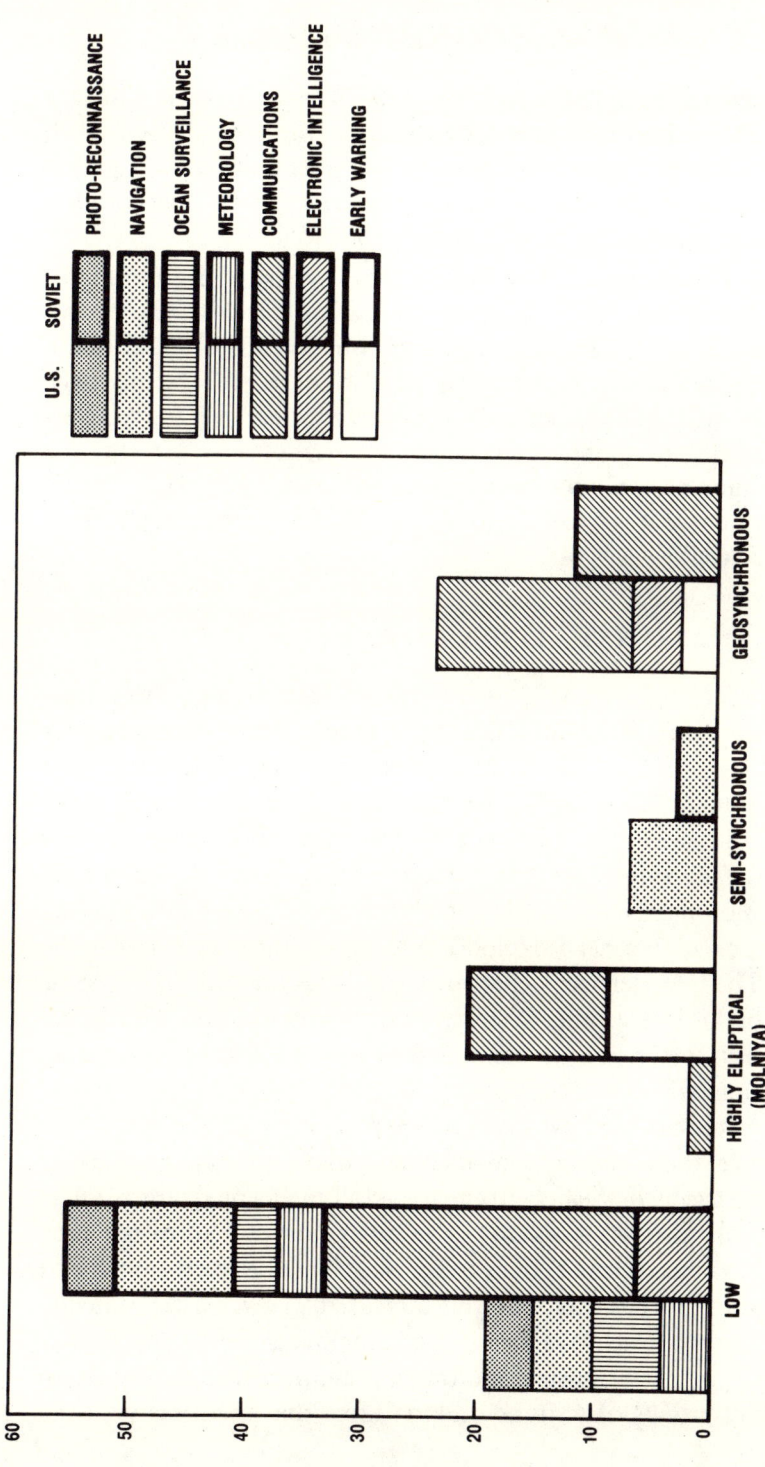

Table 1 Typical constellations of military satellites deployed by the United States and Soviet Union at any given time (includes some civilian satellites with military capabilities). The current generation of anti-satellite weapons threatens only those satellites in low orbits.

be operational in this decade. The U.S. Navstar system, which is expected to be fully operational by 1987, is now being placed in such orbits, as is the similar Soviet GLONASS system. These navigation satellites will enable military forces to determine their own location and deliver their weapons with remarkable accuracy.

Geosynchronous orbits. Satellites in these orbits, at an altitude of 36,000 kilometers, circle the earth at the same rate as the earth's rotation and remain above a fixed point on the equator. Satellites in such high orbits are invulnerable to current ASAT systems. The United States exploits such orbits for strategic early warning, electronic intelligence, and military communication. At present, the Soviet Union has only communication satellites in such orbits, but by the next decade it is expected to have a wider array of geosynchronous satellites.[1]

CURRENT ANTI-SATELLITE WEAPONS

The current Soviet ASAT is a massive device weighing more than 2000 kilograms. It is lofted into orbit atop a liquid-fueled booster derived from the old SS-9 ICBM. In all past tests, the Soviets have launched the ASAT from two pads at the Tyuratam test range. After launch, the ASAT completes one or two trips around the earth, its orbit is changed to cross that of the target, the ASAT conducts final interception maneuvers under the guidance of an on-board homing device, and its warhead explodes into pellets directed towards the target. The U.S. Defense Department estimates that the Soviet ASAT is capable of intercepting satellites in orbits up to 5000 kilometers; the highest intercept ever attempted in a Soviet test has been only 1600 kilometers.

The first test of the Soviet ASAT took place in 1968, and the twenty tests since then can be classified according to the interception technique used.[2] In the first group, the ASAT intercepted targets on its second orbital pass, using an active radar homer. Out of ten such tests, seven succeeded. About three hours elapse between launch and interception in these second-orbit attacks; this gives a U.S. satellite opportunity for evasive maneuvering or jamming of the ASAT's radar homer. It was these considerations that presumably led the Soviets, in two other groups of tests, to attempt first-orbit intercepts with the

Anti-Satellite Weapons: Weighing the Risks   151

radar homer, and second-orbit intercepts with an optical/infrared homer that detects the target's heat radiation and is not so easily foiled. In the first-orbit intercepts, four tests produced two successes; in the optical/IR homing tests, six tests yielded not a single success.

In addition to its poor test performance, the Soviet ASAT suffers from several inherent limitations. First, it can attack a satellite only when the target's ground track runs close to the ASAT launch site, which happens just twice a day; this means waiting an average of six hours to attack a given satellite. Second, the ASAT's heavy weight requires a massive booster; it can therefore only be launched from a few Soviet facilities. Third, it is difficult to fire such massive liquid-fuel boosters in rapid succession from a single launch site.

Taken together, these limitations lead informed observers in the U.S. to estimate that it would take the Soviet ASAT force a week or more to destroy those American low-altitude satellites that are theoretically within its reach, although it may take only several days to eliminate essential ELINT and photo-satellites. Attacking satellites at 20,000 kilometers or higher would require a much more massive booster rocket, and probably modifications and improvements to the interceptor as well. Futhermore, the ASAT would give ample warning of its attack, since it would take several hours to reach such altitudes.

An assessment of the threat posed by the Soviet ASAT was given several years ago by General Lew Allen, then Air Force chief of staff: "I think our general opinion is that we give it a very questionable operational capability for a few launches. In other words it is a threat we are worried about, but they have not had a test program that would cause us to believe it is a very credible threat."[3] Nothing that has happened in the Soviet test program since that assessment would suggest that the threat has increased.

The current U.S. ASAT program was begun by the Air Force in 1977. The ASAT's mission, as announced by the Carter administration, was to deter the Soviets from using their ASAT by threat of retaliation, to protect U.S. military units from observation by Soviet satellites, and to induce the Soviet Union to reach an ASAT arms-control agreement.

The U.S. Air Force ASAT weapon is launched by a two-stage rocket small enough to be carried to high altitude by an F-15 fighter. This rocket boosts a device known as a miniature homing vehicle into the path of the target satellite, using data on the target's path supplied by the ground-based satellite tracking network. The homing vehicle itself is a squat cylindrical object weighing about 15 kilograms. It has infrared sensors able to "see" the target against the cold background of space, and a set of small thrusters that can be fired to bring the cylinder into the path of its target. The resulting high-speed collision destroys the target. The maximum altitude of the U.S. ASAT has not been revealed, but the joint chiefs of staff have stated that the highest priority targets are Soviet satellites that can locate American forces—presumably photo-reconnaissance, ELINT, and ocean surveillance satellites at altitudes of 500 kilometers or less.

If it performs as it is supposed to, the U.S. ASAT will be more versatile and capable than its Soviet counterpart. In principle, any F-15 can be adapted to carry the ASAT. Furthermore, carrier-based aircraft, or mid-air refueled F-15s, can carry the ASAT to an attack position almost anywhere in the world. In contrast to the Soviet ASAT, the U.S. weapon does not go into orbit but ascends directly towards its quarry, thereby giving it little opportunity for evasion. If the United States were to invest in extensive tracking, control, and communication facilities, it could acquire an ASAT force able to destroy all Soviet low-orbit satellites in a matter of hours.

Although the Air Force is charged with the anti-satellite mission, the Army's Ballistic Missile Defense Program also contributes to the United States' ASAT capability. In the Army's Homing Overlay Experiment of 1984, an infrared-homing interceptor launched from Kwajalein destroyed a dummy ICBM warhead at an altitude of more than 150 kilometers. Since a satellite is much larger and more fragile than a warhead, there is no question but that this weapon has an ASAT potential. Indeed, this experiment indicates that the United States and Soviet Union have now tested rather similar ASAT weapons. Both weapons have the handicap of being launched by ground-based ICBM boosters, though the U.S. homing overlay interceptor exploits an infrared homing device that the Soviets have not yet mastered.

FUTURE ANTI-SATELLITE WEAPONS

In addition to these existing weapons, other ASAT techniques have been proposed, including space mines and directed energy ASATs. A space mine would be a satellite containing explosives, placed in orbit well in advance of a conflict. When detonated by remote command (or on being tampered with), it would destroy its target. In principle, space mines could threaten satellites at all altitudes. However, if the target satellite had on-board sensors and could maneuver effectively, the space mine would have to be a sophisticated device, perhaps more complicated than current ASATs, in order to stay near its target. A possible drawback of the space mine is that its placement alongside another superpower's satellite could, in and of itself, trigger a confrontation.

There has also been much interest in anti-satellite weapons that exploit directed energy (laser or neutral particle) beams. Both the United States and the Soviet Union now have laser demonstration projects that could be turned to ASAT purposes. Directed energy concepts primarily intended for space-based ballistic missiles defense would, if pursued, also have a significant ASAT potential. Even though the energy deposited on a target by a directed energy weapon decreases sharply with distance, a laser or neutral particle BMD weapon able to destroy a missile in a matter of seconds could also deposit enough energy, at a high enough rate, to damage unprotected satellites in geosynchronous orbits. In this regard, it is essential to recognize that even an ineffective space-based BMD system could be a potent ASAT, while an effective BMD system would, as a matter of course, be an ASAT system par excellence.

SATELLITE PROTECTION

There are various ways of making satellites less vulnerable to ASATs. First, satellites could be promptly replaced when they are destroyed. At present, U.S. satellites are more sophisticated, reliable, and longer lasting than their Soviet counterparts. For example, the newest Soviet photo-satellites have a lifetime of six weeks, whereas that of the U.S. KH-11 exceeds two years. U.S. communications satellites have lifespans five to ten times longer than those of the Soviets. Hence, the Soviets need to replace satellites far more frequently, and this higher Soviet launch rate may grant them a potent wartime reconstitution capability. The United States has now begun to devote greater attention to satellite replacement: for instance, by placing spare satellites in orbit, to be activated when needed.

There are other ways, varying in efficacy and cost, to protect satellites. Attack by a homing ASAT could be frustrated by a satellite that can maneuver, or that can deceive the ASAT's homing device. If there is ample warning of attack, as is true with the current Soviet ASAT, evasion could be commanded from the ground. Against the swiftly approaching U.S. ASAT, evasion would be difficult if not impossible. Radar homing devices of the sort used by the Soviet ASAT could be foiled with jammers placed on board satellites. Infrared homing devices, such as the one on the U.S. ASAT, cannot readily be jammed but could be deceived by decoys deployed by the target satellite. With the exception of photo-satellites, most current satellites have little or no ability to maneuver. But manned spacecraft such as the space shuttle could be used to refuel low orbit satellites, if they were properly designed. This would allow the satellites to carry enough fuel for evasive maneuvering.

It is also possible to achieve some degree of protection against directed energy beams and radiation. Photo-satellites, for example, could be protected against blinding from low-intensity lasers by automated shutters. Electronics can also be hardened against neutral particle beams and nuclear explosions.

If the development of ASATs continues unrestrained, we can expect to see competition between techniques of destroying satellites, and methods of protecting them. Since satellites are inherently fragile and complex, they are virtually certain to lose this competition to simpler weapons whose sole mission is to destroy them. Satellites that have mission lifetimes of many years are particularly vulnerable since they must be equipped with defenses against future, unpredictable, ASAT threats.

THE MILITARY UTILITY OF ASATS

ASATs could be used by the superpowers in a variety of confrontations ranging from wars between their client states, to full-scale nuclear war. Various scenarios can be examined, but because of the number of unknown factors they are necessarily speculative. Indeed, the most serious threats to security are likely to arise from unexpected and unwise actions that may be taken simply because of the existence of an ASAT capability.[4]

Anti-Satellite Weapons: Weighing the Risks   155

Our analysis will distinguish between possible near-term and more distant scenarios. This is a natural division to make because the current generation of ASATs does not threaten higher orbit strategic satellites, while those of the future are likely to do so.

Regional Conflicts

In the past, both superpowers have been very cautious about sharing intelligence data from satellites with even their closest allies, although they have occasionally provided allies with battlefield intelligence drawn from other sources. In 1973, for example, the Soviet Union gave Egypt aerial photographs of Israeli forces on the West Bank of the Suez Canal.[5] This superpower satellite reconnaissance of Third World battlefields is on its way to becoming a regular feature of international politics, however. Would either superpower deem it in its interest to use ASATs against the other's satellites?

The use of ASATs in such cases seems rather implausible because of the risks of escalation it would entail. Damaging or destroying the other superpower's satellite would almost certainly provoke a serious crisis. Moreover, there are less provocative, more direct, and long-accepted ways to show commitment and offer military assistance—increased logistics support, sending of advisers or even troops, intelligence support from aircraft or other traditional means—if it proved necessary to aid an ally or encourage caution in the other superpower. Unlike the destruction of a satellite, none of these actions involve a direct attack against a military asset of the other superpower.

Superpower War in Europe

That a major war between NATO and the Warsaw Pact would long remain conventional appears to us an illusion. Nevertheless, the superpowers devote the lion's share of their military budgets and training programs to preparations for conventional war. For this reason alone, we must consider the possible utility of ASATs in a conventional superpower war.

Our findings are disquieting. In the near term, neither side is likely to use ASATs because they offer no significant military advantage. In the long term, the temptation to use ASATs will be much greater. The combination of more capable ASATs with greater dependence upon satellites will create strong pressures to use these weapons. Attacks against satellites would significantly raise the nuclear alert level, producing new and complex interactions between the warning and response mechanisms of the two superpowers. They would also be likely to provoke retaliation in kind, an event that would aggravate the potential for loss of control. It seems that advanced ASATs are a military Pandora's box: at first glance, their use may be appealing, but the consequences may be disastrous.

The characteristics of existing ASATs and those that could be deployed in the next few years severely limit their utility. They cannot make possible a surprise conventional attack in Europe. Any systematic effort to destroy or interfere with intelligence satellites would itself provide relatively unambiguous warning of impending attack. Moreover, since the superpowers also use ground-based and airborne surveillance systems in Europe for early warning and intelligence gathering, simply destroying satellites would not eliminate surveillance capabilities.

ASATs could not deprive either side of theater-wide communications, either. At first sight, the United States has the greater vulnerability in this regard, because it has assigned satellites a primary responsibility for important communications missions, using ground-based systems as a backup. For the Pact, the situation is generally the reverse. Both superpowers have nevertheless built up extensive cable, microwave, and radio communication networks to meet most of their immediate military needs in the absence of satellites. Battlefield communication between different sectors, more heavily satellite dependent in NATO, could also be provided by airborne relays.

Another possible use of ASATs, after the onset of fighting, would be to destroy low-orbit photographic, electronic, and radar surveillance satellites that can enable the other side to locate military units behind the battlelines. Perhaps the most likely scenario for Soviet attacks on American satellites, for instance, would be in the aftermath of a stalled Warsaw Pact offensive. The Soviet high command might be desperate to buy a "window of darkness" behind which to regroup forces and shift the axis of advance, and so be willing to risk the repercussions of initiating ASAT warfare. An action such as this would undoubtedly be part of a coordinated attempt to deny NATO battlefield intelligence from a variety of other sources.

In this connection, it is important to recognize that satellites are likely to play an increasingly important role as sources of battlefield intelligence the longer any war lasts, due to the likely attrition of more conventional intelligence assets. Even so, ASAT operations are unlikely to deprive NATO entirely of critical military intelligence, even though they could seriously interfere with real-time intelligence gathering. In the absence of low-orbit satellites, NATO would have to rely on intelligence from aircraft, drones, ground-based sensors, and ground observers. Intelligence from such a large number of discrete sources, operated by a variety of different and only partially integrated commands, would take considerable time to collate, interpret, and relay. The resulting loss of real-time capability could severely curtail the effectiveness of NATO air defenses, and significantly affect the course of ground operations as well.

"Sweeping the skies" of low orbit satellites so as to deny surveillance of areas far behind the front would appear to be a militarily attractive objective. But there are other, less risky, ways to hinder satellites. Electronic intelligence satellites could be thwarted by jamming or spoofing, or through encryption of communications. Photo-satellites are often thwarted by the cloud cover in Central Europe, which prevails about 60 percent of the time. In clear weather, aircraft (manned and unmanned) could supply much of the photographic reconnaissance needs, even though such operations could prove costly. Radar surveillance satellites would be the most irreplaceable low-orbit satellite capability. Alternatives, such as airborne radars, must be flown in well-protected airspace and would thus gain only a limited view behind the battlelines. Radar satellites, on the other hand, are likely to be more vulnerable to jamming and spoofing.

The most commonly proposed U.S. anti-satellite use also involves shielding rear area forces: preventing Soviet satellites from tracking NATO's carrier battlegroups that protect the Airland bridge, which is vital to the reinforcement of Europe. But since the Soviets rely heavily on submarines and trawlers for continuous tracking of NATO naval forces, denying satellite coverage would not protect the ships. In any case, the Soviet ocean radar reconnaissance (RORSAT), ocean ELINT (EORSAT), and photographic satellites used to track NATO naval movements all appear to have significant operational limitations; the Soviets have experienced several failures of the RORSAT and have only once maintained a full constellation of RORSATs and EORSATs.[6] Given the technical shortcomings of Soviet ocean reconnaissance satellites, these systems could be defeated by jamming and spoofing, without the need for an ASAT. Yet for NATO to deny ocean surveillance to the Soviets entirely, it would have to clear the seas of Soviet submarines and ships, and also successfully interdict the large number of Backfire bombers used by Soviet Long Range Naval Aviation to provide additional coverage. This would be a formidable task.

Overall, the marginal military utility of the current generation of low-altitude ASATs seems to us unlikely to offset the hazards of their use. The destruction by one superpower of the other's low-orbit satellites could well provoke retaliation in kind, raising the prospect of undesired and perhaps uncontrollable escalation. It is unclear just how much loss of satellite-borne theater command-and-control either side will tolerate, especially if important ground-based facilities are being damaged or destroyed at the same time, before it resorts to graver measures. It is for these reasons that we feel the use of ASAT in any war fought in Europe in the coming years would be counterproductive—a view that we feel would probably be shared by the highest military authorities in both Washington and Moscow.

But if we look ahead to the next decade, the picture is rather different. NATO's vulnerability to Soviet ASATs is likely to increase in proportion to its growing reliance on satellite-based systems for battle management. In coming decades, NATO may have radar satellites for theater-wide surveillance of aircraft. The U.S. Navy may also become increasingly dependent upon satellites to counter stand-off threats to its carrier groups posed by Soviet aircraft and submarines. NATO forces, down to the platoon level, are expected to use Navstar satellites to improve the delivery of firepower. The follow-on-force attack (FOFA) and other deep strike strategies, if they are implemented, would make NATO even more dependent upon low-orbit satellites for locating Soviet second- and third-echelon forces behind the line of battle.

The Warsaw Pact may also become more dependent upon satellites. Some Soviet military writers envisage satellites as the solution to the fundamental dilemma of Soviet conventional strategy: the battlefield responsiveness so essential to the kind of offensives the Soviets contemplate is in conflict with their rigid and hierarchical command structure. A real-time battle management system would permit the Soviet general staff to direct operations with much more flexibility and responsiveness. If the Soviet Union develops such a capability in the course of the next decade, it could become even more dependent than NATO is upon satellites.

By the 1990s, both superpowers can also be expected to have deployed more capable ASATs, perhaps ones able to destroy satellites in geosynchronous orbits. This would bring about a highly unstable and dangerous situation because both the superpowers would be extremely dependent upon vulnerable satellite systems.

It is conceivable that mutual vulnerability would enhance mutual deterrence and restraint. Fear of escalation could also encourage restraint. An historical parallel might be found during the Korean War when, for both these reasons, the two sides tacitly respected critical sanctuaries throughout the fighting. The United States refrained from attacking the air bases in China from which MiGs sallied forth daily to strike at United Nations' positions. China and North Korea for their part never attacked the American carriers or other warships in Korean waters that provided vital air and naval support for ground operations.

Yet it is also possible that one or both of the superpowers would attack the other's satellites because of their military utility. NATO dependence upon radar satellites would give the Soviets a strong incentive to take out those satellites at the start of their offensive; likewise, NATO will be sorely tempted to attack satellites essential to the Soviets' battle management capability. The need to conduct ASAT operations in order to guarantee victory or to stave off defeat could be seen as so vital at the time that it would seem to offset the risks of escalation. The historical analogy in this instance is to Germany, which during the First World War decided to wage unrestricted submarine warfare. Both the Kaiser and his military chiefs were fully aware of the magnitude of their decision. If the U-boat campaign succeeded, Germany would win the war, and the opprobrium of using such a weapon would be of little consequence. If it failed, as indeed it did, American intervention was expected to lead, in the German chancellor's words, to *"finis Germaniae."*

Which outcome is the more likely, restraint or attack? The greater the asymmetries between satellite dependence and ASAT capability, the more prospects for mutual deterrence will decline. In the opinion of many experts, the United States will maintain a significant advantage in ASAT capability for the foreseeable future. But the U.S.

is also likely to become more dependent upon satellites, and this should reduce the temptation to exploit its ASAT advantage.

The most destabilizing situation would be one in which the superpower with the greater ASAT capability also had the least dependence on satellites. This seems unlikely to occur but it is not entirely inconceivable. By the late 1980s, the Soviet Union could have an operational ASAT against low-orbit satellites as effective as the American ASAT. At the same time, the United States could have become much more dependent upon satellites for the range of military purposes we have described. In such a circumstance, the Soviets might see few drawbacks to using their ASATs.

Nuclear War

Satellites would play a more important role in intercontinental nuclear war than in regional or conventional conflicts because they provide eyes and ears for early warning of nuclear attack, battle management of nuclear forces, and post-attack assessment of nuclear strikes.

American satellites that perform these functions are almost all in geosynchronous or semi-synchronous orbits where they cannot be attacked by the existing Soviet ASAT. Soviet satellites are now somewhat more vulnerable because the Molniya orbits of Soviet early-warning and many communications satellites are, in principle, within range of the current U.S. ASAT weapon. In the future, the development of directed energy ASATs, whether as part of ballistic missile defenses or as dedicated ASATs, could threaten satellites at all altitudes with prompt and simultaneous destruction, carried out at the same time as an offensive first strike.

To some extent, such an ASAT attack could impair the other side's ability to launch a coordinated counterforce strike in response. Nonetheless, our analysis persuades us that directed energy ASATs would not significantly effect the course of a nuclear war *once it has begun*. The real danger is that advanced ASATs would compound the difficulty of *terminating* a nuclear war, and in a crisis they would undoubtedly heighten the chance that nuclear war would break out.

Our judgment about the contribution ASATs would make in a nuclear war stems from several considerations. First, even advanced ASATs would, for the most part, only marginally degrade early-warning capabilities. For the United States, the increment of additional warning provided by satellites is important but insufficient to protect the national command authority, submarines in port, and bombers on the ground from destruction by Soviet missiles fired from off-shore submarines. The United States is more vulnerable in this regard than the Soviet Union because of its greater reliance on bombers and submarines as strategic launch platforms. It is nevertheless reasonable to assume that if an intercontinental nuclear war occurs, it will arise out of a serious crisis or war in Europe or elsewhere. American strategic forces would already be on alert and therefore less vulnerable. Contingency arrangements would also have been put into effect to shift the command center away from Washington, thus reducing the likelihood of strategic paralysis in the aftermath of an attack on command authorities.

Given the present situation, loss of early-warning satellites would be more easily borne by the Soviet Union, if only because their systems are less reliable and effective than their American counterparts. The Soviets continue to experience serious difficulties in maintaining an operational constellation of early warning satellites. Even if the Soviets established effective satellite coverage of North America, they would remain vulnerable to nuclear attacks by missiles deployed in Western Europe, and on submarines in the Mediterranean, Arctic Ocean, and, with the advent of the Trident submarine, the Indian Ocean as well. Submarine-launched Trident II missiles, when they become operational, will be accurate enough to destroy Soviet missile silos. Threats of this kind constitute a growing problem for the Soviets, as 75 percent of their MIRVed missiles are land-based, stationary, and increasingly vulnerable. Yet there is nothing to indicate that a U.S. ASAT attack on Soviet early-warning satellites or radars would significantly increase our ability to destroy their ICBMs. Destruction of these systems would in itself constitute early warning and could prompt Soviet leaders to launch their missiles.

In the aftermath of ASAT attacks, both superpowers would also lose important wartime communication capabilities. Here, the United States is probably more dependent on satellites because it uses them to communicate with submarines and airborne bombers. Although these functions could be assumed by ground stations and airborne relays, these systems are themselves highly vulnerable, and cannot be expected to endure very long in a nuclear war.

Both superpowers would presumably also use satellites for post-attack assessment. Since bombers and cruise missiles would reach their targets hours after warheads delivered by missiles, it would be most efficient to assess the result of a missile-delivered attack before following up with other forces. The U.S. government has placed great stress on post-attack assessment in recent years, in keeping with its overall notion of how an intercontinental nuclear war should be fought. The recent emphasis has been on developing the means to fight limited or protracted nuclear wars, despite the widespread conviction among strategists and senior military officers that such scenarios are unrealistic. The Soviets give every indication of sharing this skepticism and, for this reason, may judge post-attack assessment to be relatively unimportant. It also appears alien to their approach to nuclear combat, which emphasizes a massive initial strike, as opposed to a series of strikes, each followed by assessment.[7] In the case of full-scale nuclear war, both sides' satellites will, in any case, become useless after the first exchange, since satellite ground stations and launch pads will almost certainly be destroyed.

ASATs could both increase the likelihood of nuclear war and complicate its termination. Widespread destruction of communication satellites would make it difficult to restrain one's own forces or to transmit a cease-fire order, reducing still further the possibility of limiting a nuclear war once started. And if there were negotiations to end the war, it would be nearly impossible, without surveillance satellites, to monitor the adversary's behavior and gauge the extent of his compliance with any agreement.

Advanced ASATs could have a dramatic impact on crisis stability. Each side must anticipate that in the event of attack, it would have very little time to assess the threat and select an appropriate response. Laser ASATs, for example, could destroy an entire constellation of satellites almost instantly. Such an ASAT attack could coincide with a first-strike, increasing the risk and fear of "decapitation"—the destruction of a country's political-military leadership. A country could develop contingency procedures to be adopted during such crises, but even though this might reduce the probability of decapitation, it would never eliminate the fears. Policy makers could be stampeded into making these fears of war self-fulfilling.

At high levels of strategic alert, advanced ASATs would exacerbate the inherent instability of the superpowers' command-and-control systems.[8] When the alert level of strategic forces is raised, it is necessary to diffuse nuclear launch authority within the military command structure as a precaution against decapitation. However, this heightened alert status would quickly be recognized by the early-warning and response system of the other superpower, which could be pressured into taking similar precautionary action. This feedback loop could progressively dilute centralized control and result in unauthorized actions and inadvertent war. This flaw in command-and-control structures would appear to be an inevitable consequence of the desire to deter a decapitating attack—a situation that would be aggravated by advanced anti-satellite weapons, whose very existence would require a very considerable acceleration of the response tempo.

ASAT ARMS CONTROL

The hazards of escalation or preemption posed by ASATs could be averted by constraining the growth of ASAT capabilities on both sides through negotiated arms control. Such a constraint would also close an important loophole in the ABM treaty.

Unfortunately, the record on ASAT negotiations thus far has been mixed. The United States induced the Soviets to join in several rounds of talks, beginning in 1978, but formal negotiations ceased in 1979 for reasons unrelated to ASAT. Since 1981, the Soviets have repeatedly asserted an interest in resuming negotiations, and from 1983, they have observed a test moratorium. The United States has recently shown interest in negotiating, but insists on continuing the development and testing of its ASAT.

The argument against ASAT arms control was set forth in a report to Congress from the Reagan administration in March 1984. The report contended that the large variety of space systems and activities that have some ASAT capability makes it very difficult to define the term "space weapon" for arms-control purposes; that a truly comprehensive ban on tests of all methods for countering satellites could not be verified; that test bans limited to weapons dedicated to ASAT activities, though perhaps verifiable, could induce the Soviets to develop covertly an ASAT capability that relies on other systems, such as ICBMs or maneuvering spacecraft; that cooperative verification measures might require the United States to divulge sensitive information; and that ASAT arms control would not provide protection for satellite ground stations and launch facilities.

The report also argued that there are compelling considerations in favor of a U.S. ASAT capability, quite apart from these arms-control problems. In particular, U.S. ASATs would deter use of Soviet ASATs, and prevent space from being a sanctuary for Soviet satellites that would enhance the capability of their military forces, especially in naval warfare. The report concluded that "no arrangements or agreements beyond those already governing military activities in outer space have been found to date that are judged to be in the overall interest of the United States and its Allies."[9]

We disagree. A viable ASAT arms-control regime that serves U.S. security interests can be identified. In principle, there are a number of forms that an ASAT agreement might take. The most rigorous would forbid the possession, testing, and use of all techniques that can threaten the functioning or survival of satellites. The least restrictive would not limit ASAT technology, but would only expand the rudimentary body of law that governs space activities. Among the latter are "rules of the road" that would restrict close encounters between space objects of one nation and that of another, or agreements that would outlaw the use of weapons against satellites—not unlike the laws that now apply on the high seas. While such constraints would still leave ASAT weapons deployed on both sides, they merit serious consideration for their ability to prevent inadvertent incidents that might otherwise trigger a crisis or conflict. These provisions could be negotiated in their own right or be included as articles in the treaties we shall discuss below.

A vexing problem for ASAT arms control is that a wide variety of space activities have some ASAT potential. For instance, manned or unmanned maneuvering spacecraft could approach an adversary's satellite and interfere with it or destroy it. But such "spin-off" ASAT techniques would not pose a prompt threat to a set of satellites dispersed in space, and approaching several satellites simultaneously would constitute warning in itself. From a military viewpoint, this kind of "residual" ASAT capability is likely to be of marginal utility. Furthermore, whatever threat it poses could be reduced (though not eliminated) by formal "rules of the road," and by measures that allow satellites to assess such threats and evade them by maneuvering.

Another argument against ASAT arms control is that a very potent ASAT capability already exists in nuclear-armed ICBMs, SLBMs, and the Soviets' Galosh ABM interceptors. These, it is asserted, could be used to attack satellite ground stations and spacecraft at virtually any altitude with nuclear warheads. However, using nuclear warheads, even in outer space, is a risky and provocative act, and it is most unlikely that such acts would be committed unless a nuclear war were on the verge of breaking out, or had just done so. Treaties cannot offer protection once the nuclear threshold is crossed; ASAT arms control should seek to make that initial step less likely.

In sum, it is important to distinguish between *dedicated* ASAT systems that can promptly attack a whole set of satellites without resort to nuclear weapons, and *residual* ASAT capabilities in other weapons or systems. We therefore turn to possible restraints on dedicated ASATs.

A treaty that bans possession of all ASATs would be desirable, but compliance would be difficult to verify. The dismantling of readily visible facilities such as the launch sites used for the Soviet ASAT and the U.S. Army's Homing Overlay Experiment could, in principle, be verified. That would, however, be of limited utility, since ASATs could be launched from other Soviet space facilities, and Homing Overlay interceptors could presumably be launched from many U.S. ICBM silos. Even highly intrusive on-site inspection could fail to verify whether these ground-based interceptors were being covertly stockpiled, and possession of the very small F-15 ASAT would be especially difficult to establish.

A more realistic treaty would confine itself to those ASAT activities that are readily verified: ASAT tests and the development and deployment of new ASAT technologies. Such a treaty could still prevent the development of highly capable ASAT systems. An ASAT test ban treaty could be couched in a number of ways: it could forbid the testing of any technique that could prevent the functioning of a satellite, even if it caused no permanent damage. Or it could confine itself to techniques that actually damage the satellite, cause it to change its orbit, or destroy it. In the latter instance, the test ban could either cover all ASATs, regardless of their altitude capability, or only high altitude ASATs.

Activities that could hinder or prevent the proper functioning of a satellite, without causing any damage, are legion. Deception, camouflage, encryption, spoofing—all could prevent a satellite from "seeing" or "hearing" what it was intended to see or hear. These activities, however, could not be outlawed effectively. Jamming at an intensity that produced no damage could perhaps be forbidden, though it would be difficult to define unambiguously. Futhermore, jamming, encryption, spoofing, and so on are also benign and valuable countermeasures. For this reason, treaties restricted to acts that produce well defined damage or malfunctioning are to be preferred.

A high-altitude ASAT test ban treaty—one banning tests above 5000 kilometers, for example—at first glance appears very attractive. No dedicated ASAT weapons have yet been tested at such altitudes. The satellites essential to U.S. strategic forces are all at high altitudes, while the Soviet Union could presumably use a combination of geosynchronous and semi-synchronous satellites for such purposes. It would therefore appear that a high-altitude ASAT test ban could "grandfather" the current Soviet and U.S. systems and start with a clean slate, while protecting the most essential military satellites. ASAT tests at high altitudes are also more amenable to verification since they are less likely to be confused with other space activities.

This case for a high-altitude treaty is undercut, though, by the movement toward directed energy weapons, whether for ASAT or ballistic missile defense. A variety of beam weapons could, once they had been tested against low altitude satellites or missile warheads, pose a grave and prompt threat to geosynchronous satellites *without* any testing against high-altitude targets. To be effective, a high-altitude ASAT test ban would therefore have to be accompanied by a ban on directed energy weapons at low altitudes.

An ASAT test ban that covers *all* altitudes therefore provides the most viable and valuable ASAT constraint. By including all ASAT weapons, such a ban would also close the most important loophole in the ABM treaty. The severe strictures on BMD development imposed by that treaty could be evaded by dressing the first phase of a BMD program in ASAT clothing. Indeed, quite apart from any treaty, the kinship between ASAT and BMD techniques, and the great vulnerability of satellites as compared to missiles, make the ASAT route a natural path towards BMD. By the same token, a program that is only intended to produce an ASAT capability could, under many circumstances, be construed by the opponent as a budding and illegal BMD program. Since the ABM treaty is the foundation on which all strategic arms limitation rests, the prevention of further ASAT competition would strengthen the shaky arms-control edifice.

The Reagan administration has contended that an ASAT test ban treaty is not verifiable. Yet that judgment is disputed by a number of officials who have held high-level positions in intelligence and military space programs.[10] These officials stress that while verification cannot guarantee the total absence of covert testing of single weapons, it would be virtually impossible to test surreptitiously a system of the complexity required for a prompt and reliable ASAT capability. Extensive experience in observing Soviet programs has convinced these experts that the Soviet military would have little confidence in a system that had not undergone realistic tests, and that such realistic tests could not escape detection. Their statements also underline a point that is not often appreciated outside the intelligence community, to wit, that a properly drawn treaty eases the task of intelligence assessment since it dramatically reduces the pace of technological change.

The Reagan administration's pessimism regarding ASAT verification stands in sharp contrast to its optimism over ballistic missile defense. The space surveillance complex that would mastermind the BMD system envisaged by the SDI is supposed to track vast numbers of space objects, discriminate among them, assign interceptors to each, assess the success of intercept, and reassign weapons—all in a matter of minutes. Moreover, it would have to perform these automated functions while itself under direct attack, and in the environment created by nuclear explosions. If such a formidable task can even be contemplated, it is presumably possible to develop a peacetime space surveillance system that would monitor the relatively infrequent and isolated events that are candidates for clandestine ASAT tests, especially as that system need not come to prompt decisions and would involve human beings.

A viable and adequately verifiable test ban treaty *can* be constructed, provided it is focused on the objective of preventing the growth of a *prompt* and *dedicated* Soviet ASAT capability. If the objective is enlarged to encompass a host of secondary concerns raised by systems that have only *residual* ASAT capability, a viable treaty will no longer be possible. One should also recognize that ASAT arms control would be cost-effective. Under unrestrained ASAT arms competition, we would have to buy far more elaborate space surveillance facilities and satellite defenses—not to mention the ASATs themselves—than under ASAT treaty constraints.

CONCLUSION

Both superpowers are about to make fundamental decisions concerning the military uses of space, decisions that will have far-reaching implications for their security, in the widest sense of the term. Without doubt, their most fateful decision will be whether and how to pursue a space-based ballistic missile defense. Another critical choice they face is whether to continue placing ever greater reliance on military satellites. As satellites grow in military importance they also become more attractive targets. If, at the same time, space-based BMD is allowed to evolve, the capability for satellite destruction will increase dramatically. Unless these conflicting trends are resolved, the superpowers' unending quests for greater individual security will, once again, result instead in a mutual loss of security.

How are we to avoid this outcome, especially given the technologies that can be envisaged? We can do so only by maintaining and reinforcing the goals and stipulations of the ABM treaty. We should accompany this reinforcement with a comprehensive ban on the testing of space weapons, and more ambitious programs to protect satellites. As we have argued above, this is the prudent policy choice, for it recognizes that the space weapons technologies presently under consideration have a considerably greater potential for exacerbating tensions and enlarging conflicts, than they do for bolstering deterrence or improving actual military missions. Under such treaty constraints, the United States would possess much more secure space-based intelligence, early-warning, and command-and-control facilities at considerably lower cost than it would in an environment in which both superpowers were continually refining their ability to destroy satellites.

It must be admitted that a side effect of these constraints would be the creation of a sanctuary in space that will be used to deploy satellites with ever greater military capabilities. Some would argue that such a development is intolerable and therefore unstable. But the alternative—unrestrained militarization of space—is far less tolerable. The superpowers must learn to live with the hazards posed by relatively secure satellites, for in the nuclear age the prevention of war must take precedence over the ability to wage it efficiently.

ENDNOTES

[1] For data on Soviet satellite programs, see Nicholas L. Johnson, *The Soviet Year in Space: 1982* and *The Soviet Year in Space: 1983* (Colorado Springs, Co.: Teledyne Brown Engineering, 1983 and 1984); Stephen M. Meyer, "Soviet Military Programmes and the 'New High Ground,'" *Survival*, Sept.-Oct. 1983, pp. 204–15.

[2] For more on the Soviet ASAT test program, see Richard L. Garwin, Kurt Gottfried, and Donald L. Hafner, "Anti-Satellite Weapons," *Scientific American*, June 1984, pp. 45–55.

[3] Testimony by Gen. Lew Allen, Senate Foreign Relations Committee, July 11, 1979.

[4] Assessments of future ASAT capabilities are further complicated by the fact that they will depend as much on political choices as they do on technology. In the course of the next few years, both superpowers will have to decide how dependent they wish to become on space-based systems for their military needs, how much money and effort they are willing to spend to protect their satellites or to make them redundant, and of course, what kinds of ASAT weapons they want to develop and deploy. These decisions cannot be predicted with any accuracy.

[5] Anwar L. Sadat, *In Search of Identity: An Autobiography* (New York: Harper & Row, 1977), p. 260; Mohammed Heikal, in *The Road to Ramadan* (New York: Ballantine Books, 1975), p. 241, insists that these were not satellite photographs.

[6] Naval authorities nevertheless maintain that these satellites give the Soviets an additional important surveillance capability. Louise Hodgden, in "Satellites at Sea: Space and Naval Warfare," *Naval War College Review*, July–Aug. 1984, pp. 31–45, reviews official statements to this effect.

[7] Stephen M. Meyer, "Soviet Military Programs and the 'New High Ground,'" *Survival*, Sept.-Oct. 1983, pp. 204–15.

[8] Paul Bracken, *The Command and Control of Nuclear Forces* (New Haven, Conn.: Yale University Press, 1983), makes a good case for the likelihood of the national command authority becoming so fragmented in the course of a nuclear war that important strategic decisions might well be made, or even have to be made, at levels several steps or more down from the top.

[9] President's Report to Congress on U.S. Policy on ASAT Arms Control, March 31, 1984. According to Leslie H. Gelb in the *New York Times* (Aug. 3, 1984), "important segments of the Navy" do not concur with the report's claim that ASATs are required to counter the threat posed by Soviet ocean reconnaissance satellites, and instead believe that the Navy has more to lose if American communication satellites are attacked than it has to gain by ASAT attacks on Soviet ocean reconnaissance satellites. Furthermore, Gelb reports that "civilian leaders in the Pentagon and the Arms Control and Disarmament Agency have rejected these [military] views and have sustained the [ASAT] program," and that the March 31 report represents a compromise between these conflicting positions of the civilian and military leaderships.

[10] See R. Jeffrey Smith, *Science;* May 18, 1984, pp. 693–695; the individuals interviewed include Leslie Dirks, until 1982 CIA deputy director for research and technology; Walter B. Slocombe, formerly deputy under secretary of defense;

James Reynolds, former manager of the Navstar program; and Michael May, associate director of Livermore National Laboratory. See also Robert W. Buchheim's May 2, 1984, testimony to the House Foreign Affairs Committee. There has been some concern that ground-based or airborne optical lasers could be developed clandestinely for ASAT purposes. Low-intensity lasers that could swamp a photo reconnaissance satellite, or even damage it, are feasible and would probably evade detection, but it would be a simple matter to equip such satellites with shutters triggered by light sensors. According to Michael May, covert installation of high-intensity ASAT lasers would not be possible.

Ashton B. Carter

The Relationship of ASAT and BMD Systems

PUBLIC OPINION AND POLICY LEADERSHIP today insist on merging anti-satellite weapons and ballistic missile defenses under the general rubric of "Star Wars." But how accurate is it to treat ASAT and BMD as but two facets of a single security problem? At their earliest beginnings more than thirty years ago, ASAT and BMD were treated as naturally related security concerns, for the simple reason that missiles and satellites were both viewed as extensions of the long-range aircraft as a means of strategic attack. But in the intervening decades, ASAT and BMD programs have gone their separate ways. Satellites evolved to carry sensors and radio transponders, not nuclear warheads, while ICBMs have remained faithful descendants of the strategic bomber. ASAT and BMD therefore came to respond to distinct military missions and to pose distinct arms-control problems. There are sound reasons for public policy to treat them separately.

But while their military roles diverged, ASAT and BMD systems continued to share certain technical characteristics, and the argument for merging the two issues focuses precisely on this technological overlap. Because of this overlap, certain types of ASAT systems would possess marginal BMD capability, and many types of BMD systems would have quite substantial ASAT capability. Though this overlap must not obscure the substantial differences between ASAT and BMD, it does give rise to several genuine complications for BMD and ASAT policy. On the one hand, the overlap poses an "ASAT upgrade" problem: unrestrained development or deployment of ASATs having some BMD capability could circumvent the strict limits on missile defenses in the ABM treaty. On the other hand, imposing restraints on ASAT development could also impede development of advanced BMD technologies (or countermeasures to BMD) at an earlier stage than the terms of the ABM treaty alone would require; this might render comprehensive ASAT arms control incompatible with the Reagan administration's Strategic Defense Initiative (SDI). Proponents of the SDI could, on this ground, be led to oppose ASAT arms control, while critics of the SDI could use ASAT arms control as a political tool to suppress BMD.[1]

There are potential hazards in making ASAT and BMD policies hostage to each other in this way. If the policy debate over ASATs is wrongly transformed into a symbolic battle over BMD, we may lose the opportunity to judge the military utility of ASATs and proposals for ASAT arms control on their intrinsic merits. At the same time, some near-term BMD deployment options do not require space-based components, and the merits of these deployments may be overlooked if they are wrongly swept up in the drama over the "militarization of space." One should hope instead for a policy debate that addresses the overlap between ASAT and BMD, while respecting the basic differences between their roles in military security.

DIFFERENCES BETWEEN ASAT AND BMD

In the broadest strategic and military sense, ASAT and BMD have little connection. Ballistic missile defense is concerned with providing protection to the nation in the midst of nuclear war, and its affinities in overall military planning are therefore to air defense, civil defense, anti-submarine warfare, anti-tactical ballistic missile defenses, and other activities that have little or nothing to do with space or with ASATs.

Anti-satellite weapons, on the other hand, might well play a role in nuclear war, but many of the ASAT scenarios cited as worthy of concern have little or nothing to do with nuclear war; they range all the way from peacetime tampering with reconnaissance satellites through varying levels of use in crisis and conventional war. Indeed, ASAT use during nuclear war may not be regarded as the most worrisome use of ASAT, nor, for that matter, one that can be

The Relationship of ASAT and BMD Systems 173

easily guarded against through satellite defense programs or arms-control agreements.

From the point of view of arms control, ASAT and BMD could scarcely be more different. The ABM treaty imposes firm and comprehensive restraints on BMD, and few question the ability of the United States to detect militarily significant violations of its provisions. Virtually no arms-control limits exist on ASAT activities, however, and many doubt the verifiability of agreements that would seek to eliminate all or most threats to satellites.

The argument that technological advances are blurring the distinction between ASAT and BMD and thus compel us to consider them as a single issue is in many ways overdrawn. It is true that both missions may involve intercepting objects at roughly comparable altitudes and speed. But the visibility and vulnerability of satellites, boosters, post-boost vehicles, and warheads differ somewhat; and satellites may orbit at altitudes many times higher than those reached by ICBMs. Future advances in directed-energy technologies might well bring greater commonality to ASAT and BMD weapons. But some very significant methods for disrupting satellite functions—space mines, attacks on ground stations, deception—have no BMD analogues. And important terminal and mid-course BMD systems now under study have no space-based components and would work at low-orbital or sub-orbital altitudes.

Above all, the performance requirements of effective ASAT and BMD systems are almost incomparable. ASAT attack can be mounted from a friendly country, but BMD interception of ICBM boosters must take place from outer space and over enemy territory. BMDs must handle many targets in a short time, whereas ASATs need engage relatively few targets at a leisurely pace. The ASAT operator picks the timing of attack; BMD must react to the initiative of the offense. BMD must operate in the most hostile circumstances imaginable—in the midst of nuclear war—whereas ASATs are just as likely to be used during crises or conventional conflicts.

In short, while there is little technical difference between the act of destroying a satellite in orbit and the act of intercepting an ICBM in flight, there is an enormous difference between the effective performance of a BMD *system* and an ASAT *system*.

EXAMPLES OF ASAT-BMD OVERLAP

To gain some sense of the balance between their technological similarities and performance differences, it will be useful to compare the capabilities of a few illustrative ASAT and BMD systems that share the same technologies.

Current Soviet ASATs and ABMs

The current Soviet ASAT interceptor is a multi-ton device launched by an ICBM-sized missile into an orbit that crosses the orbit of its target. It is guided in the final moments of intercept by a radar or optical homing sensor. The ASAT bears little or no resemblance to Soviet BMD systems under development, in either hardware or interception technique, and it has no BMD capability.

The Soviet Galosh ABM interceptor missiles deployed around Moscow, and currently being upgraded, can fly to altitudes of up to several hundred kilometers and laterally to comparable distances. Ground radars guide Galosh's ascent until it is close enough to its target, at which point its nuclear warhead detonates. Galosh's multi-megaton nuclear warhead could destroy satellites of today's design at ranges of many hundreds of kilometers. In upgrading the Moscow ABM system, the Soviets are deploying one hundred interceptors in all, some portion of them modified Galosh missiles. Galosh missiles therefore constitute, in principle at least, a potent nuclear ASAT system with several dozen low-altitude interceptors. It is even conceivable that with proper radar and guidance, Galosh could use a non-nuclear warhead for ASAT attacks. (The remainder of the upgraded Moscow ABM interceptors will be shorter-range missiles with little or no ASAT capacity.)

Current U.S. ASATs and ABMs

The U.S. homing vehicle ASAT, which would be launched by an F-15 and is undergoing testing, and the Army's Homing Overlay BMD device both involve guiding a small space vehicle to the vicinity of its target with a long-range sensor, activating an infrared homing sensor as the vehicle draws nearer to its target, and destroying the target by direct impact. The Army's Homing Overlay test of 1984, in which an ICBM warhead was intercepted and destroyed, demonstrated that objects at altitudes of several hundred kilometers and at near-orbital

The Relationship of ASAT and BMD Systems

speeds can be intercepted by such homing devices. This homing technology is therefore clearly applicable to both ASAT and BMD, if matched to the proper boosters and integrated into an overall system.

If the United States were to deploy a BMD site consisting of a hundred Homing Overlay interceptors (the number permitted under the ABM treaty), together with appropriate radars or other long-range sensors, such a system would be capable (generically, at least) of making large numbers of low-altitude ASAT intercepts within a short time. Since the Overlay interceptor missile's range is envisioned to be long enough to allow BMD coverage of much of the continental U.S. from a central launch site, the Overlay system would not have to wait for satellite ground tracks to pass very near to the launch site to make ASAT intercepts. The system would therefore make intercepts within at most a few hours of receiving an attack order. On the other hand, this BMD system would threaten only low-orbit satellites.

The U.S. homing vehicle ASAT, as currently conceived, probably has no capability whatsoever to intercept missile warheads launched from the Soviet Union. Among other reasons, the long-range sensors that provide guidance for the ASAT mission (the spacetrack network of ground-based radars, which include the early-warning radars along U.S. borders) are probably inadequate for ICBM intercept, and communicating their data to the ASAT interceptor's guidance system in a timely manner is also probably impractical.

But let us suppose that the United States or the Soviet Union were to deploy an ASAT system consisting of homing vehicles such as those now being tested by the U.S., based not on F-15s but on ground-launched missiles capable of reaching high altitudes. Suppose appropriate long-range sensors for these ASAT interceptors were also provided. The small number of interceptors deployed for an ASAT-only mission—perhaps less than a hundred—would clearly not constitute a major BMD system. A system designed as an ASAT, and not adapted to the stringent requirements of BMD, would also be susceptible to the usual panoply of BMD penetration aids and countermeasures, including nuclear attack on its components. Seen from this point of view, such an ASAT system would seem to be an insignificant missile defense.

From another point of view, however, even this limited BMD potential could not be entirely neglected. A hundred long-range ASAT interceptors of this sort could defend the entire country against small attacks. Even against large attacks, preferential defense tactics would allow either superpower to ensure with high probability the survival of a small number of selected targets. It is precisely the possibility of employing limited ASAT systems in such ways that gives rise to the "ASAT upgrade" issue. This issue will be treated in further detail below, but from this example it is already clear that one's concern over ASAT upgrade is determined more by strategic outlook—whether the survival of a few selected targets makes a critical difference in strategic stability—than by technical details.

Space-based lasers

This discussion will confine itself to orbiting chemical lasers, but most remarks apply as well to orbiting particle-beam weapons. Here, the ASAT potential of a good BMD system is quite clear. If a space-based laser BMD system had the beam brightness and number of orbiting battle stations required for effective BMD, and if its sensors could be adapted to the task of finding satellites as well as missiles (probably a minor matter), it would be capable of instantaneous attack on satellites at all altitudes. Even against heavily shielded satellites, the laser beam could presumably dwell on the target long enough to cause damage.

A space-based laser system designed specifically for the ASAT mission, on the other hand, would probably consist of a smaller number of orbiting stations of lesser beam brightness and smaller fuel supply, since the total number of targets to be attacked would be fewer, and the time available longer, than for the BMD task. Moreover, an ASAT system would presumably not be designed to withstand very vigorous BMD penetration tactics. Both sides could easily calculate the potential of such an ASAT system to intercept ICBMs and SLBMs. The space-based laser ASATs would be able to intercept a small number (let us say fifty) of simultaneously launched ICBMs; it could intercept a larger number if the ICBM launches were spread out in time, permitting each laser to turn to new targets when finished with the first wave. In contrast with the previous case, a space-based laser ASAT of this sort would *not* have the capability for preferential defense of a few selected sites, since it would intercept ICBMs in the early stages of flight before they revealed their intended targets. However, a space-based laser ASAT could completely interdict small nuclear attacks (fifty missiles, in this case), thus forcing an attacker to mount far larger attacks or to employ special penetration aids for which the ASAT was unprepared. Once again there is a potential ASAT upgrade issue, depending on one's strategic outlook.

Ground-based lasers.

A ground-based laser BMD system would require much greater beam power than its ASAT counterpart, since its energy would have to be transmitted over longer ranges (up to an orbiting mirror and then back down toward the earth) and destroy its target as rapidly as possible. Such a BMD system would be able to attack satellites directly if they passed over its ground site, or destroy them over the horizon if its orbiting mirrors could be oriented for ASAT attacks.

A ground-based laser ASAT system would not require orbiting mirrors to redirect the beam, though such mirrors might in fact enable the system to attack a satellite without waiting for it to pass overhead. (An orbiting mirror would also permit use of adaptive optics to compensate for degradation of the laser beam as it passed through the atmosphere.[2]) Without mirrors to bend its beam around the earth, such an ASAT would clearly have no BMD capability, and even with mirrors its beam would not be powerful enough for BMD if originally designed only for the ASAT mission. A ground-based laser ASAT could be a way station en route to a BMD capability, however, allowing tests of basic techniques that could later be scaled up to BMD requirements. The U.S. SDI program will reportedly follow this route in its investigation of excimer lasers.[3]

X-ray lasers

X-ray lasers, powered by nuclear explosions, could be stationed in space, but they are also light and compact enough to be employed in a "pop-up" mode in which the weapons would be rapidly launched into space when needed. An orbiting X-ray laser would display much of the same ASAT-BMD interaction as the space-based chemical lasers described above.

A pop-up X-ray laser BMD system deployed by the U.S. near Soviet borders in order to intercept ICBMs (probably a quite impractical approach), or along U.S. coasts to intercept Soviet SLBMs, could clearly double as an ASAT system. The missile carrying the laser device would merely need to climb above the atmosphere (or to an altitude where the X-ray beam could "bleach" its way through the upper atmosphere) and detonate, destroying any satellite within its—potentially long—range and line of sight.

An X-ray laser would have an advantage over an ordinary ground-launched direct-ascent ASAT interceptor since it would not need to ascend all the way to its target to destroy it. Because X-ray lasers would be powered by nuclear explosives detonated in space, however, they would be appropriate for ASAT use only in extreme circumstances when breaching the nuclear threshold was thought acceptable. Such a pop-up laser ASAT deployed on U.S. soil would have no capability to intercept ICBMs ascending from the Soviet Union, since it could not attain line-of-sight altitude to the ICBMs before they burned out. On the other hand, such a system might have limited use in intercepting SLBMs launched from near U.S. coasts, thus raising again, in a very constrained form, the ASAT upgrade issue.

Sensors

Like the weapons themselves, the sensor systems that support BMD and ASAT systems can serve a dual purpose. In the past, the tracking of ICBM warheads for BMD has been performed by ground-based radars, at least some of which would also be capable of tracking satellites at lower altitudes. This tracking data would allow the ASAT to plan its attack and to respond to the launch of a new satellite within several orbital revolutions. Conversely, radars built for space-tracking can also track ICBMs. However, to avoid ambiguity, a nation can locate its space-track radars in places inappropriate for BMD (for instance, beyond national borders or along the nation's periphery and facing outward).

The U.S. has considered replacing or augmenting its space-track radar network with a Space Surveillance and Tracking System (SSTS) consisting of satellites with infrared sensors able to distinguish space objects as warm spots against the cold background of space. Such sensors would be able to track not only satellites but also ICBM warheads and post-boost vehicles, and they are in fact under investigation in the SDI. For the BMD task, such sensors would have to be able to detect and keep track of thousands or tens of thousands of targets instantaneously and to discriminate missile warheads from decoys. For the ASAT task, it would be sufficient to be able to track

The Relationship of ASAT and BMD Systems 179

fewer targets and to discriminate over a longer period of time. But it may be difficult for either side to judge, when looking at the other side's sensor systems, precisely what their capacities are.

If such sensors are able to track ICBMs "in flight trajectory" with enough precision to support a BMD mission, they might qualify as "ABM components" under the terms of the ABM treaty. If so, deployment of SSTS's for ASAT support would violate the treaty's ban on space-basing of ABM components. A similar problem to ASAT-BMD overlap has already arisen with the case of the Soviet radar under construction at Krasnoyarsk in the central Soviet Union. The Soviets claim the Krasnoyarsk radar is for space-track only; the U.S. points out that the radar's inherent BMD capability (however limited) was precisely the reason the ABM treaty required that such radars be deployed only on the Soviet periphery. Further issues of ASAT-BMD overlap could emerge with the use of other types of sensors, such as space-based radars, laser radars, and satellites with short-wave infrared detection sensors.

A number of general principles emerge from this brief review of ASAT and BMD technologies. An efficient BMD system could, with little or no effort, be made a potent ASAT, though in some cases with only limited altitude capability. A system designed specifically for the ASAT mission, on the other hand, generally would be a poor BMD. However, in many instances the technology itself would not distinguish between the two missions in a way that would make possible a clear judgment that an ASAT system had *no* BMD capability. Actual ASAT and BMD systems might in fact have particular limitations resulting from details of their design that would prevent or constrain dual use. But such detailed limitations might not be visible to another nation's intelligence analysts. Finally, even the rather limited BMD capacity of dedicated ASAT systems might be viewed as strategically significant by one side or the other. It is from these complications that the ASAT upgrade issue arises.

THE ASAT UPGRADE ISSUE

Early BMDs evolved out of surface-to-air missile (SAM) systems used for intercepting bombers. From the start, there was concern that SAM systems might be upgraded to enable them to intercept missile warheads as well as aircraft. Since missile warheads travel twenty or thirty times faster than aircraft and have much smaller radar visibility, missile defense requires better radars, faster data processing, and faster interceptor missiles than air defense.[4] The SAM systems use the same type of components—radars and interceptors—as BMD systems, and it is possible that these interceptors might, in the right circumstances (such as being equipped with nuclear warheads), be capable of intercepting missile warheads. Particularly susceptible to interception are those warheads with low reentry speeds, such as older models and some submarine-launched models. By deploying SAM systems with BMD potential, a superpower could circumvent the limits of the ABM treaty. The treaty's litmus test for SAM upgrade is whether the SAM system has ever been tested against strategic missile warheads. Such testing "in an ABM mode" would qualify the SAM as a BMD for treaty purposes.

American concern with SAM upgrade began with the Soviet SA–5 in the 1960s and continues today with the SA–10 and SA–12. Over the years, the Soviets have improved the capabilities of their SAMs to keep up with improved U.S. bomber capabilities. At the same time, U.S. missile warheads have grown faster and harder to detect. Despite their ability to intercept certain U.S. missile warheads under some conditions, Soviet SAMs are clearly not designed as BMDs, and *as systems* their BMD potential is extremely limited. In general, their radars cannot search for missiles at long range over wide sectors of the sky, do not seem to have the data processing capability to handle large numbers of warheads quickly, cannot discriminate warheads from decoys, and their interceptors cannot handle modern fast warheads and certain reentry angles. All in all, the effectiveness of Soviet SAMs against U.S. missile attacks would be marginal.

Nonetheless, there have sometimes been reasons why SAM upgrading has seemed more plausible than it needed to be. In some early U.S. ICBMs, for example, the warhead and upper booster stage flew close to one another, allowing a relatively unsophisticated Soviet radar to use the large upper stage as a locator.[5] Also, it is of little comfort to point up a Soviet SAM's susceptibility to penetration aids such as decoys if the United States has not incorporated such aids in its offensive planning. What is more, there are circumstances (at least according to some analysts) in which the ability to intercept a few U.S.

warheads would be significant if it could give the Soviets a virtual guarantee that a few selected targets in their country would survive.

The analogous ASAT upgrade issue stems from U.S. fears that the Soviet Union might deploy a system nominally intended for ASAT use, but in fact capable of BMD. Even if they did not deploy such a system, the Soviets might develop advanced BMD technology in the guise of ASAT development, preparing for "leakout" or "breakout" from the ABM treaty. By testing against orbiting objects but not against missiles "in flight trajectory," the Soviets could seek to evade the treaty's strictures on BMD development. ASAT activities that pressed upon the edges of the ABM treaty would cause both sides to hedge their bets against surprises by pushing forward their own programs. A spiral of mutual fear and suspicion might ultimately erode whatever arms-control relationship the two sides wished to preserve.

A future ASAT upgrade issue would probably recapitulate the SAM upgrade problem. As in the SAM case, no one would claim that if one set out to design ASAT-only and BMD-only systems from scratch that the two *system* designs would have much in common, although individual components might use similar technologies. At issue instead would be the strategic significance of an ASAT's limited BMD capability, and this in turn would depend more on the observer's broad strategic outlook than on technical details. To those convinced that the ability to deliver large numbers of nuclear warheads to the Soviet Union is sufficient to assure deterrence, a Soviet ASAT's modest BMD capability would be meaningless, even in a case where U.S. retaliatory forces had already been reduced by a Soviet first strike. To those convinced that the ability to assure destruction of selected military targets is essential to deterrence, however, even minor Soviet interference with U.S. missile attack plans would be unwelcome. Opinion would differ even more over the extent to which Soviet ASAT work would prepare them to develop truly effective BMD and place them in a position to circumvent or abrogate the ABM treaty. Anxious observers would point to common components—weapons and sensors—while doubters would point to the very different system requirements of ASAT and BMD.

To illustrate these ambiguities, we will consider two hypothetical examples of ASAT upgrading, one of U.S. and one of Soviet deployment. These examples are purely illustrative, and are not judgments or predictions about the actual future of ASAT upgrading.

Example 1. The United States deploys an ASAT system consisting of fifty land-based missiles carrying infrared homing vehicles. The ASAT system uses data from ground-based radars and from a Space Surveillance and Tracking System (SSTS) to select targets and to guide the homing vehicles to their vicinity. The United States tests its system extensively against a variety of orbiting targets, but never against sub-orbital objects.

The Soviets worry that these fifty interceptors could be used against missile warheads. Fifty warheads is a small fraction of a Soviet arsenal of thousands. But with its interceptors, the United States can try to preserve a handful of targets selected secretly from the large number targeted by the Soviet Union—a few airfields capable of servicing U.S. airborne command posts and bombers, a ground-command post, an antenna for communication with strategic submarines, a satellite ground station, a radar, a few Minuteman missiles outfitted with radios for broadcast of launch orders. This preferential defense is a substantial annoyance to Soviet planners, since they must now target several warheads at every target they wish to be confident of destroying, and the United States can virtually assure the survival of a small number of carefully chosen targets (by using all fifty interceptors to defend ten sites, for instance).[6] In small or selective Soviet attacks, the relative effect of a small U.S. BMD capability would be greater still.[7]

Adding another twist to this example, suppose the United States now tests from the Kwajalein Atoll (the terminus of the western ICBM missile test range) a BMD interceptor using a different booster from that used by the ASAT system, but with a very similar homing vehicle. Fixed, ground-based radars provide long-range data to guide the BMD interceptor to its target, but the SSTS gathers data in an adjunct mode during the tests. The Soviets protest that the fifty-missile ASAT system has, in effect, now been tested "in an ABM mode" as well. They assert that since the SSTS might be used to replace, rather than supplement, the ground-based radars, this test violates the ABM treaty's ban on space-based sensors. The United States rejects this interpretation, and deploys 150 more ASAT inter-

The Relationship of ASAT and BMD Systems 183

ceptors. The United States also maintains the industrial capability to produce more homing vehicles, prompting Soviet fear of a U.S. BMD breakout. The Soviets, hedging their bets against this possibility, consider deploying more ICBMs, penetration aids, and their own BMD.

Example 2. The Soviets launch a space structure combining a moderately powerful chemical laser with a large mirror. They test destruction of satellite targets, using a long-wave infrared SSTS and possibly laser radars to direct the laser beam. Eventually a total of ten such lasers appear in orbit.

The United States realizes that these laser satellites could attack ICBM and SLBM boosters and post-boost vehicles, even though the lasers are a hundred times less capable than would be required to handle simultaneous launch of hundreds of U.S. ICBMs. The Soviet lasers can therefore destroy, let us say, between ten and one hundred U.S. ICBMs boosters in a U.S. missile attack. The United States does not wish to go to the cost and trouble of countering the Soviet lasers with specialized penetration tactics. The Soviet lasers threaten a larger fraction of U.S. SLBMs, since SLBMs are spread over wide ocean areas where they are subject to attack by many orbiting lasers. Against a U.S. ICBM force reduced by a Soviet first strike, the Soviet protection is greater still. However, the Soviet lasers cannot accomplish the preferential defense that the U.S. ASAT in the previous example was able to accomplish, since the precise destination of U.S. missiles is not apparent during their boost phase. The Soviet lasers can also attack British, French, and Chinese missiles, creating great concern in these nations.

Over the years, the Soviets replace their early versions of these laser satellites with new models. U.S. intelligence cannot estimate precisely the output of the factories that make the lasers, the unit costs, or the characteristics of future models. U.S. intelligence also cannot determine whether inactive Soviet laser satellites that remain in orbit are capable of reactivation. The Soviet lasers have never been tested against strategic ballistic missiles "in flight trajectory," but they have been tested against a variety of orbiting targets. The Soviet Union has also performed, in other space programs, pointing and tracking experiments on ballistic missiles.

184   Ashton B. Carter

The United States considers survival of the Soviet lasers in wartime intolerable and makes contingency plans to attack them with ground-based lasers.

These examples, though somewhat fanciful, demonstrate the potential for a real ASAT upgrade dilemma. Doubters can always claim that a nation intending to use its ASAT for BMD would only be able to intercept a fraction of opposing missiles. Yet others worry that even a modest capability could be useful against SLBMs, against third powers, and against ICBMs in some scenarios; that the defense's capability poses an especially alarming threat to a retaliatory force that is reduced and disorganized by a first-strike attack; that the offense will not, in fact, deploy penetration aids that could overcome the defense; or that breakout or further upgrading is imminent.

ASAT ARMS CONTROL AS AN IMPEDIMENT TO BMD RESEARCH

A second aspect of ASAT-BMD overlap involves possible arms-control agreements. The ABM treaty permits research and restricted development of BMDs. Both nations value this latitude to seek more effective BMD concepts and to hedge against unexpected activities by the other side. In the United States, the Reagan administration has made such research a major part of its Strategic Defense Initiative. ASAT arms control might increase the limits on BMD-related activities and impede progress of the SDI.

One can think of a few explicit cases of conflict between possible ASAT bans and useful BMD development activities. For instance, the ABM treaty permits the United States to test-launch interceptors from Kwajalein against missile warheads. Such a test system might well be considered a low-altitude ASAT, even if it were never tested against orbiting objects. An ASAT treaty intending to ban low-altitude ASAT capabilities would either have to ban such tests or draw a dubious distinction between intercept of orbital and sub-orbital objects. To take another example, suppose the United States wished to develop powerful ground-based lasers and associated adaptive optics to investigate their BMD potential; an ASAT treaty might prohibit propagation of high-energy laser beams from the ground to orbiting

The Relationship of ASAT and BMD Systems

targets. The extent to which ASAT bans would actually impede BMD research that is genuinely necessary to explore the technology's potential would obviously depend on both the details of an ASAT treaty and the degree to which BMD researchers are able to maneuver around treaty limitations.

Perhaps the main impediment that ASAT arms control presents to the SDI is political rather than technical. Public debate will very likely construe even carefully delimited ASAT negotiations as applying to all space weaponry, and will interpret U.S. participation as an implicit commitment to tone down the national initiative on BMD. Proponents of ASAT arms control will advertise their wares as a tool to combat the SDI, even though the ABM treaty already places thoroughgoing limits on deployment and on many types of development of advanced BMDs. Opponents of the SDI might be led to support ASAT arms-control negotiations they otherwise would not attach much importance to.

ASATS AS COUNTERMEASURES AGAINST BMD

If ASAT technology has much in common with that of advanced BMDs, it is even more directly linked to advanced offense, that is, to countermeasures against defensive systems. All advanced BMDs envisioned rely on weapons or crucial sensors based in space. ASAT attack on these components is probably the cheapest and most effective offensive countermeasure.

Since the signing of the ABM treaty, the United States and the Soviet Union have conducted penetration-aid research as a hedge against breakout, as a challenge to their own BMD R&D programs, and as a continuing demonstration of the treaty's underlying principle—that confident and cost-effective defense of their national territories is not possible. This research, it is widely agreed, strengthens both sides' confidence in the treaty.

If aggressive BMD research programs proceed, both sides will presumably wish to develop and test ASAT interceptor missiles, space mines, and directed energy weapons that could be used against BMD battle stations. Limitations on testing and deployment of ASAT, by depriving the two sides in some measure of the opportunity to demonstrate their ability and determination to counter defenses, might have the paradoxical result of increasing the two nations' fear of breakout, and might even tempt them to initiate deployment of BMDs that could not survive ASAT attack.

FOUR FUTURES

The importance of shared ASAT-BMD characteristics and functions depends largely on the future course of legal restrictions. There are four conceivable ways in which arms-control limits on BMD and ASAT systems could be combined: 1) the status quo, with an ABM treaty and no ASAT treaty; 2) an ABM treaty plus an ASAT treaty; 3) an ASAT treaty but no ABM treaty; and 4) no limits on either ASAT or BMD systems. Quite apart from the intrinsic merits and demerits of ASAT and BMD arms-control limitations, it would appear that in all four combinations the overlap of ASAT and BMD will give rise to problems and inconsistencies.

1) ABM treaty; no ASAT treaty.

The most important issue in this case would be whether unrestrained ASAT activity would ultimately erode the ABM treaty. ASATs tested solely against orbiting objects would not fall under the treaty's defining criterion, since they would not be tested against ballistic missiles in sub-orbital flight. Both sides could thus test space-based, mobile, and other ASAT components that are technically akin to BMD components, circumventing the treaty's ban on all but fixed, land-based, radar-plus-interceptor BMD systems. Actual deployment of ASATs that possessed the potential for upgrading to partial BMDs would create further anxieties. If the SAM upgrade precedent is any guide, debate in the United States over the threat of Soviet ASAT upgrading will turn less on estimates of the technical capability of the Soviet ASAT to intercept missiles than on the strategic significance of the modest BMD potential of the ASAT.

Though the technical requirements of effective ASAT are much less challenging than those of effective BMD, there seems to be no legal formulation that distinguishes between the two missions so as to dispel all ambiguities.[8] In an amicable period of U.S.-Soviet relations, regular discussions might temper the disruptive effects of these inevitable ambiguities. But it would not be easy in the best of climates to strike a constructive balance, in the United States at least, between

The Relationship of ASAT and BMD Systems

a willingness to tolerate some ambiguities and a determination to challenge Soviet behavior that surpasses comfortable bounds.

2) ASAT treaty and ABM treaty

Supposing the United States and the Soviet Union wished to limit ASATs, managed to satisfy verification requirements, and found enough common ground to proceed to agreement; they would still have to accept the fact that ASAT limitations will constrain their BMD research programs. If the ASAT treaty forbids test interception of orbital objects, for instance, can it tolerate the tests against suborbital objects explicitly permitted by the ABM treaty? Even if the ASAT negotiations handle the technical ambiguities in such a way so as to minimize interference with BMD research, political opinion in the United States is very likely to view the spirit of the ASAT treaty as obliging the United States to scale back its Strategic Defense Initiative.

3) ASAT treaty but no ABM treaty

For obvious political reasons, collapse of the ABM treaty seems unlikely to coincide with wide-ranging ASAT arms control. But even if compatible politically, unrestrained BMD is incompatible technically with a comprehensive ASAT ban since almost all types of truly effective BMDs would also be overwhelmingly powerful ASATs. On the other hand, some observers imagine a time when both sides cooperate in limiting ASAT activity during or after deployment of advanced space-based BMDs, peacefully ushering in a defense-dominated world. One may doubt the plausibility of this political vision, but it is logically possible.

4) No arms control

If enthusiasm for technical advances or snowballing political misunderstandings lead the two superpowers to abandon the ABM treaty, there will no longer be any question of the legal compatibility of their ASAT and BMD programs. The technical outcome of this future is uncertain, but it is possible that potent ASATs would drive BMD deployment from space, and that other military support missions heretofore conducted from space (communications, reconnaissance and surveillance, navigation, meteorology) would have to find more survivable backups. In such a world, space would be ASAT dominated, and strategic warfare would remain offense dominated.

*   *   *

Although there is as much perception as reality in the overlap between ASAT and BMD, the reality is nonetheless palpable and unavoidable. It grows out of certain inherent technical similarities between the two types of systems, to which is added a strong (if not always well-informed) popular identification of ASAT and BMD as one and the same "Star Wars" vision.

But in fact ASAT and BMD systems respond to fundamentally different security concerns. ASATs challenge the rather paradoxical proposition that space—alone among earth, sea, and air—should be a sanctuary in which satellites can conduct a wide range of military activities, such as surveillance and navigation, free from fear of destruction. BMD concerns itself with only one military contingency—the ultimate contingency of nuclear war—and with the proposition that the imperfect defenses which it is within technology's power to build can contribute to avoiding such an event.

Until public policy settles on an answer to these two propositions separately, a policy for managing their overlap will continue to elude us. Indeed, discussion of the subject both in the U.S. and in the 1985 Geneva arms-control talks seems to be taking an opposite path, considering ASAT and BMD precisely in their region of overlap. To do this risks forfeiting judgment of each on its own merits, while at the same time making resolution of one dependent on resolution of the other. Only when we have clear policies on ASAT and on BMD individually will their difficult area of overlap prove tractable.

ENDNOTES

[1] These issues have been raised previously in a similar vein by Donald M. Kerr, "Implications of Anti-Satellite Weapons for ABM Issues," paper prepared for the SIPRI Symposium on "Outer Space: Can Militarization be Checked," Stockholm, Sept. 21–23, 1983 (Los Alamos report LA–UR–83–2455 [Rev]):

> Without an ASAT Treaty, development and deployment of directed energy ASAT technology would offer a severe challenge to the present ABM Treaty regime, as well as posing potentially serious new realms for military

competition and threatening one of the key stabilizing and limiting mechanisms between the superpowers. On the other hand, the negotiation of a restrictive ASAT treaty, because of the substantial intermingling of ASAT and ABM technologies, might pose insurmountable obstacles to the development of many of the most promising BMD technologies.

[2] See Ashton B. Carter, *Directed Energy Missile Defense in Space* (U.S. Congress, Office of Technology Assessment, April 1984), Chap. 3.

[3] Clarence A. Robinson, Jr., "BMD Research Draws Strategic Focus," *Aviation Week & Space Technology*, June 18, 1984, p. 83.

[4] See Stephen Weiner, "Systems and Technology" in *Ballistic Missile Defense*, ed. Ashton B. Carter and David N. Schwartz (Washington D.C.: The Brookings Institution, 1984) pp. 73–74.

[5] See Sayre Stevens, "The Soviet BMD Program in Carter and Schwartz, op. cit., p. 206.

[6] See Ashton B. Carter, "BMD Applications: Performance and Limitations," in Carter and Schwartz, op. cit., p. 104 and 150ff.

[7] See "Defense Against Ballistic Missiles: An Assessment of Technologies and Policy Implications," (Department of Defense, April 1984), containing conclusions of the Future Security Strategies Study headed by Fred S. Hoffman.

[8] A strictly legal distinction could be based on ASAT intercept of *orbiting* objects and BMD intercept of objects on *suborbital* trajectories that intersect the earth's atmosphere. Whether a space object intercepted by one side in a test was on an orbital or suborbital trajectory would be easy for the other side to determine. The ABM treaty defines a BMD as "a system to counter strategic ballistic missiles or their elements in flight trajectory." A corresponding ASAT treaty could refer to "space objects in stable orbits." The legal distinctions would then be clear: a system never tested against suborbital objects (i.e., "never tested in an ABM mode") would not be subject to ABM treaty limitations, and a system tested only against suborbital trajectories would be immune from ASAT treaty restrictions (since it was never tested "in an ASAT mode"). While this distinction is legally neat, technically unambiguous, and easy to verify, few would accept that an extensively tested BMD had no ASAT capability and vice-versa. As always, it is the *capability* of a system, and not its nominal or declared purpose, that must form the basis of treaty definitions.

II. Implications for Security

Abram Chayes, Antonia Handler Chayes,
Eliot Spitzer

Space Weapons: The Legal Context

SINCE THE BEGINNING OF THE SPACE AGE, political leaders and international lawyers have worked to establish a legal regime to govern activities in outer space. In the following three decades, a considerable, if by no means comprehensive, body of law has evolved. Elaborate conventions now regulate the placement of satellites in geostationary orbit, the allocation of frequencies for space communications, liability for space accidents, weather reporting networks, and many other specific activities in outer space.

This paper examines the bearing of this body of law on the use of outer space for ballistic missile defense and anti-satellite weapons. It begins with a brief examination of the general orientation of the Outer Space Treaty concerning military activities in space.[1] For the most part, this takes the form of general principles and guidelines, significant more for the broad attitudes and approach they express than for the setting down of positive legal rules. Detailed regulation of space-based ballistic missile defense systems is to be found not in those documents dealing with outer space generally, but, as is so often the case, in functionally specific agreements—agreements directed expressly at arms control, the provisions of which cover outer space as well as other environments. The most important of these is the Treaty on the Limitation of Anti-Ballistic Missile Systems (ABM treaty), which the United States and the Soviet Union signed and ratified in 1972.[2]

Today, more than a dozen years later, the ABM treaty remains the only permanent and legally operative bilateral arms-control agreement fully in effect between the two superpowers—a thin legacy of detente and its hope of attaining substantial reductions of strategic arms. This single document bears an extraordinarily heavy burden in U.S.-Soviet security relations. McGeorge Bundy, George Kennan, Robert McNamara, and Gerard Smith have argued that "the ABM treaty stands at the very center of the effort to limit the strategic arms race by international agreements."[3] It is a delicate only child. If it fails, the future of any negotiated treaty regime of arms control will be thrown into question. The implementation and effectiveness of the ABM treaty therefore merit special scrutiny.

The fundamental strategic assumption underlying the treaty is that the security of the United States is best guaranteed by a relationship of mutual deterrence between itself and the Soviet Union, and that the stability of this relationship would be threatened by the deployment of defensive systems that might call into question either side's retaliatory capability. To this end, the basic provisions of the treaty sharply curtail for the indefinite future the development of such defensive systems, permitting only token deployment at a single site.

Those who drafted the treaty knew that technology would not stand still: its provisions therefore apply not only to system concepts that were current in the early 1970s, but to unforeseen "systems based on other physical principles" as well. But no treaty can anticipate every eventuality, and the ABM treaty, like any other legal instrument, is subject to varying interpretations, especially as technological change has altered the factual context. Unlike domestic legal systems, however, there is no impartial tribunal to give an authoritative or binding interpretation when disputes arise; the achievement of the treaty objective depends on the continued commitment of the two parties.

New developments have raised questions about the strength of that commitment on both sides. The president of the United States has announced an objective that is in fundamental opposition to the treaty goals: the creation of a defensive shield over the United States. He has established a powerful bureaucratic organization disposing of large resources to accomplish the strategic objective. Members of the Reagan administration, including the president himself, were opposed to the ABM treaty in 1972; the secretary of defense has, more recently, publicly questioned its utility. The Soviets are conducting their own extensive BMD research programs, though without the same public scrutiny that exists in the United States as a result of the presidential initiative and congressional review. In addition, there is a strong conviction among U.S. government officials and large segments of the population that the Soviets are in fact violating the treaty, thus demonstrating the absence of any commitment to it.[4]

Moreover, deterioration in the political relationship between the two countries is taking its toll on this fragile treaty regime. Since behavior under the treaty can only be regulated by discussion between the parties themselves, a relationship marked by incivility and petulance, accusations and counter-charges, makes it next to impossible to resolve questions that inevitably arise concerning ambiguous activity, apparent violations, or differences in interpretation.

The altered technological and political setting since 1972 has thus brought the ABM treaty under heavy pressure. As we have noted, President Reagan has challenged the fundamental strategic assumption on which the treaty was based: that anti-missile systems erode rather than enhance national security. He has backed this challenge with a $26 billion five-year research program, conceived as the first step in an effort to build and deploy a strategic defensive system. This raises the question of what will happen to the treaty in the interim. Can the treaty survive the SDI program and similar Soviet efforts? What sorts of political and legal pressures can the ABM treaty withstand?

THE OUTER SPACE TREATY

The first efforts in the field of space law reflected a spirit of international cooperation and a determination to depart sharply from previous treatment of new territories or common environments in international law. This cooperative spirit pervades the Outer Space Treaty, which was developed in the United Nations in 1967 to establish a general framework for activities in outer space, and now has approximately eighty signatories. The preamble of the Outer Space Treaty and the General Assembly resolutions leading up to it proclaim the lofty principle of "peaceful use." Unlike the continents newly discovered by Europeans from the 16th to 19th centuries, "outer space, including the moon and other celestial bodies, is not subject to national appropriation."[5] And unlike the high seas, which since Salamis and Actium have been the arena of decisive military engagements, the exploration and use of space is to be "for peaceful purposes."[6]

The specific rules embodied in this treaty, however, are rather more guarded in their restraints on national military activities. Article IV, the key provision, states that "The moon and other celestial bodies shall be used ... exclusively for peaceful purposes." As for outer space generally, the only provision restricting activities forbids the placing "in orbit around the Earth any objects carrying nuclear weapons or any other kinds of weapons of mass destruction ... or station[ing] such weapons in outer space in any other manner."[7] The "peaceful purposes" rubric applied to the moon and other celestial bodies is never defined in the treaty, but presumably comprehends more than the simple prohibition applied to outer space generally.

The reason for the different treatment of "celestial bodies" and "outer space" generally was to accomodate nuclear ballistic missiles, which were just entering the arsenals of the U.S. and the Soviet Union as the treaty was being negotiated. A major portion of the trajectory of such missiles is in outer space, but they do not go into orbit. The language of Article IV was carefully chosen to ensure that the general principle of "peaceful uses" would not interfere with the testing of these weapons.[8]

The treaty also remains silent on the use of military satellites for reconnaissance, surveillance, early warning, and communications. The United States has always taken the position that such "passive" military uses are compatible with a doctrine of peaceful purposes. The Soviets, at first, seemed to take the contrary view. An early Soviet draft of the proposed treaty, drawn up at a time when the United States had a monopoly on observation satellites, contained a provision expressly forbidding their use. The United States and its allies opposed this provision. They argued that international law did not forbid observation of a state from points outside its national territory, and that there was no sound justification for making an exception in the case of outer space. The Soviet Union eventually conceded on this point, but perhaps the change of position had as much to do with its acquisition of the relevant technology as with the force of the U.S. legal argument.

In any case, it is clear from this history that reconnaissance and other "passive" military satellites are not prohibited by the Outer

Space Treaty. This conclusion has since been confirmed by the provisions of the ABM treaty and other arms-control agreements in which the United States and the Soviet Union endorse the use of "national technical means of verification" to assure compliance, and agree not to interfere with them.

Although only a few provisions of the Outer Space Treaty deal specifically with military activities, and those that do leave much ground uncovered, the affirmation of the basic principles of peaceful purposes and international cooperation in exploration and use nevertheless remains important for the construction and application of more specific agreements governing outer space activities. The principles reflected widespread attitudes toward the new environment of space in the late 1960s, when the treaty was adopted, and there is little reason to suppose that those attitudes are different today. The principles of the treaty have remained largely intact throughout the past thirty years of outer space activity. During this time, there has been general agreement between the superpowers that the principle of peaceful use could accommodate passive military uses. And though both Soviet and U.S. military forces have increased the use of space for these purposes, and have even conducted research and development on programs that would go beyond those limits, the actual pursuit of military activities in outer space has so far all been of the passive variety. Ballistic missile defense (BMD) and anti-satellite (ASAT) systems could well represent the first significant challenge to the continued viability of the first and only international legal framework that has governed outer space.

THE ABM TREATY

The Treaty on the Limitation of Anti-Ballistic Missile Systems, which took effect in 1972, is the only bilateral agreement in full force between the United States and the Soviet Union limiting the armaments of the two countries. It is the linchpin of a thirty-year effort to limit the strategic weapons of the superpowers.

The chief purpose and effect of the treaty is to eliminate defensive—that is, anti-ballistic missile—systems from the arsenals of the two countries (with the exception of a single designated site on each side, sharply limited in area and armament). To that end, the first obligation undertaken by each government, as set forth in Article I of the Treaty, is: "not to deploy ABM systems for the defense of the territory of its country...."

It is clear that the task President Reagan has set before the American scientific community—to devise systems that will "interrupt and destroy strategic ballistic missiles before they reach our own soil"—is a task that, if accomplished, would flatly violate the solemn treaty obligations of the United States. (The express ban on deployment of ABM systems for defense of the territory of its country means that any ABM system designed to intercept missiles in the boost phase is necessarily barred by the treaty, because in that phase the targets of the incoming missiles cannot be determined.)

The plain meaning of Article I is fully corroborated by an analysis of the treaty's more detailed provisions. For example, under Article V, each country undertakes the comprehensive obligation "not to develop, test, or deploy any ABM systems or components that are sea-based, air-based, space-based, or mobile land-based." This sweeping prohibition is not limited to deployment, but expressly extends to development and testing as well; it applies not only to entire systems, but with equal force to components. The only exception to the prohibition is fixed land-based systems, and here what is permitted is highly circumscribed. Deployment is confined to a single limited site 150 kilometers in radius and with no more than one hundred launchers, as specified in Article III. Since, according to Article VII, it is permissible to modernize existing fixed land-based systems, development and testing of such systems are also allowed by the treaty. Such activities, however, can only occur on "current or additionally agreed test ranges." In the U.S., testing and development are confined to two test sites, located at White Sands, New Mexico, and at Kwajalein Atoll in the Pacific, the two sites identified by the U.S. delegation as already existing at the time the treaty was concluded. Similarly, the Soviet Union is limited to test ranges at Sary Shagan and Kamchatka. These are categorical limitations imposed by the agreement.

It has occasionally been suggested that the treaty does not apply to ABM systems based on exotic technologies such as lasers or particle beams, both of which are part of the Strategic Defense Initiative (an argument that has not, it should be said, been made by the administration). The suggestion seems to rest on the definition of ABM systems contained in Article II of the treaty:

Space Weapons: The Legal Context

An ABM system is a system to counter strategic ballistic missiles, or their elements in flight trajectory, currently consisting of:

(a) ABM interceptor missiles, which are interceptor missiles constructed and deployed for an ABM role, or of a type tested in an ABM mode;

(b) ABM launchers, which are launchers constructed and deployed for launching ABM interceptor missiles; and

(c) ABM radars, which are radars constructed and deployed for an ABM role, or of a type tested in an ABM mode.

It is contended that exotic technologies do not use interceptor missiles or launchers or radars, and thus do not fall into the categories of ABM systems banned by the treaty. The argument proceeds with reference to Agreed Statement D:

In order to insure fulfillment of the obligation not to deploy ABM systems and their components except as permitted in Article III of the Treaty, the Parties agree that in the event ABM systems based on other physical principles and including components capable of substituting for ABM interceptor missiles, ABM launchers, or ABM radars are created in the future, specific limitations on such systems and their components would be subject to discussion in accordance with Article XIII [establishing a Standing Consultative Commission] and agreement in accordance with Article XIV [providing for possible amendment] of the Treaty.

From this it is argued that the only limitation on exotic systems is a requirement of consultation with the other party to the treaty.

This argument is specious. Article II cannot be read so narrowly. Mindful of potential advances in technology, the drafters defined ABM systems in the most general and comprehensive terms: "systems designed to counter strategic ballistic missiles or their elements in flight trajectory." The use of the word "currently" is a recognition that ABM systems might not always consist of the components enumerated, and indeed were not expected to. As Dr. Raymond Garthoff, a member of the U.S. negotiating team, stated: "The word 'currently' was deliberately inserted into a previously adopted text of Article II . . . in order to have the very effect of closing a loophole to the ban [on future ABM systems]."[9]

Nor does Agreed Statement D open the door for systems based on exotic principles. On the contrary, the language "to insure fulfillment of the obligation not to deploy ABM systems and their components except as provided in Article III" makes it clear that the Statement applies only to those deployments permitted by Article III, that is, fixed land-based systems at a single designated site. Replacement of these by "systems based on other physical principles" is permitted, according to the Statement, only by amendment of the treaty after consultation between the parties. Nor does the Statement modify the prohibition in Article V against development, testing, or deployment of systems or components "which are sea-based, air-based, space-based, or mobile land-based."[10]

This straightforward meaning of the language of the Treaty is fully supported by the legislative history—including the analysis and explanation in the president's submission of the treaty to the Senate,[11] and the testimony in the Senate hearings preceding ratification. These are part and parcel of the ratification process and represent the understanding of the treaty obligations accepted by the president and by Congress. They confirm the interpretation that the treaty is broad in its prohibitions (not only against deployment, but extending to development and testing) and narrow and explicit in its exceptions for permitted activity. This view has been reinforced by executive and congressional commentary since ratification. Secretary of State Rogers during the preratification hearings before the Senate Committee on Foreign Relations confirmed that the treaty would extend even to exotic defense systems.

The treaty provides for other important qualitative limitations. The parties will undertake not to develop, test or deploy ABM systems or components which are sea-based, air-based, space-based or mobile land-based.... Perhaps of even greater importance as a qualitative limitation is that the parties have agreed that future exotic types of ABM systems, i.e., systems depending on such devices as lasers, may not be deployed, even in permitted areas.[12]

The absolute prohibition on any development, testing, or deployment of space-based ABM systems—including those dependent on exotic technology—was also explicitly recognized by Secretary of Defense Laird[13] and Ambassador Gerard C. Smith, chief U.S. negotiator of the treaty and head of the Arms Control and Disarmament Agency (ACDA) at the time the treaty was concluded.[14]

The report of the Senate Committee on Foreign Relations, recommending that the Senate approve the treaty quoted from those passages of Secretary of State Rogers's testimony stressing the absolute nature of the ban on space development, testing, and deployment.[15] In the floor debate on the treaty, the absolute ban on space activities was accepted without question, as was the ban on deployment of "exotic" ABM systems, even at permissible fixed land-based ABM sites.[16]

The Arms Control Impact Statements, prepared annually by ACDA have uniformly adopted this same interpretation. The statement of fiscal year 1984—the most recent available—represents the official position of the present administration. It says:

> The ABM Treaty bans the development, testing, and deployment of all ABM systems and components that are sea-based, air-based, space-based, or mobile land-based. In addition, although the Treaty allows the development and testing of fixed, land-based ABM systems and components based on other physical principles (such as lasers or particle beams) . . . the Treaty prohibits the deployment of such fixed, land-based systems and components unless the Parties consult and amend the Treaty.
>
> The ABM Treaty prohibition on development, testing and deployment of space-based ABM systems, or components for such systems, applies to directed energy technology (or any other technology) used for this purpose. Thus, when such DE [directed energy] programs enter the field testing phase they become constrained by these ABM Treaty obligations.[17]

Recent congressional testimony by former government officials further demonstrates the uniform acceptance of this view.[18]

ISSUES OF TREATY INTERPRETATION

The legal defense of the Strategic Defense Initiative, however, has not relied on the general claim that its programs and activities are outside the purview of the ABM treaty. For the present, it rests primarily on the claim that these are *research* activities, and that the treaty places no strictures on "research," as it does on "development" and "testing." For the future, even the proponents and managers of the SDI recognize that the program, amounting to $26 billion in the first five years, must ultimately come up against the treaty limits. Yet the likelihood of being able to negotiate satisfactory amendments or create a substitute treaty has not been seriously addressed in any public forum. Instead, the administration is relying on ambiguities in the treaty language to provide a legal rationale for program developments as they arise. Three areas of ambiguity, in particular, lend themselves to such use:

(1) What is the line between *research,* which is not prohibited by the treaty, and *development* which, except for fixed, land-based systems, is barred for all types of ABM systems, including space-based?

(2) What is the difference between a *component*, which is subject to treaty limitation on development and testing, and parts or elements of a system, which might not be characterized as components?

(3) To what extent can dual or multi-purpose technology, which might be relevant to, or even intended for use in, ABM systems, be developed and tested in connection with other systems *not* covered by the treaty—such as anti-satellite (ASAT) systems or anti-tactical ballistic missile (ATBM) systems?

The 1985 Report to Congress on the Strategic Defense Initiative includes an Appendix entitled "Compliance of the Strategic Defense Initiative with the ABM Treaty." It sets forth the legal justification for the fifteen presently programmed SDI tests and experiments. The analysis illustrates all three problems identified above. It establishes three categories of permitted experiments: *(1)* conceptual design or laboratory testing; *(2)* "field testing" of devices that are not ABM components or prototypes of ABM components; and *(3)* "field tests" of fixed land-based ABM components—presumably permitted under Articles III, IV, and VII of the treaty, dealing with modernization of fixed land-based systems. The difficulties raised under this framework will be touched on in the analysis below.

Development

ACDA Director Gerard C. Smith was questioned on this subject by Senator Henry Jackson during the Senate hearings on approval of the ABM treaty. A written response was prepared by the administration after a thorough review of the negotiating record. It states:

The prohibitions on development contained in the ABM Treaty would start at that part of the development process where field testing is initiated on either a prototype or breadboard model. It was understood by both sides that the prohibition on "development" applies to activities involved after a component moves from the laboratory development and testing stage to the field testing stage, wherever performed. The fact that early stages of the development process, such as laboratory testing, would pose problems for verification by national technical means is an important consideration in reaching this definition.[19]

The definition of "development" as any work performed outside the laboratory remains the official United States position, and has been reiterated in Arms Control Impact Statements issued since the adoption of the treaty.[20]

The line that is drawn is thus a functional one, related to the method accepted by both parties for verifying compliance with treaty provisions: "national technical means of verification" (NTM). It is fair to say that if an activity cannot be monitored by NTM, it is not prohibited by the treaty; the two parties, particularly the United States, have been unwilling to accept constraints that cannot be verified. Conversely, any test of a component is prohibited if it can be observed by national technical means (or could be observed if the country in question were complying with its treaty obligation not to use "deliberate concealment measures which impede verification by national technical means"). At least, there would be a heavy burden on it to establish that such activity was mere "research," and did not amount to development or testing within the meaning of the treaty.[21]

The Compliance Appendix seems to adopt this view. It describes Category 1 experiments as preceding "field testing" and as not verifiable by NTM. The analysis relies on the quotation from Ambassador Smith's testimony reproduced above. Two of the fifteen programmed experiments and part of a third are placed in this first category of "under-roof experiments."

Components

As has been noted, Article II defines "current" components as ABM interceptor missiles, ABM launchers, and ABM radars. In addition, Agreed Statement D, dealing with exotic systems, refers to "components capable of substituting for" ABM interceptor missiles, ABM launchers, and ABM radars.

The Presidential Communication transmitting the treaty to the Senate develops this concept. It defines a component as "a device to perform the current functions of ABM launchers, interceptors, or radars."[22] It adds that devices other than these three "could be used as adjuncts to an ABM system, providing that such devices were not capable of substituting for one or more of these components." But as new technology and system concepts move further away from those that prevailed in the early 1970s, these notions of "substitution" or functional equivalence become increasingly less helpful in interpreting the treaty. Systems currently under consideration may have no direct analogues to the "missiles," "launchers," or "radars" of an earlier technology. The functions previously performed by these "components" may be redistributed among the elements of the system in different ways. It is possible that a complete ABM system could be made up of elements no one of which would perform the specific functions of a missile, launcher, or radar of earlier technology.

The prohibition in Section V on the testing and development of components was specifically designed to prevent circumvention of the limitations on testing and development of systems by disaggregation. It would be ironic if this prohibition could be evaded simply by disaggregating the system along different axes than those of the original system.

The Compliance Appendix graphically illustrates the problem. The TRIAD program, begun in the Carter administration, is to be carried forward under the SDI program. It now consists of the ALPHA laser (a chemical laser), LODE/LAMP, (a precision segmented mirror with associated optics) and an Acquisition Tracking and Pointing system (ATP) consisting of a telescope and sensors for identifying and tracking the target. The laser is to generate a beam to be projected against the mirror and pointed at the target by the telescope. According to the Compliance Appendix, the ALPHA and LODE/LAMP experiments will be under roof, and are thus permissible "research." As for ATP, there "is a distinct possibility" of field tests in space in which the telescope and passive sensors, mounted in the space shuttle, would be used to measure booster plumes—i.e., would, in the language of the treaty, be "tested against ballistic missiles in flight trajectory" or "in an ABM mode." (The Appendix notes that further compliance review will be had when the shuttle mission is more precisely defined.)

The Compliance Appendix takes the position that the space test is permissible because "the experiments will use technologies which are only a part of the set of technologies ultimately required for an ABM component." Even ATP alone, however, would amount, if perfected,

Space Weapons: The Legal Context 205

to a very significant portion of an ABM system. No doubt it could not "substitute for" a traditional ABM missile, launcher, or radar. But if so, that is because the basic system concept is different. Moreover, the three technologies together certainly seem large enough to be a "component." And they have been linked from the beginning as parts of the same program. Can it be said that because they are tested separately the treaty remains inviolate? It would seem that this is just the kind of development and testing process that the "component" provisions of the treaty were designed to bring to a halt at an early stage. In any case, even if the new system concepts seem to provide some flexibility, as the technologies become promising enough to move out from "under-roof," it will be increasingly difficult to argue that a major element of an ABM system is something less than a component just because all the elements are not tested simultaneously or do not mature at the same rate.

For a number of the projected experiments, the Appendix advances a different argument to meet the prohibition against testing "components." It asserts that the tests will be conducted at power levels or with other parameters below what is required for an ABM weapon. For instance, the Boost Surveillance and Tracking System (BSTS) will be used to measure the signature of booster plumes, a necessary aspect of boost-phase target acquisition. But the tracking satellite will not be given the computational hardware necessary to do so "in real time." Similarly, the Space Surveillance and Tracking System (SSTS) will be capable of performing functions relevant to an ABM system, but its capabilities "will be significantly less than those necessary to achieve ABM performance levels." Certainly, more is needed to avoid the strictures against "testing" an ABM component than simply turning down the power or deliberately limiting some other parameter of the device being tested. But even if the failure to reach levels of performance required for ABM capability is due to insufficiently developed technology, it is hard to see how this can avoid the treaty prohibition on testing *and development* of ABM components. The whole purpose of experiments with immature technology, after all, is to *develop* a component capable of performing the ABM mission.

Dual-purpose technologies

In the case of the Triad, legal analysis is much assisted by relatively unambiguous indications, verbal and otherwise, of the ultimate object of the exercise. But purpose or intention are subjective criteria, neither observable by national technical means, nor easy to prove persuasively by other means. Thus, in the case of dual-purpose technologies that *might* achieve, but do not yet have, ABM capability, the intention of the party conducting the development will always be in doubt. This is especially so for the USSR, where weapons decisions are not required to undergo public evaluation and justification. For this reason, dual-purpose technologies present the most difficult problem of treaty interpretation, and ultimately pose the most serious threat to the existing ABM treaty.

As is discussed by Ashton Carter elsewhere in this collection, the technology for ABM systems and ASAT systems may be closely interrelated. An ABM system designed to intercept ballistic missiles in outer space will almost necessarily have an ASAT capability at some altitudes and regions in space, since the task of locating, tracking, and destroying a single satellite in orbit is much less demanding than defending against multiple missiles. Likewise, much ASAT technology has ABM implications since the basic functions of tracking, pointing, and destroying objects in space are broadly similar. Yet there is no treaty banning the testing and development of anti-satellite weapons as such. While Article XII of the ABM treaty—as well as Article XV of SALT II—prohibits interference with the other party's national means of verification, it is a prohibition, not against the development and testing of ASATs, but only against their *use,* specifically, against satellites performing treaty verification functions.

Anti-tactical ballistic missile (ATBM) systems were also deliberately omitted from the ABM treaty at the instance of the United States, apparently to protect the SAM-D program then under way.[23] The definition of an ABM system in Article II only includes "systems to counter *strategic* ballistic missiles." There are significant differences in trajectory, approach angle, and terminal velocity between tactical missiles (having less than intercontinental range) and ICBMs (and to a much lesser extent, SLBMs). Nevertheless, there is a good deal of overlap between the missions and functions of ABM and ATBM system components.

To the extent that these non-ABM systems pursue traditional configurations, Article VI of the treaty provides some constraints. It prevents giving "missiles, launchers or radars other than ABM missiles, launchers or radars capabilities to counter strategic ballistic missiles or their elements in flight trajectory." The United States insisted on including this provision because of concern over the possible upgrading of the Soviet SAM-5 air defense system to ABM levels.[24] The U.S. Homing Overlay Experiment (HOE) of June 1984 raised questions under this provision. The test was criticized both by the Soviets and within the United States for using a Minuteman ICBM ("a missile other than an ABM interceptor missile") to bring down a reentry vehicle from another Minuteman ("to counter [a] strategic ballistic missile ... in flight trajectory"); the experiment may also have violated the Article VI prohibition against testing non-ABM components in an ABM mode.

Defenders of the experiment argue that it was permissible under Article IV of the treaty: the system was land-based; the test was conducted at Kwajelein, a designated test site; and the interceptor was a specially modified missile using two stages of the Minuteman I plus a new third stage. Nevertheless, since Minuteman I had never been regarded as an ABM interceptor, it was not unreasonable to claim that it fell within the definition of a missile "other than an ABM interceptor missile" set forth in Article VI.

Several of the most problematic experiments in the current SDI program are rationalized on the basis that the device in question will be tested against satellites rather than ballistic missiles. Among these are the Space Surveillance and Tracking System experiment discussed above, as well as the Kinetic Kill Vehicle (a rocket-propelled projectile launched from space) and the space-based Railgun Experiment.

The basic purpose of Article VI—to prevent the upgrading of non-ABM systems—could arguably cover "exotic" as well as conventional systems. The application of the Article to systems incorporating exotic technologies is problematic, however, because its express language deals only with familiar elements (missiles, launchers, and radars) and does not use the general term "components" found elsewhere in the treaty. Even if we were to accept such an inferential extension of Article VI to newer technologies, it would prohibit only the final act of "upgrading"—giving the system "capabilities to counter strategic ballistic missiles ... in flight trajectory." Development, testing, and even deployment of such sub-ABM systems would not be barred so long as they do not possess the prohibited capabilities. SDI program experiments conducted against satellites with power levels or other performance criteria below that required for

ABM missions are apparently justified on this basis. Nevertheless, if such a system were deployed, the potential for upgrade might make or appear to make a sudden breakout from treaty limitations feasible, and even, in certain political circumstances, more likely.

A useful litmus test in these cases of exotic dual-purpose technologies would be whether the United States would consider the Soviets in violation if they conducted the same experiments. It seems likely that, in the past, the United States would have raised serious questions as to the compliance of such activities with the treaty. Thus, it has taken a "strict constructionist" position in suggesting that the Soviet SAM–12 might have or easily be given ABM capability, or that certain radars were "mobile" because of the short installation time. In the context of the current SDI program, however, the United States might be willing to accede to a broader range of Soviet experiments as a way of validating its own. That would come close, in effect, to a tacit amendment of the treaty to eliminate the prohibition against the development of ABM systems—an amendment that would be operative *before* there was a basis for deciding whether the new systems would ultimately provide more security than the existing treaty.

It should be noted here that there is one particular exotic technology not subject to these vagaries of interpretation: that is a space-based X-ray laser powered by a nuclear explosion. The nuclear component of such a system would run into legal constraints quite apart from the ABM treaty. Testing of the system in space would be prohibited by the Limited Test Ban Treaty.[25] Article I of that treaty prohibits "any nuclear weapons test explosion, *or any other nuclear explosion* . . . in . . . outer space. . . ." The testing prohibition applies regardless of whether the nuclear component is characterized as a "weapon." Moreover, if it *is* regarded as a "weapon" and is to remain stationed in space for a period before it is used, it would also run afoul of the Outer Space Treaty's undertaking "not to place in orbit around the Earth any objects carrying nuclear weapons."[26]

As for ATBM systems, there would be room for doubt about the true purpose of a U.S. program justified under that rubric. There is no military threat to the United States from tactical ballistic missiles; ATBM systems would only be developed for use in defense of Western Europe. Since no such program could be seriously undertaken without extensive consultation and advance agreement from the NATO allies—something that has not yet taken place—any extensive

U.S. ATBM activity undertaken *without* NATO approval would most likely be a subterfuge for ABM work.

The Soviet situation is different. The Pershing II deployment in Western Europe furnishes the Soviets with a military justification for ATBM development. On this analysis, Soviet ATBM activity would not be prohibited so long as it did not achieve the capability to intercept strategic ballistic missiles in flight trajectory. Nevertheless, it hardly needs saying that the United States is unlikely to accept a reading of the treaty in which ATBM development is forbidden to the United States but not the Soviet Union.

Issues of treaty interpretation—such as those discussed above—can never be resolved definitively. No court or other third party tribunal sits with jurisdiction to answer such questions authoritatively. At the same time, the interpretation of treaties, like other legal instruments, consists of more than playing games with words to see what stretched constructions they can be made to bear. The enterprise is subject to the universal maxim that cases of genuine doubt are to be resolved so as to further, not frustrate, the basic purposes of the instrument. In the case of the ABM treaty, the basic purpose is clear and appears in the opening words: "Each party undertakes to limit anti-ballistic missile systems. . . ." In light of this fundamental objective, there is little doubt how the questions that have been raised about the coverage of the treaty should be answered.

Yet neither the United States nor the Soviet Union seems to have adopted a strict constructionist approach to questions that arise under the treaty. Instead, each seems to be embarking unilaterally on an expanding series of programs, more or less defensible on technical legal grounds. Individually and cumulatively, however, such activities may have considerable ABM potential. The likely outcome of this behavior is that even without unequivocal treaty violations, both parties will be deprived of the assurance they had sought from the ABM treaty: that the other party would not prepare a "break-out" from the basic prohibition against deployment of ABM systems. The resulting insecurity will itself contribute to the erosion of the treaty, and to the deterioration of the relationship between the parties.

It would clarify the existing legal and political situation if the two nations were to negotiate new agreements specifically addressed to ASAT and ATBM systems. The agreements might prohibit outright the development, testing, and deployment of such systems. At the very least, they could define with greater precision the scope and capabilities of ASAT and ATBM programs and deployments both sides are willing to live with. It may not be too much to say that only such a clarifying agreement can save the ABM treaty from eroding away as a result of gradual encroachments on the part of the two countries.

COMPLIANCE

Although no courts have jurisdiction over the interpretation or enforcement of the ABM treaty, and no formal sanctions exist for breach, the treaty does contain an important innovation in the area of compliance. Article XIII establishes a Standing Consultative Commission (SCC) "to promote the objectives and implementation of the provisions of this treaty. . . ." The principal function of the commission is "to consider questions concerning compliance with the obligations assumed and related situations which may be considered ambiguous"; the SALT I interim agreement, SALT II, and the 1971 Accident Measures Agreement give the SCC similar functions with respect to their provisions.

The Standing Consultative Commission consists of a commissioner and deputy commissioner from each side, supported by appropriate staff and advisory personnel. It meets twice a year, and its proceedings are confidential. It has no permanent secretariat or offices and no independent authority to make decisions or treaty interpretations. Commissioners act only on instructions from their governments. On important matters, the U.S. commissioner's instructions are approved by the president.

The commission is essentially a continuing and expert body for regular and confidential consultation between the parties. Issues concerning compliance and interpretation that may arise under the treaties may be referred to the commission, which attempts to clarify uncertainties and ambiguities and thus to maintain the confidence of the parties in the continued integrity of the agreement. Although discussion of compliance issues presumably generates pressure for adherence to treaty requirements, the commission is not, strictly speaking, an enforcement agency. It is not well-suited to deal with accusation or confrontation. It operates in a setting where each side, through NTM and other intelligence, possesses extensive but incomplete information about the other, but neither knows exactly what the other knows. The commission is thus an agency through which a party, by providing "on a voluntary basis such information as either party considers necessary to assure confidence in compliance,"[27] can persuasively assure the other that it is in fact carrying out its obligations.

The experience with the SCC may be roughly divided into two periods, the first lasting from 1972, when the commission was organized, until 1980; the second, from 1980 to the present. Public information about the first period is derived primarily from accounts published by the State Department in 1979 in connection with Senate consideration of SALT II.[28] During that period, eight compliance-related issues were raised by the United States, five by the Soviet Union. Only a few of these involved the ABM treaty.

For the most part, the issues were minor, turning on technical questions of treaty interpretation or minor uncertainties about the factual situation observed. The most important question raised by the United States was whether a Soviet SA-5 air defense radar had been "tested in an ABM mode"—by being used to track an incoming missile during a test flight—in violation of Article VI of the ABM treaty. The United States acknowledged that the activity was ambiguous and the Soviet Union maintained that it had not violated the treaty. Nevertheless, according to the State Department report, shortly after the issue was raised the Soviets stopped this tracking activity, although subsequently the United States has complained of sporadic resumption. Discussion in the SCC over a period of several years has resulted in some clarification concerning the meaning of "testing in an ABM mode"—a clarification that the U.S. had sought since the initial negotiation of the treaty. Other complaints by the United States in this period were, on the whole, resolved satisfactorily, either by cessation of the questionable activity or by the supply of information that, in connection with NTM, satisfied the United States that the activity was not prohibited by the treaty.

The most serious Soviet complaint against the U.S. concerned the use, beginning in 1973, of 2700-square-foot environmental shelters to shield Minuteman silos that were being hardened at Malmstrom Air Force Base. The Soviets claimed that this practice violated the obligation of Article V of the Interim Agreement "not to use deliberate concealment measures which impede verification by national technical means. . . ." In 1977, the United States reduced the size of these shelters, but they were not completely removed until the conclusion of the SALT II negotiations in 1979. The United States has maintained that other actions questioned by the Soviets were in compliance with the treaty, and therefore did not require modification of conduct.

The experience during the second period of the SCC's history is far less satisfactory. On February 1, 1985 a White House Compliance Report to Congress stated: "The U.S. Government judges . . . that the new large phased-array radar under construction at Krasnoyarsk constitutes a violation of legal obligations under the Anti-Ballistic Missile Treaty of 1972."[29] The report also cited, with varying degrees of certainty, Soviet violation of a number of other treaties and commitments in the security field. The USSR has responded in kind, making similar wide-ranging accusations against the United States, including the charge that U.S. radar deployments were in violation of commitments under the ABM treaty. Some of these charges reopened issues the United States thought had been settled in the earlier period. The most recent barrage of accusations has continued for more than a year now, with most of the conflicts having been formally raised within the commission before being publicly aired.

Unlike the complaints brought before the SCC in the earlier period, current charges raise questions of compliance with major substantive obligations under the treaty. Since much of the supporting evidence is classified, it is difficult to make a judgment on the merits of the claims. As to the charges made by the United States, it is undisputed that the Soviet Union is constructing a large phased-array radar near Krasnoyarsk in Siberia. Since it is sited more than 700 kilometers from the nearest frontier and oriented northeast, it cannot be said to be located "along the periphery of its national territory and oriented outward," as is required for early-warning radars under Article VI. Agreed Statement F prohibits large phased-array radars in other locations "except for the purposes of tracking objects in outer space or for use as national technical means of verification." The Soviets claim that the radar is for space tracking, but the United States says its technical features are not consistent with a space-track mission. The Soviets have countered that once the radar is completed it will be apparent that it is designed to track objects in outer space. The great weight of expert opinion in and out of government at this writing is that the device at Krasnoyarsk is an early-warning radar not distinguishable from other early-warning radars. Its position and orientation has very limited utility for satellite tracking. There has been some suggestion that the radar may be designed for use in conjunction with the Soviet ASAT program.[30] If so, it might be characterized as a space-tracking radar and therefore be permitted by the treaty. In that case, since the radar would also have early-warning capability, this would be another illustration of the problems raised by overlapping and multiple purpose technologies.

The Soviet Union, for its part, challenges U.S. phased-array radar deployments, which it says can be used as a basis for a nationwide ABM defense, thus violating Article I of the treaty.[31] In addition to installations in Massachusetts and California, examples that had previously been raised in the SCC, the Soviets pointed to new U.S. deployments "in a southern direction." This presumably is a reference to large phased-array radars now being constructed in Georgia and Texas. These are perhaps close enough to the continental boundary to be "on the periphery" of the United States, but since their angle of coverage is said to be 240 degrees, they are at least arguably not "oriented outward."

Despite charges by both superpowers of serious breaches of the treaty, neither party has yet withdrawn from the treaty or indicated any intention of doing so. Perhaps this reticence reflects the remaining uncertainties in the parties' knowledge of the facts. But it is also surely a recognition of the huge political costs, both domestic and international, that would be involved in formal abrogation of the treaty.

The resulting situation is very unsatisfactory: the charges made by both sides remain on the record, and provide a basis for deprecating the importance of complying with the treaty in discussions of the SDI and other programs. Meanwhile, both the United States and the Soviet Union are proceeding with programs, particularly in the ASAT and ATBM area, that are of dubious validity under the treaty, and are justifying them with increasingly casuistic arguments. All of this combines to erode confidence in the treaty, and may perhaps reach a point where the political costs of withdrawal would become acceptable.

This is not the process contemplated by the treaty. Questions of compliance and disputes about treaty interpretation were to be taken to the SCC and discussed there in good faith, in confidence, and with a view to resolving them on a mutually agreeable basis. It is hard to be satisfied today that either party is following this injunction, or availing itself of the opportunity provided in the SCC to resolve ambiguities and provide assurance of compliance with treaty obligations. Instead, the dominant political tone of confrontation between the United States and the Soviet Union has saturated the proceedings of the SCC and reduced its utility for the central questions on its agenda.

SUMMARY AND CONCLUSIONS

There is a framework of legal principles and treaty obligations—fragmentary and incomplete, perhaps—governing military activities in outer space. As we have seen, the Outer Space Treaty enunciates principles of international cooperation and the use of space for peaceful purposes. We have also seen that, from the beginning, this goal was broadly understood to accommodate passive military uses such as reconnaissance and communications. The positive rules of law laid down by the Outer Space Treaty, on the other hand, are much narrower in scope. One of them prohibits nuclear weapons in orbit. This, taken together with the ban in the Limited Test Ban Treaty on nuclear explosions in outer space, has a decisive bearing on a narrow range of ABM and ASAT technology, specifically, the use of nuclear explosions in space as a power source for lasers.

The principal legal rules relevant to SDI activities are articulated in the amended ABM treaty. These rules call for:

(1) prohibition of the development, testing, and deployment of space-based ABMs and ABM components, whether dependent on existing or exotic technologies. This prohibition also applies to air-based, sea-based, and mobile land-based ABMs and ABM components, that is, to everything but fixed land-based systems and components;

(2) prohibition on the deployment of fixed, land-based ABM systems and components, except at a single designated site not more than 150 kilometers in radius, centered on the national capitol or a missile silo field, and containing not more than one hundred launchers;

Space Weapons: The Legal Context

(3) no prohibition on the development and testing of fixed, land-based ABMs and ABM components of traditional technologies at existing test ranges;

(4) prohibition of deployment of exotic ABM technologies even if fixed and land-based, and even at the existing test ranges, except after discussion and agreement with the other party;

(5) prohibition on upgrading non-ABM missiles, launchers, and radars to an ABM capability, or on testing them in an ABM mode, and, arguably, by inference, restriction of upgrading components (that are not missiles, launchers, or radars) of sub-ABM systems to ABM capability;

(6) no prohibition against research.

It is apparent that the SDI enterprise as a whole, its objectives and philosophy, are simply at odds with the purposes and objectives of this treaty. Moreover, whatever the exact technical limits on testing and development may be—the difference between "research" and "development," between a "component" and something else—it is inevitable that under the current presidential mandate and Defense Department response, these limits will be breached and the treaty violated outright within a period of time that is relatively short compared to the time it would take to develop, evaluate, and deploy SDI technologies and systems. Attempts to develop ABM technologies under the label of ASAT or ATBM programs would be legally disingenuous, technically costly, and in any event could only extend arguably permissible development a few years. Meanwhile, the interrelation of ASAT and ABM technology will continue to generate disputes over the legality of particular actions under the treaty. While these issues could be addressed and perhaps resolved ad hoc by a well-functioning SCC, it would be far better—indeed, it may be essential to the continued viability of the treaty—to eliminate the source of these disputes by calling a prompt mutual moratorium on ASAT testing, followed by serious negotiation of a treaty to define the limits of ASAT activity. The successful conclusion of such a treaty would have intrinsic merit apart from its clarifying impact on the ABM treaty.

The ABM treaty is not the embodiment of ultimate truth. It represents a judgment, at least by the United States, that its security is enhanced and the stability of the strategic balance strengthened if both sides forswear defensive systems. Like other legal arrangements, this treaty may be subject to revision in the light of changing technology or reconsideration of the values at stake. It has been suggested, for example, that even within the general framework of deterrence theory, point defense of silos might be an effective way of protecting ICBMs during a period of strategic arms reduction. This would be permissible to a limited extent under the current ABM treaty. The scope of permitted terminal defenses could be extended by relatively modest amendments to the treaty (subject, of course, to Senate approval).

The SDI embodies a much more far-reaching, indeed, a fundamental challenge to the policy assumptions of the treaty. It may be that the government and people of the United States will decide to make such a drastic change in direction. If so, the way to do it is not by nibbling away at the ABM treaty. By engaging in legalistic—not to say sophistic—interpretations that bend the treaty's language and torture its basic meaning, we do not "preserve" or "comply with" a treaty. And in the process, we are not only undermining the ABM treaty, we are severely compromising the possibility of confident reliance on all present or future arms-control regimes. It would be more straightforward to invoke the process prescribed by the treaty to amend it or withdraw from it in accordance with its terms. Until this is done, however, the United States is bound by the treaty as it stands, and thus not only by constraints under international law, but by obligations that are, under the Constitution, "the supreme law of the land."

ENDNOTES

[1] *Treaty on Principles Governing the Activities of States in the Exploration and Use of Outer Space, Including the Moon and Other Celestial Bodies*, January 27, 1967, 18 U.S.T. 2410, T.I.A.S. No. 6347, 610 U.N.T.S. 205 (hereafter cited as Outer Space Treaty).

[2] *Treaty Between the United States and the U.S.S.R. on the Limitation of Anti-Ballistic Missile Systems*, October 3, 1972, 23 U.S.T. 3435, T.I.A.S. No. 7503 (hereafter cited as ABM treaty).

[3] McGeorge Bundy, George Kennan, Robert McNamara, Gerard Smith, "The President's Choice: Star Wars or Arms Control," *Foreign Affairs*, Winter 1984–85.

[4] *Soviet Compliance with Arms Control Agreements*, State Department Special Report no. 122, Feb. 1, 1985.

[5] Outer Space Treaty, Article II.

[6] Outer Space Treaty, Preamble, Article IV.
[7] Outer Space Treaty, Article IV.
[8] Article IX of the SALT II accords did forbid the development, testing, or deployment of fractional orbital missiles (FOBS). FOBS must be distinguished from ICBMs, however, because ordinary ICBMs are not considered to be in orbit.
[9] Raymond Garthoff, correspondence in *International Security*, Summer 1977, pp. 107–09.
[10] ABM Treaty, Article V (1).
[11] The Communication of the President transmitting the treaty and recommending that the Senate advise and consent to ratification, explains the interaction of the clauses of the treaty:

> *Future ABM Systems*
> A potential problem dealt with by the Treaty is that which would be created if an ABM system were developed in the future which did not consist of interceptor missiles, launchers, and radars. The treaty would not permit the deployment of such a system or of components thereof capable of substituting for ABM interceptor missiles, launchers or radars: Article II (1) defines an ABM system in terms of its function as "a system to counter strategic ballistic missiles or their elements in flight trajectory," noting that such systems "currently" consist of ABM interceptor missiles, ABM launchers, and ABM radars. Article III contains a prohibition on the deployment of ABM systems or their components except as specified therein, and it permits deployment only of ABM interceptor missiles, ABM launchers, and ABM radars. . . .

> The presidential statement makes it clear that development of even these components is forbidden for any other than fixed land-based systems:

> *Development, Testing, and Other Limitations*
> Article V limits development and testing, as well as deployment of certain types of ABM systems and components. Paragraph V (1) limits such activities to fixed, land-based ABM systems and components by prohibiting the development, testing, or deployment of ABM systems or components which are sea-based, air-based, or mobile land-based. . . .

> See the *Communication from the President of the United States, Transmitting Copies of the Treaty on the Limitation of Anti-Ballistic Missile Systems and the Interim Agreement on Certain Measures with Respect to the Limitation of Strategic Offensive Arms*, 92nd Congress, 2nd sess., 1972, House Doc. 311, pp. 9–10.

[12] *Strategic Arms Limitation Agreements, Hearings Before the Senate Committee on Foreign Relations on Executive L. 92-2*, pp. 5–6 (testimony of Secy. of State Rogers); see also, *Agreement on Limitation of Strategic Offensive Weapons, Hearing Before the House Committee on Foreign Affairs*, 92nd Congress, 2nd sess., 1972, p. 5 (statement of Secy. of State Rogers, reemphasizing the absolute ban on development, testing, and deployment of space-based systems and the ban on deployment of "exotic" ABM systems even on fixed, land-based sites).
[13] *Military Implications of the Treaty on Limitation of Anti-Ballistic Missile Systems and the Interim Agreement on Limitation of Strategic Offensive Arms, Hearing before the Senate Committee on Armed Services*, 92nd Congress, 2nd sess., 1972,

pp. 40–41 (answers to prepared questions supplied by Secy. of Defense Laird): "There is, however, a prohibition on the development, testing, or deployment of ABM systems which are space-based.... There are no restrictions on the development of lasers for fixed, land-based ABM systems. The sides have agreed, however, that deployment of such systems which would be capable of substituting for current ABM components ... shall be subject to discussion...."

[14] *United States-Soviet Relations, Hearings before the Senate Committee on Foreign Relations*, 98th Congress, 1st sess., 1983, p. 180 (prepared statement by Hon. Gerard Smith): "Very briefly, my understanding of the ABM Treaty is that ... [d]evelopment, testing, and deployment of sea, air, space, or mobile land-based systems was banned; [i]f land-based systems are developed using so-called 'exotic' components—lasers, particle accelerators, etc.—they could not be deployed unless the treaty was amended."

[15] See *Treaty on Limitation of Anti-Ballistic Missile Systems, Report of the Senate Committee on Foreign Relations*, 92nd Congress, 2nd sess., 1972, p. 3.

[16] See *Congressional Record*, August 3, 1972, 92nd Congress, 2nd sess., S. 26703 (statement of Sen. Buckley asserting that the treaty banned all space development of laser ABMs).

[17] *Fiscal Year 1984 Control Impact Statements*, 98th Congress, 1st sess., 1983, pp. 266–67, as well as the relevant sections of the Arms Control Impact Statements for fiscal years 1979 through 1983.

[18] See, e.g., *Controlling Space Weapons, Hearings before the Senate Committee on Foreign Relations*, 98th Congress, 1st sess., 1983; *United States-Soviet Relations*, loc. cit. (testimony of Hon. Gerard Smith).

[19] *Military Implications of the Treaty on the Limitations of Anti-Ballistic Missile Systems*, op. cit., p. 377.

[20] *Fiscal Year 1984 Arms Control Impact Statements*, op. cit.

[21] The Russian text of Article V uses the word *sozdavat*, which translates more nearly as "create" rather than "develop." This may lay the basis for an even narrower reading of permitted research.

[22] *Communication from the President*, op. cit.

[23] Interview with Dean Albert Carnesale, John F. Kennedy School of Government, Harvard University, formerly Senior Advisor to Head of SALT I delegation.

[24] Ibid.

[25] *Treaty Banning Nuclear Weapon Tests in the Atmosphere, in Outer Space and Under Water*, Oct. 10, 1963, 14 U.S.T. 1313, T.I.A.S. No. 5433, 480 U.N.T.S. 43.

[26] Outer Space Treaty, Article IV, op. cit.

[27] ABM treaty, Article XIII, op. cit.

[28] See, e.g., *Compliance with I Agreements*, U.S. Dept. of State, Special Report no. 55, July 1979.

[29] *Soviet Non-Compliance with Arms Control Agreements*, op. cit.

[30] *Arms Control Treaty Compliance*, Federation of American Scientists Public Interest Report, *Journal of the Federation of American Scientists*, March 1984.

[31] USSR Aide Memoire to the U.S. on Arms Violations, *Foreign Broadcast Information Service: Soviet Union*, Jan. 30, 1984, pp. AA1–AA5.

John C. Toomay

The Case for Ballistic Missile Defense

WITH HIS STRATEGIC DEFENSE INITIATIVE, President Reagan has called on American scientists and engineers to devise a defense against nuclear ballistic missiles. In proposing that the United States switch from an offense-oriented to a defense-oriented strategic policy, the president has sparked a lively public debate about the merits of such a policy. As he noted in his March 1983 address, our current strategic policy is to deter aggression through the promise of retaliation. Deterrence has worked, and we have averted a nuclear exchange of any kind, for over forty years. But I believe we cannot afford to be complacent about changes that have occurred in the strategic balance over these years. As a member of the Defensive Technologies Study Team (Fletcher panel) that advised the president, I am persuaded that an increased effort to develop ballistic missile defenses is a wise course for the United States.

THE DETERRENCE EQUATION

Deterrence has become a complex and subtle matter. The fine points of how to maintain deterrence in peacetime, in crisis, even during war, have been the subject of much discussion. But the basic concept of deterrence is relatively simple. The word itself comes from the Latin, meaning "from fear": deterrence is the state of not acting for fear of the consequences. Gen. Russell Dougherty, former head of the Strategic Air Command, has described it as the product of military capability and the will to use it.[1] In the case of nuclear deterrence, we must pay far more attention to our opponent's capabilities—the weapons and the means of using them—than to his will. We cannot gamble that he might be lacking in will, when the consequences of being wrong are so grave. Therefore, our current strategic posture is based substantially on the power of our offensive forces. It emphasizes the *capabilities* of the Soviets rather than their intent. We do not question that the Soviets have the will to act; instead, we make conservative estimates of the capabilities of Soviet strategic forces and assume they will use them in the way most destructive to our interests.

Our deterrence equation must also take into account our estimate of what the Soviets believe is unacceptable damage to their country. Because we have little confidence in our abilities to probe the psyches of Soviet leaders, our estimate is again very conservative. Our strategic force requirements are determined by working backward from the level of potential damage that we are confident will ensure deterrence. We believe we can deter the Soviets if their calculations always show that we are capable of inflicting unacceptable damage on them in retaliation, even after they have executed an all-out surprise attack against us.

Our ability to gauge what deters the Soviets has been sufficient to avert a nuclear exchange of any kind for over forty years. I assert that this is evidence that deterrence has worked. We seem to accept these decades of deterrence quite casually, even complacently, for two reasons. The first is manifest: human nature comes to accept the status quo as natural. The second is vital: until recently, the margin of our advantage over the Soviets in strategic forces has been wide. Even very large errors in our estimates of Soviet intent or Soviet capability would not have significantly changed the equation.

Today, however, the situation is less sanguine: the relation is one that some call "rough parity." There is much debate over the significance of this shifted balance. Devotees of minimum deterrence argue that when both superpowers have thousands of nuclear weapons only a few hundred of which are capable of inflicting massive damage, an increase in numbers, no matter how asymmetrical the result, is irrelevant and therefore cannot jeopardize deterrence. Advocates of stronger strategic forces, on the other hand, warn that any imbalance encourages Soviet intransigence and adventurism, and a gross imbalance might invite a limited nuclear attack, which could easily escalate into a massive exchange.

The Case for BMD   221

We have no way of knowing, before the fact, which of these two interpretations is correct. Yet we can reflect on the experiences we have had. The Cuban Missile Crisis of 1962 is often cited as an example of Soviet accommodation to our nuclear superiority; in contrast, Soviet stubbornness during the intermediate nuclear force (INF) negotiations of 1981–83 is seen as an example of Soviet implacability in our present world of "parity." Such evidence is anecdotal and hence not altogether convincing. Perhaps we need to address the matter in a different way: what have we gained by letting our strategic superiority erode, and what have we risked? The benefits have presumably included reduced expenditures on strategic forces, reduced numbers of nuclear weapons, and an international reputation as a peace-seeking nation. The world has not been made a safer place, however, nor has nuclear war been made less likely by such erosion. The potential risks from an eroded strategic posture are far more onerous. They include new wars of national liberation, major incursions into, for example, Europe, or actual nuclear exchanges and loss of sovereignty and freedom.

We have tried various stratagems since the mid-1960s to sustain our strategic position. At first, we sought to preserve superiority; this was followed by "balance"; then came, in rapid succession, "essential equivalence" and "countervailing forces"—all euphemisms for a declining strategic margin. During much of the 1960s and 1970s, we cut the budget for nuclear forces almost every year. We have made some efforts to bolster our strategic forces, though with mixed results. The B-1 bomber, at first cancelled, has been revived, though in drastically reduced numbers. The MX has been approved, though again in reduced numbers. Minuteman's accuracy and hardening have been improved, cruise missiles are being deployed on bombers, and Trident submarines are operational.

Nevertheless, our efforts pale beside those of the Soviets, whose massive buildups have now spanned over twenty years and have brought them parity in numbers of strategic warheads and superiority in total explosive power. Moreover, the Soviets have devoted themselves to the full range of strategic forces, including defenses. They have maintained and are now improving their ABM system around Moscow; their air defense network is extensive; Soviet civil defense preparations are vastly greater than our own; and they are pressing ahead with advanced ballistic missile defense technologies.

The United States has also tried detente and arms control as methods of stabilizing the strategic balance. The SALT negotiations may have diverted, but they did not halt, the Soviet buildup of forces; our efforts at limiting strategic arms by negotiations have proved disappointing. As things now stand, we are frustrated in our dealings with the Soviets, we have no consensus at home about what our policy should be, and our media's recent focus on the horrors of nuclear war has unsettled us. What we are in need of is a new approach, one that might allow us to escape our current dilemmas while still preserving viable strategic forces. The addition of strategic defenses to our national policy serves just this function.

To discuss the promise of ballistic missile defense, we must first consider what the Strategic Defense Initiative refers to as "intermediate capabilities."[2] I will begin by examining those types of defense that raise relatively clear issues—defense of valuable hard targets, local area defense, and light area defense—and move gradually to consider full, "perfect" defense of the nation, the most contentious issue. (It will be useful, throughout the discussion, to bear in mind that these defenses could also be useful for our allies and in some cases would demand deployment by them.) We will trace how these defenses would carry us from the offense-oriented deterrence we now know to a potentially more powerful, but certainly less familiar, defense-oriented strategy.

HARD-POINT DEFENSE

One dimension of the Soviet strategic buildup that must concern us is the increase in their number of highly accurate, high-yield ICBM warheads appropriate for attacking U.S. deterrence forces and other hardened military targets. Carrying out such an attack would by no means be an easy task, as the Scowcroft Commission has pointed out. Yet when the Soviets devote so much effort and resources to acquiring a military capability, we must take seriously the potential threat that it poses. While hard-point defenses of our retaliatory forces were not explicitly addressed by the Fletcher panel (they had, after all, been studied almost continuously since the early 1960s), the possibility was examined by the Hoffman panel, which came out in support of such a policy.

The Case for BMD

Hard-point defenses would complement deterrence in a straightforward manner. Our present retaliatory forces achieve survivability in many ways: bombers are launched on warning (mobility); submarines patrol under the oceans (mobility and concealment); and land-based ICBMs are deployed in silos (hardening). As the accuracy of Soviet ICBMs has improved, however, the ability of our hardened silos to withstand attack has become a matter of acute concern. Of the several ways to preserve or reestablish the survivability of our ICBMs, defense shows considerable promise. Hard-point defenses are an alternative to deploying more offensive forces in order to maintain the strategic balance. Instead of installing an offensive missile in yet another silo, a defensive interceptor could be deployed to destroy an attacking warhead before it reaches the silo. Conceptually, either of these methods increases the survivability of our retaliatory forces. The choice a country makes would depend on the cost-effectiveness of offensive versus defensive systems and their political and diplomatic ramifications. Three factors so far seem to have prevented the United States from deploying hard-point defenses: the existence of the ABM treaty, uncertainties about the importance of ICBM "vulnerability," and disagreements about what should be done to rectify it.

It is difficult for most people to grasp that the time to act on the survivability of our land-based ICBMs is while their effectiveness is still intact. This point deserves emphasis. If we wait until the threat to our retaliatory forces is unambiguously clear, we will have plunged ourselves into a situation of great strategic instability. Since it takes a decade or more to design, develop, and deploy any major weapons innovation, the time to begin improving our forces is long before the danger actually confronts us.

In our past calculations of what we needed for effective deterrence, we have always presumed that the Soviets would make a massive attack against our Minuteman ICBMs—as well as many other targets—and that only a fraction of them would survive. The role of active defenses would be to assure survival of at least this fraction in the face of an increasing Soviet threat. One tactic available to us is preferential defense. In preferential defense, the defender marshals his resources (in this case BMD interceptors) to defend only a limited number of points he has secretly selected. If the defender concentrates his defense effort to protect a few targets, the attacker must also increase the size of his attack on those targets if he wishes to destroy them. But if the attacker does not know in advance which targets will be defended, he must increase the magnitude of his attack on *all* targets. Just a few BMD interceptors on the part of the defense, therefore, forces the offense to multiply his forces manyfold—or develop a wholly new tactic.[3] To be effective, a preferential defense system must be able both to defend itself and to give the *appearance* of being able to defend an area much wider than the few targets actually selected.

Some systems being considered for defense of future land-based ICBM systems would have even greater leverage. For example, a system could combine defense with deceptive missile basing, which would move a relatively small number of ICBMs among a very large number of shelters. Because the location of each missile would be kept secret, the enemy would be compelled to attack all the shelters to be sure of destroying the missiles hidden within some of them. One defensive interceptor accompanying each ICBM would approximately double the number of warheads required to destroy all the missiles. If there were two hundred ICBMs and four thousand shelters, for instance, the enemy might consider an attack with four thousand warheads to be adequate. But if each ICBM were protected by an interceptor, the enemy would have to target *two* warheads on each shelter—or eight thousand warheads in all—in order to be successful.[4]

The small, hard-mobile ICBM—or Midgetman—system that is recommended by the president's Commission on Strategic Forces has some inherent suitability for such high-leverage defense. In this system, the missiles would be moved randomly on mobile launchers over government land in the Southwest. Since the Soviets would not know where each launcher was at any moment, but would presumably know the bounds of the overall deployment area, they would have to barrage the entire area with warheads to destroy the missiles. A mobile launcher is inherently a softer target than a silo and may potentially be damaged even by warheads that explode a considerable distance away, thus making it harder for a limited defense to protect the missile. Still, if an interceptor (and its associated equipment) accompanied each missile and shot only at those warheads that would destroy the missile, the warhead requirements for the Soviets would be doubled.

There is little doubt of the cost-effectiveness of such defenses. The Soviets would be hard-pressed to add, say, four thousand warheads to their strategic arsenal for every two hundred interceptors and associated equipment we deploy. The marginal cost to the attacker of adding another warhead to his missile force and modifying his post-boost vehicles seems to be about $3 million apiece—or $12 billion for four thousand additional warheads. A modern BMD interceptor would cost somewhat less than $3 million, although the two hundred small radars and the complex system integration equipment would be expensive, perhaps $10 million per missile—or $2 to $3 billion for the whole system. These ratios favor the defense by about five to one. Adding a second round of interceptors would increase the leverage even more.

Hard-site defenses also have very favorable attributes from the perspective of strategic arms control. They add to deterrence because they protect retaliatory forces while leaving urban-industrial targets unprotected, and they add to stability because they do not threaten either the cities or the forces of the other side. They have the additional advantage of providing the command authority with more time to make crucial decisions. Of course, the ABM treaty, as amended, allows each side to defend either an area containing missile silos, or its national capital. The forces permitted by the treaty (only one hundred interceptors) are insufficient for creating an effective overall defense, however. Thus, even a defense that is almost wholly in accord with the principles of strategic arms control would require renegotiation of the ABM treaty for deployment. Verification of limitations on such defenses would be feasible in several ways, although on-site inspection would probably be required.

It is inappropriate to leave off discussion of these high-leverage defense systems without noting their technical difficulties. Unless the ICBM launchers or shelters can be made as hard as their advocates claim, and the exact location of the missiles at any moment kept secret, preferential defense will lose its effectiveness. And unless the defense radars (or other sensors) are as hardened as the missiles they are protecting, or are rapidly replaceable if damaged, an attacker could degrade or nullify the defense by attacking its weakest part. The offense might try other clever tactics: shoot-look-shoot, nuclear blackout of the defense radars, or maneuvering warheads. Shoot-look-shoot entails risks, because while the attacker is looking, we could be retaliating with the ICBMs that survived the first strike. Nuclear blackout is not so effective against a preferential defense that is able to intercept warheads in the last seconds of flight. And maneuvering vehicles lose velocity and accuracy very rapidly when making hard turns at low altitudes. All these countermeasures deserve further investigation as part of a BMD program.

In sum, there are important factors favoring ballistic missile defense of our land-based ICBMs. Furthermore, we foresee no apparent detrimental effects if the Soviets should also deploy such defenses, within reasonable constraints. The technical and political problems do not seem insurmountable; it is rather an inchoate public suspicion of defense, and apprehensions about tampering with the ABM treaty, that are the obstacles to be overcome.

LOCAL AREA DEFENSES

Another "intermediate" defense option would be defense of valuable, but not necessarily hardened, targets or clumps of targets such as air bases, command-and-control centers, and seaports. This was addressed explicitly by the Hoffman study and implicitly by the Fletcher panel and so is a legitimate option within the SDI.

The merits of defense for such local area targets are not so easy to state as for land-based ICBM forces. Many of these targets are in or near cities and most are soft. The technical problems of defense are greater because incoming warheads have to be intercepted much farther away. And where such targets are few in number and the survival of each quite important, preferential defense may not be feasible—any or all of them might be destroyed by a concentrated attack. Given these target characteristics, it would usually be more cost effective as a first step to try more conventional means of protection: proliferation, mobility, secrecy, hardening, or some combination of these.

Over the years, studies have been made of ballistic missile defense for some valuable targets such as strategic bomber bases. Currently, we ensure that a significant fraction of our bombers will survive attack by putting them on twenty-four-hour alert, ready to take off on warning of attack. During crises, the fraction on alert can be raised and the bombers dispersed to other bases. Defending bomber bases arguably offers two advantages over these procedures: it would also protect that portion of the bomber force not on alert, and it would increase the survivability of airfields after a nuclear exchange. Yet defending the bases would probably be the most expensive way of guaranteeing the survival of bombers. A full defense of airfields would have to intercept large numbers of SLBMs, ICBMs, and cruise missiles—all of which the Soviets could place on target almost at their leisure, because once our alert forces have departed, the remaining planes are slow to respond. Moreover, defense of bomber bases presumes nuclear war-fighting and protracted nuclear war. It thus raises the question whether money spent on defending air bases might be better spent on forces that would increase deterrence and prevent nuclear exchanges in the first place.

As for command-and-control centers, increasing their numbers seems a more efficient way to ensure continued operations. Suggestions that we defend them seem to stem more from a sense that it would be easier to obtain funding for defense than for adequate command-and-control facilities themselves. But if we think of the national command authority (NCA) as people, rather than facilities, the issue is more easily resolved. The whereabouts of the president and secretary of defense at the start of an attack can never be anticipated, so their safety can never be fully guaranteed. Short of an almost leak-proof whole-country defense, the most effective tactic for preserving command authorities is to equip and train the NCA hierarchy and staff.

For similar reasons, defense of seaports probably does not make sense. The number of major seaports is small, they are soft targets, and they tend to be far apart. The Soviets could simply overwhelm the defenses if they choose to make a large enough attack. More important, however, seaports are more of a military convenience than a necessity. The Soviets could not stop the flow of goods to our allies by attacking seaports. There are far too many ways to transport material of all kinds: over the beaches; by air; through improvised ports; or through the small ports, marinas, and piers that dot our 23,000 miles of coastline.

Defending various valuable but soft targets with local area defenses, then, does not seem promising.

LIGHT AREA DEFENSES

So-called "light area defenses" would be a step beyond local area defenses. They would protect the country as a whole from a light attack, that is, a level of nuclear attack substantially less than what would be expected in a major nuclear exchange. The United States started to deploy a light area defense, the Sentinel system, in the 1960s. In 1968, Sentinel's name was changed to Safeguard, its mission was changed to protecting Minuteman ICBMs, and its light area defense role was markedly reduced. After the ABM treaty was signed, the few sites that had been built were dismantled. Studies of light area defenses continued, however, on the assumption that the United States would eventually need them. One rationale was that we needed to raise the threshold of any small nuclear attacks made on us by the Soviets. It cannot be denied that as the commitment of weapons required by an attacker increases, and the intensity of probable responses therefore becomes larger, the likelihood of his engaging in impulsive acts declines dramatically. It was further argued that the whole country needs protection—with increasing urgency as nuclear weapons proliferate—from attacks by nations with small nuclear arsenals, from accidental or unauthorized attacks by any source, and from attacks by terrorist groups. Supporters of light area defense reason that although such attacks may be unlikely, there is a finite probability that one or more will eventually occur.

Until recently, light area defenses had to rely on strings of radars along the coasts, as ground-based radars could not survey beyond the horizon. Short-range surveillance begat numerous (and expensive) short-range missile sites. Several developments since the early 1970s have altered that situation. Air- and space-based surveillance of huge areas is now feasible. Small, accurate, hypervelocity BMD warheads are practical, making the interceptors that carry them much faster given the same overall weight. The huge signal and data processing requirements of a central site can probably be accommodated now because of vast improvements in computers. Space-based directed-energy weapons could provide light area protection even if they could not achieve whole-country ballistic missile defense.

Light area defenses would also have growth potential as the threat became more sophisticated or the potential number of attacking warheads increased; more interceptors could simply be added. The sensors and computers are the key components for upgrading; they need higher capacities and improved software to solve the complex problem of finding warheads among a variety of penetration aids. As this upgrading is achieved, the light area defense could gradually become the terminal portion of a whole-country defense.

An upgraded light area defense system deployed overseas might also provide a dense defense against attacks on NATO by Soviet theater nuclear forces. Our European friends might be comforted by the realization that the United States would no longer be constrained in its diplomatic support by the concern that the Soviets might engage in nuclear adventurism against Europe as a response. If the Soviets also deployed a defense, it might neutralize the retaliatory forces of both Great Britain and France, but both could field sophisticated penetration aids that would add to the Soviet Union's uncertainties about its defense effectiveness. Of course, with the nuclear threshold raised on both sides by the presence of defenses, NATO readiness at the conventional level would become even more important than it is now.[5]

While effective light area defenses would require modification of the ABM treaty if we decided to deploy them, they seem to violate none of the precepts of strategic arms-control theory. They would simply force the Soviets to increase the scale and risk of a successful nuclear attack, thereby deterring an attack and increasing stability. Small numbers of interceptors would not significantly undermine assured destruction. They would not increase the level of nuclear weapons in the world. Even if they drove the Soviets to counter with an increase in offensive weapons, that increase would presumably be nullified by the number of weapons that would be intercepted and rendered useless. The world should be a somewhat safer place with light area defenses, particularly since they could evolve to meet an increasing threat.

In the cases of hard-point and local area defenses, it was noted that alternatives to defense could be used to achieve the same objectives, sometimes at lower costs. There are no alternatives, though, to light area defense, so we must think quite differently about the costs involved. Cost comparisons must be tied to the damage that expenditures on light area defense could prevent, including the loss of human life. Tens of thousands of human deaths and billions of dollars of property damage would be a tragic price to pay when they could be avoided by deploying an efficient light area defense.

One point should not be forgotten, however: none of these hard-point, local area, or light area defenses against ballistic missiles can provide effective protection unless augmented by air defenses against bombers and cruise missiles. Although BMD systems have substantial capability against airborne threats, they are insufficient. Air defense would also be required for whole-country BMD systems.

WHOLE-COUNTRY DEFENSE

Whole-country defense is the subject of greatest contention in discussions of ballistic missile defense. President Reagan's March 1983 speech set the protection of all Americans from the threat of all nuclear weapons as the ultimate goal. The Fletcher panel also focused on defense-in-depth of the whole nation as the heart of its study, acknowledging lesser deployments as potentially useful components of the whole.

When ICBMs first appeared in 1957, whole-country defense was an immediate concern. But in the following decade, the outlook for a successful city defense became bleaker. When a 1968 Defense Department study showed that even a very expensive city defense system would not reduce the casualties sustained in a heavy attack, the major funding for city defense R&D was cut back and substantial funds were diverted to hard-point defense.

What were the technical deficiencies that spelled the demise of city defense? Boost-phase defenses had been examined in the early 1960s and found impractical—surveillance requirements could not be met with existing technology, and interceptor velocities (even from a space platform) were too slow to allow a sensible system design. The effectiveness of penetration aids in space, where the drag of the atmosphere is not a factor, made an effective mid-course defense unlikely. Although atmospheric drag would separate warheads from decoys during the terminal phase, the altitudes at which this occurred were too low for city defense. And the capacities of computers required to handle an intense attack were far beyond the state of the art.

But if whole-country defense was laid to rest in the late 1960s, why resurrect it now? Technological advancements have occurred that promise order-of-magnitude improvements in every area previously deficient. Multi-spectral infrared sensor technology allows surveillance of the whole earth from deep in space; computer processing capacity has increased a thousandfold. The most exciting new technologies are speed-of-light weapons, which could substitute for slower interceptors and make space-based boost-phase intercept feasible. Regardless of current criticisms, such weapons might prove cost-effective. After all, the chemical energy equivalent of ten kilograms of fuel would be sufficient to kill a hardened Soviet missile weighing hundreds of thousands of kilograms. Furthermore, protection and survivability of space platforms can in principle be enhanced, so there is some prospect that they can win a long-term offense-defense competition.

There are still a good number of problems to be addressed. Intercepts occurring earlier in the ICBM's trajectory (as early as the boost-phase, in fact) must cope with a panoply of penetration aids, and it is important to determine whether new technologies will meet that requirement. It is also important to assess whether the systems that emerge will be lethal enough to perform the mission, and robust enough to survive a massive effort by the Soviets to nullify or counter them.

POSSIBLE SOVIET ACTIONS AND REACTIONS

The Soviets will doubtless regard the various tiers of a BMD system differently, reacting more strongly to space-based weapons systems than to the more familiar terminal or mid-course systems. How, then, might the Soviets respond to our deployment of a broad-coverage, terminal/mid-course ballistic missile defense that contained airborne and space-based sensors, but no space-based weapons?

First, the Soviets would have to assess the overall current and potential effectiveness of our defense and build up their offensive forces accordingly. In order to attack those targets they believe are critical, they would have to take into account that our defense can range over a large area, and they would therefore have to use large multiples of the actual number of targets. In a democracy, preferential defense of regions is obviously inappropriate. Yet if an attack occurs, our defenses, although deployed for broad protection, must surely first be employed to protect those sites critical to our survival as a nation. The worst aspect for the Soviets will be uncertainty; not knowing which targets we are determined to save, they cannot be confident that their attack will succeed. To counter the obstacles posed to a successful preemptive strike by the survivability of our retaliatory forces and the existence of our defenses, the Soviets will be driven to building defenses of their own. To be uncertain of their capabilities against us, while certain that we could annihilate them should be intolerable for the Soviets.

The result would be a world in which both superpowers have deployed offensive *and* defensive strategic systems. Would such a situation merely reestablish, but at a greater cost to both sides, the status quo?[6] The answer is no, because of two new, key factors. Defenses will have raised the threshold for effective attacks so high that only major nuclear strikes would have an effect, and the uncertainty of outcomes would be great. Thus, the likelihood of a nuclear exchange would be substantially reduced, the spectrum of conditions under which it might occur greatly narrowed, and both sides would have means of protection against accidental launches and third party attacks.

The Soviets may react more vigorously to a U.S. defense system if it is deployed in space. Since space-based weapons have to be deployed gradually, on orbital paths the Soviets can readily predict, we should expect that the Soviets, fully aware of the threat such weapons pose, would make every effort to destroy them. The Soviets may calculate that speed-of-light weapons in space would grant global hegemony to the first country that deploys them successfully. They may reason that once either side gets a space-based system in place, its survivability may depend on preventing the other side from undertaking unauthorized launches into space. They might also reason that even if *both* sides deployed space defenses, the co-presence in space of Soviet and U.S. directed energy weapons of approximately equal capability might be extremely destabilizing, as an attack by either side on the other's defense could grant a country global hegemony in a matter of minutes. Motivated by such fears, the Soviets may consider attacking our defense satellites and launch pads.

Accommodating such extreme possibilities would require heroic measures. Our launch sites might be hardened against all but nuclear attacks, they might be defended, or their security might be part of a negotiated arrangement with the Soviets covering defenses. As for security in orbit, defensive satellites (DSATs) could be launched into orbit with each directed energy battle station.

Because of its obvious expense and complexity, a space-based defense will probably not be deployed unless it is a low-leakage system. How would such BMD further affect the strategic balance set by terminal/mid-course defenses? First, it would remove us from the era of deterrence by mutual assured destruction. An all-out Soviet attack may still wreak havoc on our country, but it could not annihilate us. Second, it would place increased demands on the Soviet defense system, which would be driven to achieve balance with our own. Our inability to destroy each other would establish a new deterrence, one based not on the terror of offensive forces, but on respect for defensive systems. This situation has been characterized by some as a move from assured vulnerability to assured survivability. This is not exactly the case, since no imaginable set of defenses can prevent a determined and resourceful enemy from detonating nuclear weapons in our country. But the attacker's certainty that his objectives (other than naked terrorism) cannot be achieved, provides a quantum change in strategic perspective.

How does whole-country defense fit in with strategic arms-control theory? It would certainly be consistent with important goals of arms control. It would improve stability by raising the nuclear threshold and introducing a greater measure of uncertainty into an attacker's plan. There are those who disagree with this: they argue that defense is *de*stabilizing because deployment of defenses may alarm the other side by signaling a changing strategic balance, thereby encouraging a preemptive nuclear attack; and that defenses, known to be imperfect, will only be used by the *attacker*—against his opponent's ragged response to a massive first strike. The first of these objections would be sound only if there existed a huge imbalance in superpower forces—which is not at all the case today. The second objection would be valid only if it were the case that reducing the number of attacking warheads would allow poor defenses to become nearly perfect. But a defense of acknowledged mediocrity can be penetrated in a variety of ways by forces whose survival from an all-out first strike can be assured. A ragged retaliatory strike, therefore, is inexcusable militarily; neither side should allow the balance or the quality of offensive forces to become so degraded.

Arms control would still have an important role in a world of whole-country defenses. Even though the defense systems may themselves be non-nuclear, the total number of nuclear weapons in the world would be increased if the Soviets responded to our defense deployments (or we respond to theirs) by deploying a larger number of offensive forces. Moreover, in the absence of mutual restraints, either country may feel compelled to continue augmenting its offensive forces without limit, even if cost-exchange ratios favor defenses. That a truly massive effort in offensive deployment could overcome any defense is a certainty. Thus, even in a defense-oriented world, agreements on acceptable levels of offensive nuclear arms would be imperative.

Whole-country BMD does violate one of the precepts of current strategic arms-control theory: it renders retaliatory forces less effective. But a defense orientation would rely on a different form of deterrence: the would-be aggressor would be disuaded not by the fear of annihilation so much as by the recognition of futility. Particularly if the cost-exchange ratios favor defense, the motivation would be to negotiate. In the end, the level of nuclear weapons required by the superpowers should become just sufficient so as to maintain an adequate margin over any combination of forces of lesser nuclear powers.

NEAR-TERM COST ESTIMATES

The last question to ask about the move toward a BMD-oriented world might be whether the amount of money earmarked for the president's Strategic Defense Initiative is too much. The plan was to spend about $26 billion over the first five years (1984–89), then to decide in the early 1990s which technologies are suitable for incorporation into systems. This would give future administrations the option to proceed with engineering development of those technologies and to deploy them—if feasible—in the first decade of the twenty-first century. Some technologies, and therefore some systems, can be ready for engineering development (building of prototypes) before 1993 and deployment before 2000. But the $26 billion SDI estimate includes no full-scale engineering development for these systems, and does not provide funds beyond 1989.

Estimated expenditures for the SDI through the 1980s average around $5 billion per year. Historically, the Pentagon has allotted about 10 percent of its budget to research and development of all kinds. The current amount is about $30 billion per year. By 1989, the SDI will require about 17 percent of the Pentagon's R&D budget—not insignificant, but certainly affordable. Government officials have estimated that $15 to $20 billion of the projected $26 billion would have been spent anyway in research and development on projects that are technologically similar to potential defense systems, on work to prevent technological surprise, or on efforts to keep penetration-aids technology at the state of the art.[7] The whole SDI budget will amount to far less than what Americans will spend on cigarettes or cosmetics during the same period.

The costs of possible future systems development can only be guessed at. Major systems such as submarine-launched ballistic missiles and bombers cost several billion of today's dollars to develop. A complete BMD system might consist of as many as ten separate systems, costing perhaps $40 billion for full engineering development. Estimates of the full-scale engineering development costs, like the ultimate systems costs, should be given little attention and still less credence now, especially since they seem to be within the broad bounds of affordability.

In considering these costs, we need to take a broader view than is usual in the formulation of public policy. Both the promise and the problems of the new defense technologies being investigated in the SDI will require a long time to clarify. This urges us to seek a new perspective on our long-term strategic posture. Because we elect our politicians to terms of two, four, and six years, and make five-year guidance plans for the Defense Department, we find it hard to think in terms of the twenty-, fifty-, or one hundred-year future. But the challenge we face is a massive one: averting nuclear holocaust for centuries. Our ambitions to create an effective whole-country defense using futuristic weapons should be viewed in that light. And the costs of developing such a defense should be considered in those terms as well.

THE SDI AND ARMS CONTROL

Perhaps the most delicate issue raised by the SDI concerns the coordination of defense research with the development of a rational strategy for strategic arms reductions. As I noted above, arms control will remain important even in a defense-oriented world. Therefore, we must not blunder into situations that would be disastrous to our prospects for strategic arms reductions. Neither do we wish to cling to tenets of arms control that are flawed or outdated. Each defense discussed above would require renegotiation of the ABM treaty before its deployment, and testing many of the components in the ABM mode might also require treaty modifications. Our existing treaties are the product of honest and diligent efforts, and reflect the views, ideals, and technologies of the time. Nevertheless, they should not be viewed as sacrosanct. These treaties were all originally devised for security purposes, and if those purposes can be shown to be better served in new ways, the treaties should be adapted to changing circumstances.

As I have argued above, there are a variety of defense systems beneficial to American security that appear consistent with arms-control goals implicitly endorsed by the Soviets in their signing of past agreements. In many instances, the Soviets came to accept arms-control measures they initially condemned—the ABM treaty is an excellent example of this change in Soviet posture. We should not be surprised if the Soviets react to the SDI in much the same way.

SUMMARY

I have argued that many new technologies show promise of making a spectrum of BMD systems practical; that a strategic posture based on defense can contribute to deterrence, increase stability, and reduce—and perhaps eventually eliminate—the threat of nuclear extermination; and that we ought to support a vigorous development of BMD technologies to discover their true potential.

In saying these new technologies show promise, I am not assuming, or offering a guarantee of, success. The Fletcher panel concluded only that the technologies were sufficiently understood for us to sketch out broad systems concepts for a multilayered defense. The crucial technological issues will take years of research, engineering, and analysis to resolve; they certainly cannot be resolved now by polemics. Yet in our open society, any large and technically ambitious initiative like the SDI is subject to immediate and widespread attention from the media and the citizenry. Temptations are irresistible for intellectuals, who deal in words rather than hardware, to attack broad system concepts as if they were actual system designs, and to predict confidently the outcome of decades of research. Any broad-based technology program such as the Strategic Defense Initiative must be guided by a set of conceptual systems that is admittedly incomplete and imperfect. Unless the public is discriminating, it may find itself unable to distinguish issues from quibbles in the debates over the direction of the SDI. Thus, the seeds of good ideas may be lost before any facts can be determined. For now, the question should not be whether conceptual systems will work, but whether the technology is worth pursuing to find out what can be achieved: let us stick to that question.

ENDNOTES

[1] Gen. Russell E. Dougherty, "Capability × Will = Deterrence," *Air Force Magazine*, June 1984, pp. 7–8.

[2] Texts of the Fletcher and Hoffman reports, as well as the Defense Secretary's summary report, can be found in "Strategic Defense and Anti-Satellite Weapons," Senate Committee on Foreign Relations, 98th Cong., 2nd sess., April 25, 1984, pp. 94–175.

[3] One such tactic is shoot-look-shoot, in which the attacker fires a first wave of missiles, watches to see which sites are defended successfully, and then fires a second wave concentrated on those targets. In the time required for shoot-look-shoot, however, the defender may retaliate with his surviving forces.

[4] See Harold Brown, *Thinking About National Security: Defense and Foreign Policy in a Dangerous World* (Boulder, Colo.: Westview Press, 1983), p. 71. Brown describes the concept of preferential defense for ICBMs, but notes that unless the ICBMs are deployed in deceptive basing, adding ballistic missile defenses will not have a significant effect on ICBM vulnerability.

[5] Colin S. Gray, *Nuclear Strategy and Strategic Planning* (Philadelphia: Foreign Policy Research Institute, 1984), pp. 86–92, provides a point-by-point discussion of the effects that BMD might have on the NATO alliance.

[6] In the late 1960s, when the United States first considered deploying a ballistic missile defense, Secretary of Defense Robert McNamara argued, for instance, that "all we would accomplish by deploying ABM systems against one another would be to increase greatly our respective defense expenditures without any gain in real security for either side." *Department of Defense Annual Report, FY 1968* (Washington, D.C.: Government Printing Office, 1967), p. 40.

[7] "SDI Hearings Focus on Financial Aspects," *Aerospace Daily*, May 10, 1984, pp. 57–58.

George Rathjens and Jack Ruina

BMD and Strategic Instability

IN HIS "STAR WARS" SPEECH of March 23, 1983, in which he launched the Strategic Defense Initiative (SDI), President Reagan called for an effort to develop a defense against ballistic missiles which would change the basis of national security policy in a fundamental way. Instead of relying on the threat of punitive damage as a means of deterring the use of nuclear weapons against the United States and its allies, the SDI aims at preventing their delivery, through the development and deployment of active defenses. In calling for a defense that would render nuclear weapons "impotent and obsolete," the president was wittingly or unwittingly implying not only a 100-percent effective defense against ballistic missiles, but complementary and equally effective defenses against all other means of delivery of nuclear weapons as well—a point that was subsequently made explicit by Secretary of Defense Caspar Weinberger.

President Reagan's goal of a perfect defense stirred the interest of the general public and of many public figures who have long sought an alternative to deterrence based on mutual assured destruction (MAD). Yet MAD is not a policy capable of being changed by political will; it is rather the inevitable consequence of the superpowers having the nuclear arsenals they have. Recalling the Manhattan and Apollo projects, which overcame seemingly difficult technological problems, many argue in support of the president's initiative. They do so, however, not fully appreciating the difference between the SDI and these two earlier efforts, i.e., that the SDI involves competing against a determined and resourceful adversary *as well* as unlocking nature's secrets and harnessing technology.

A substantial amount of support for the SDI has also developed within the community of military technologists and strategists. This has occurred less from a belief that a perfect defense against nuclear weapons can be realized than from a general interest in defense missions far less demanding than that proposed by the president. The tension between the president's goal of a perfect defense and many of his supporters' more modest goals of limited defenses was reflected in the presidential study groups (especially the Fletcher and Hoffman panels) that were formed after the president's speech, as well as in the administration's January 1985 special SDI report.

More limited roles for BMD include those that were considered when BMD was a major national issue in the 1960s: minimizing damage in the event of an all-out ballistic missile attack; defense of hardened military installations such as ICBM silos and command-and-control facilities; defense against unauthorized or accidental attack; and defense against lesser nuclear powers. Additional justifications for BMD in the 1980s include limiting the possibility of a Soviet "light" attack with nuclear weapons, and inducing the Soviets to reduce their commitment to land-based missiles (now their dominant nuclear force) in favor of other systems such as submarine-launched missiles (SLBMs), which to some seem less worrisome.

The counter-arguments to these more modest goals are also similar to those invoked in the past. The first argument is that BMD technology is not up to some of the tasks being considered. The second is that these objectives, if they need be met at all, can be better realized through means other than active defense. The third is that in reaction to U.S. BMD deployment, the Soviets will not only increase the size and capability of their offensive nuclear forces but will also develop and deploy a BMD system of their own that we must assume will be about as effective as ours. The result of this competition could be a net loss in the security of both countries. The fourth argument seeks to counter the claim that deploying a BMD of even limited capabilities would provide valuable operational experience for subsequent deployment of a system intended for population defense. Whatever fears a country might have about its adversary's "breakout" capability (rapid deployment of an operational system) would certainly be exacerbated by its development, training, production, and testing of a limited but operational BMD.

Concerns about deployment aside, the prospect of a greatly expanded SDI R&D program is troublesome. First, it will likely increase incentives for each side to improve its offense because of the enormous uncertainty in estimating the effectiveness of defenses during the R&D phase, and because of the inherent conservatism that leads military planners to base their R&D and procurement on worst-case analyses. In addition, the effect of the SDI program, and similar Soviet efforts, will be to undermine U.S.-Soviet arms-control negotiations, as neither side will be willing to make deep cuts in its offensive forces if work on strategic defenses is intensified. The reality of mutual assured destruction may be unsettling, but neither country is likely to give up, or even significantly reduce, its offensive retaliatory capability in the face of possible defensive deployments by the other. Indeed, the history of the nuclear weapons competition between the U.S. and the Soviet Union demonstrates that each country will do whatever it feels is necessary to maintain this capability for deterring nuclear attack.

We therefore see serious harm resulting from the current SDI program and similar Soviet BMD efforts. This includes:

1) an increase in U.S.-Soviet tensions and intensification of the arms race;

2) the possibility that one of the superpowers would withdraw from the ABM treaty and deploy some type of BMD, even one of doubtful effectiveness;

3) erosion of the confidence of our European allies who are deeply troubled by the implications an SDI program and possible BMD deployment might pose for their security;

4) a false sense that technology and new weapons systems can eliminate the threat of nuclear destruction.

We want to emphasize that we find the pursuit of strategic defenses worrisome for these reasons, and not—as some maintain—because of concern over *crisis instability*. In our view, fears that deployment of a (purportedly) effective population defense could lead to substantially increased risk-taking during a crisis, or possibly even to a preemptive first strike, on the grounds that such a defense would be more effective against a degraded retaliatory force, are unwarranted.

The very possibility of such a defense, and its consequences for crisis instability, can be almost totally discounted given the offense-dominant nature of nuclear weapons, and the technical realities facing strategic defenses.

Nevertheless, we shall treat the possibilities of *perfect* and also *very good* defenses analytically. We are paying more attention than we believe is warranted on technical and cost-effectiveness grounds simply because it is the promise of an effective shield against nuclear weapons that has captured the imagination and hopes of the president and the American public. We will also discuss more realistic defense possibilities, including two special cases: defense of selected military assets, particularly strategic forces; and defense against unauthorized, accidental, or third country attack.

COST-EXCHANGE RATIOS

One important consideration in judging the merits of any type of defensive system is the cost-exchange ratio involved, that is, the incremental cost to the offense to ensure the ability of its warheads to reach their targets, compared to the added cost to the defense to intercept those warheads.

As an illustration, consider whether it would make sense for the U.S. to invest in defense of its industry and population from nuclear attack using technology presently available. We can hypothesize a variety of defensive measures, including air defense, civil defense, and different BMD deployment schemes, that might save a million lives in the event of a nuclear attack, and then estimate the minimum cost to the Soviet Union of upgrading its offensive capabilities just enough to negate each of these hypothetical American investments.

If the Soviet Union could negate such defensive measures at a cost much lower than the incremental American investment in defense, then U.S. investment in defense would likely be of little use. Assuming the Soviet Union wanted to maintain its current ability to inflict damage on the U.S., it would presumably make whatever investment would be required to offset ours. If we, on the other hand, could develop a defense system that could be negated only by relatively costly Soviet improvements, investment by the U.S. in defense would seem desirable. In this case, the Soviet Union presumably would not react, because any improvement in its offensive capabilities could be negated by a much smaller American investment in defense.

There is a third possibility. If the cost-exchange ratio for the best defense option we could identify were near unity—if the marginal costs to the offense and defense were about equal—then deciding whether to implement the defense option would be more difficult. Would the Soviets be less or more determined to threaten additional destruction than we would be to negate that possibility? If both nations were to attach the same weight to that increment of destruction, could the U.S., with its stronger economy, more easily afford to reduce the expected damage than the Soviet Union could afford to offset that reduction?

In any event, estimates of cost-exchange ratios are of limited value since they generally reflect the likelihood that each country overestimates the effectiveness of the adversary's weapons programs, while underestimating his own. Nonetheless, ratios either highly favorable to the offense or highly favorable to the defense do indicate, respectively, offense and defense dominance. A ratio near unity may well imply a vigorous offense/defense arms race, with each side straining to offset the improvements made by the other.

It should be remembered that calculations of cost-exchange ratios will necessarily be highly tentative in the absence of concrete formulations about what the defense will look like. It is true, however, that the more nearly "perfect" a defense strives to be, the more likely it is that additional investments will be relatively unattractive. The reason for this is the adversary's option of *preferential offense targeting*. An attacker can selectively choose which targets to attack, and concentrate its forces on them, while the defender must be prepared to defend all targets against such efforts. This challenge is especially acute when the defense is seeking to protect cities and population.

On the other hand, a limited defense that seeks to protect a fraction of its targets may benefit from *preferential defense*. The defender can concentrate his defenses to protect a fraction of his targets, but the attacker, not knowing which targets are to be defended, will be driven to expend more weapons than needed against undefended targets if he is to have much hope of destroying defended ones. Such limited defense is likely to be of use particularly in assuring the survival of retaliatory capabilities. Even if the defender is left with only 20 percent of his one thousand ICBMs, that would constitute a potent retaliatory force.

The cost-exchange ratio will, of course, depend sensitively on the technological options available to the two sides, and it is conceivable that it could change dramatically if new options become available. The current interest of the SDI in boost-phase defense illustrates this point. One of its great conceptual attractions as a "perfect" or highly effective defense—as opposed to the terminal defenses of the 1960s—is that it avoids the problem of offensive preferential targeting: each increment in defense capability provides the same degree of added protection to all targets. However, as we will elaborate below, whether concentrating on boost-phase defense can *in fact* yield a favorable cost-exchange ratio for the defense is very doubtful given the many technical challenges involved, the relative ease of countering a boost-phase defensive system, and the difficulty of protecting a space-based BMD system against direct attack.

PERFECT DEFENSE

There is something unprecedented about a defense that would make nuclear weapons "impotent and obsolete": it *really would* have to be perfect, not just very good. Non-nuclear offensive systems have been "rendered impotent and obsolete" by defenses that were less than perfect (armored knights by archers and the musket, for example), but the challenge is much more difficult when the issue is defense against nuclear weapons, with their individual destructive power.

In non-nuclear combat, the survival of some percentage of the loser's military forces has rarely been militarily or politically important. But with defense against nuclear weapons there is a world of difference between 100 and even 99 percent effectiveness. The former would, indeed, make nuclear weapons "impotent and obsolete," except perhaps for coercive purposes against nations not having defenses. A defense that is 99 percent effective would, on the other hand, result in unacceptable leakage in a war between the superpowers. The detonation of even as few as five or ten Soviet warheads (less than one-tenth of one percent of the Soviet arsenal) on U.S. cities would cause unparalleled destruction, something McGeorge Bundy has rightly characterized as "a disaster beyond history."[1]

To relieve all public concern about nuclear weapons, to assure political leaders that a defensive system would eliminate the possibility of a nuclear explosion on their own or allied territory, and to convince one's adversaries that nuclear weapons would not be at all useful to them—and "impotent and obsolete" would have to entail all these things—a defense would have to be virtually 100 percent effective, and perceived as such by both sides. This means that the system would have to perform flawlessly the very first time it is called upon, that it be invulnerable to direct attack, and that it be effective against *all* means of nuclear weapons delivery, including aircraft and cruise missiles as well as ballistic missiles.

In summary, we do not believe perfect defense to be a realistic possibility, particularly if the Soviet Union reacts by improving its offensive capabilities, as the Hoffman panel has recognized is likely[2] (but as the manager of the SDI program does not).[3]

VERY GOOD—BUT LESS THAN PERFECT—DEFENSE

It is our conviction that even a very good defense is not a possibility, assuming one is dealing with a determined adversary with substantial resources at his disposal. By "very good," we mean a defense capable of affecting the behavior of nations in times of crisis. Defenses that would be 99 percent or more effective against a "first strike" would meet this criterion. Such defenses might well affect decisions by political leaders in the event of crises in the Middle East, the Persian Gulf, or Europe. If one or both superpowers were confident that it could reduce the number of attacking warheads to a small number, say ten or less, then the U.S. and/or Soviet Union would presumably be willing to take larger risks in escalating a conflict. This might be the case even if it posited a somewhat poorer defense performance, e.g., if it faced the possibility of ten to fifty warheads leaking through. We would be doubtful, though, about its taking such risks if the expected leakage exceeded a hundred. This suggests that in order to be of political-military worth, defenses would have to be effective at the 99 percent-plus level.

Such defenses would have a terribly—and we use this adverb literally—troublesome effect on crisis instability. However effective defenses in this range might be against an adversary's "first strike," they would be still more effective against a ragged retaliatory force.

Expected leakage could be sufficiently reduced to justify political/military decisions that would be judged unacceptable in the absence of defense. If technology and cost-effectiveness considerations warranted such defenses, a country might proceed with their development and deployment on the assumption that a situation might develop in which it would be willing to launch a first strike; once a defense system is deployed, the incentive to strike first in a crisis situation could be overwhelming. If an opponent's defenses were vulnerable, the attractiveness of preemption would increase. Country A might strike first at B's defenses, so degrading them that a significant number of A's warheads might then be delivered against B's offensive force, reducing it to the point where A's defenses could reduce the leakage from a retaliatory attack by B to "acceptable" levels. This creates quite an incentive, and a reciprocally reinforcing one, for both sides to strike first during a crisis.

We feel it necessary to stress again that, in our view, this type of very good defense is illusory, at least as it applies to confrontation between the superpowers. The ongoing competition between defensive developments on the one hand, and offensive countermeasures and defense suppression techniques on the other, will mean that even if some BMD systems are deployed, the goal of a perfect, or even very good, defense will be unattainable. For this reason, longer-term considerations of crisis instability should be of little real concern.

Concerns about arms-race instability, on the other hand, are indeed justified, inasmuch as arms races seem to be driven as much—or even more—by fear of what *might* happen as by what is likely. If there were even a remote prospect that either superpower, or both, could deploy defenses that would be 99 percent effective, there would be enormous pressure for the would-be defender to improve the figure and for its adversary to reduce it. From an arms-race perspective, it is difficult to imagine a prospect more destabilizing.

FAIR-TO-GOOD DEFENSES

Few advocates of the SDI, with the possible exception of President Reagan and Defense Secretary Weinberger, believe that perfect or very good defense, as we have used the term, is realizable. Some, however, do envisage the possibility of such dramatic improvements in defenses that the cost-exchange ratio may approach unity at, say, a 50-percent effectiveness level. That is, they envisage an offense-defense technological balance that would permit, if not lead to, the deployment of defenses and offensive forces such that the expected level of damage in the event of conflict would be around 50 percent, or anywhere in the range of 1–90 percent (our arguments are invariant over this range). Efforts to decrease the damage level to below 50 percent through additional increments of defense could be offset by incremental improvements in offenses at less cost, while efforts by the offense to increase the level of destruction beyond 50 percent could be negated less expensively by incremental improvements in defense.

Were technological developments to permit such defenses, we would expect to see strong interest in deployment, for if the cost-exchange ratio were unity at the 50 percent destruction level, it would surely be much more favorable to the defense at higher levels of expected destruction. Each side would presumably have an impetus to deploy defenses sufficient to hold damage to some level of destruction, perhaps 70 percent, that would clearly be cost effective, while its adversary would not be strongly inclined to try to increase it beyond that level. This would follow given the unfavorable cost-exchange ratio to the offense and the fact that it would gain little, if any, political-military utility from increments in prospective damage level in this range.

In such circumstances, one would expect a continuing technological competition that might shift the balance one way or the other. From the perspective of crisis stability, however, the situation would be much less worrisome than if the cost-exchange ratio were unity at a level characterizing very good defenses—such as 99.9 percent—where small changes in capabilities would be politically and militarily exploitable, and where there would likely be enormous advantage in launching a "first strike."

OFFENSE DOMINANCE

We now turn to the current situation, in which the offense is dominant over virtually the whole range of destruction levels. This condition has obtained throughout the nuclear era, and is likely to persist between the superpowers for the indefinite future. As long as it does, we can be reasonably sure that the superpowers will invest the resources necessary to overcome each other's defenses. Since the 1950s, for example, the Soviets have spent billions of dollars upgrading their extensive air defense systems, yet the U.S. has continued its efforts to be able to penetrate them. Over the last decade, the U.S. has spent billions of dollars to modernize the B-52 bomber and equip it with short-range attack missiles and later with cruise missiles, to procure the advanced B-1 bomber, and to develop Stealth techniques that would allow U.S. bombers in the 1990s to evade radar detection and reach targets in the USSR. What makes this example particularly compelling is that, for both the U.S. and USSR, manned bombers are far less important for retaliatory deterrence than either ICBMs or SLBMs, yet the competition has continued. And, of course, there has been competition between ICBMs and ABMs, even though technology and cost-exchange considerations have always militated against large-scale defensive deployment, at least of a sort intended to protect cities and industry. ICBM penetration aids were developed in anticipation of the possibility of defenses being deployed, and the prospect of ABM deployments was also a—if not the—factor in the impetus to develop and deploy multiple warheads (MIRVs). Thus, almost every effort by one superpower to improve its defense capabilities can be expected to provoke improvement in its adversary's offense capabilities for as long as defensive developments do not significantly alter the current offense-dominated situation.

DEFENSE OF SELECTED MILITARY ASSETS

Up to now, we have been addressing defense of cities and populations, a fundamentally difficult problem, given the enormous destructive capability of nuclear weapons, the vulnerability of these targets, and the fact that loss of even a small fraction of them would be catastrophic. Defense of selected military targets, however (ICBM silos are usually cited), is a much easier task. It is quite possible that, unlike defense of population and social infrastructure, cost-exchange considerations will favor this type of defense. There are two reasons for this.

We have already mentioned that preferential defense can give great leverage to the defense in any situation where the same components (interceptor missiles, aircraft, etc.) can be used to defend any of several targets *and* where the loss of a large fraction of the targets is tolerable.

Second, the defense of military targets is often easier because it is possible to protect them against the effects of nuclear attack by passive means such as "hardening" (something not possible for cities), which enables them to survive near-misses. This is particularly important since the firing of defensive interceptors can be delayed until late enough in a warhead's trajectory for the defense to discriminate between warheads likely to destroy a target unless intercepted and decoys and warheads likely to miss their targets, or warheads aimed at targets already destroyed.

But cost-exchange considerations are not the only ones that bear on the decision whether to deploy defenses. One must also decide if the targets are worth defending, and whether there is indeed a threat. Placing a monetary value on the loss of a city and its population is clearly a formidable—some might argue, impossible—task; it is clearly easier to make judgments about the worth of missile silos or other military targets. One should also raise the question whether there are alternatives to defense that could accomplish the same military mission, e.g., providing for the delivery of a desired level of retaliatory attack. When such comparisons are made—and this must involve not just dollar costs, but considerations of stability and other factors—it could be that retaliatory deterrence would be better served by buying larger numbers of offensive missiles or by placing greater reliance on mobile systems.

However, in the absence of a considerably increased Soviet threat, we would argue against either buying defenses to protect U.S. strategic forces or increasing offensive forces. We would argue this on the grounds that the so-called "window of vulnerability" has been much overrated. As President Reagan's Commission on Strategic Forces, chaired by General Brent Scowcroft, recognized, the enormous uncertainties over whether a Soviet attack on U.S. strategic forces would be successful, and the tremendous risk that such an attack could lead to general nuclear war, make the threat of such an attack implausible, at least in the near term.

As a final note, we should point out that it is possible to develop and deploy a respectable terminal BMD system for hardened military targets with state-of-the-art technology. The Department of Defense has pursued relevant technologies for hard-target defense since the signing of the ABM treaty in 1972. Yet it is not at all clear that the massive SDI program now underway can or will lead to the development of hard-site defense systems that are any more useful or cost effective than those that might emerge from the continued orderly pursuit of BMD technology as it existed prior to the SDI. The SDI program therefore seems irrelevant to decisions about whether to pursue hard-site defense deployment; indeed, it may even be diversionary.

DEFENSE AGAINST UNAUTHORIZED, ACCIDENTAL, OR THIRD COUNTRY ATTACK

The case against defense in general is not quite as strong as the foregoing suggests, for we have not dealt with the possible role of defense in coping with threats to either of the superpowers from unauthorized or accidental launches and nuclear attack by third countries. Because such defenses would not have to meet the same stringent technical, cost-effective, and survivability criteria as other BMD systems, they could well provide an effective hedge against limited attacks. For example, deployment of defenses may be attractive to the superpowers in dealing with a threat from a third party, as its limited resources could make it difficult for it to counter heavy investment in defenses by a superpower. It is conceivable that a superpower could deploy a defense that would be very effective against a third power, but relatively ineffective against its main adversary, because of the much greater initial offensive capability of the adversary superpower and because it would more likely improve its offensive capabilities to negate any increment in investment in defense.

If the Soviet Union deployed such a defense, against China for example, how ought the United States to react? From a narrow cost-benefit/game-theory perspective, it would only make sense for the U.S. to improve its offensive capabilities (and/or deploy defenses) if it seemed plausible that the Soviet initiative might lead to acquisition of a "very good" defense against the U.S. In the real world, however, the propensity for worst-case analysis and the pressures of domestic politics and international image-making would make a U.S. response likely, even if the Soviet defense in question seemed quite incapable of dealing with American offensive forces. As for possible nuclear threats to the U.S. from Third World nations or terrorist groups, their nuclear weapons are more likely to be smuggled in by ship or plane than to be delivered by ballistic missiles.

A decision by either superpower to deploy defenses designed to protect against unauthorized or accidental launch of ballistic missiles could also provoke concern. A light, nationwide defense of the U.S. might not call into question Soviet retaliatory capabilities, even after a U.S. preemptive strike. Yet it would be provocative to the extent that it could conceivably be rapidly upgraded to a more robust defense capable of dealing with a ragged Soviet second strike.

At present, even though the ABM treaty places very strict limits on the deployment, testing, and procurement of BMD systems and components, many critics of the treaty claim that the Soviets, through a combination of legal loopholes and illegal activity, are striving for a breakout capability. If the treaty is amended to allow even limited nationwide defenses, fears about upgrade and breakout capabilities will become still more serious. Even if the ABM treaty could be modified to permit deployment of limited defenses, such deployment could prove counterproductive if the end result is increased arms-race instabilities and a new round of offensive and defensive competition. And withdrawal from the treaty for the purpose of deploying a limited, thin area defense is likely to have even more serious consequences for the arms race, since there would be no defined limits to how much defense could be deployed. Treaty withdrawal would also likely have very negative consequences for arms control generally, and for our relationship with the Soviet Union and our allies.

As we noted in the case of defense of hardened targets, a reasonable thin area defense intended to protect against some unauthorized, accidental, or third-country attacks can be developed with existing technology. With time, such systems can surely be improved, but we need not await breakthroughs in technology resulting from the SDI program to address realistically the question of the net benefits of such a defense and to decide whether it should be deployed.

POLITICAL EFFECTS OF THE SDI

We have argued that there is virtually no prospect of realizing President Reagan's objective of "rendering nuclear weapons impotent and obsolete," or of achieving even a very good defense of population and industry—at least if the threat is from the Soviet Union. This opinion seems to be shared by most of those involved in the SDI program. It is necessary, then, to ask: what is likely to come of the SDI, and should we be concerned about it?

In all likelihood, a superpower competition in strategic defenses will lead to an increased Soviet-American offensive arms race. We expect that U.S. defensive developments will lead to an acceleration of Soviet efforts both to develop similar capabilities for defense and to improve offensive capabilities, and we can expect corresponding U.S. reactions to Soviet efforts. If this process continues for some time—and clearly, the SDI has a very distant time horizon—pressures will develop for some kind of BMD deployment in both countries. Indeed, support is already developing in this country for a defense of ICBM silos and other military assets.

Here, perhaps, the American experience of the late 1960s is relevant. After years of R&D, and confronted with the beginnings of a limited Soviet BMD deployment, President Johnson could not resist pressures to go ahead with *some* kind of BMD deployment. Since there was no possibility of deploying a system that could be very effective against a massive, determined Soviet attack, a decision was made to deploy a system (Sentinel) that might cope with a much more limited attack, e.g., missiles launched by accident or by China. The system's objective and mode of deployment were later changed (as was the name, to Safeguard) to emphasize defense of Minuteman missiles. Something like this could well happen again, perhaps driven more by domestic political and institutional considerations—as was the case with the Sentinel and Safeguard—than by military arguments. Such a situation could further accelerate the arms race, particularly if the system in question even *appeared* to have the potential for upgrading to a nationwide defense of some effectiveness.

In this context, a greatly accelerated U.S. R&D effort on strategic defenses might introduce an insurmountable obstacle to arms-control negotiations. As the Soviets have made clear, they will not be willing to accept reductions in offensive forces until strict limitations are placed on defense efforts. Yet the Reagan administration has been equally adamant that it does not consider the SDI to be a negotiable bargaining chip.

Moreover, the SDI poses a real threat to the ABM treaty, which is probably the most significant arms-control agreement of the post–World War II period. By introducing some measure of certainty into U.S. and Soviet military planning, the treaty has weakened a major justification for upgrading offensive capabilities and developing defenses. It is difficult to believe that any deployment of new defenses would be so limited as to fall within the constraints of the treaty. Indeed, we are more concerned that there will be great pressure—perhaps successful—to scrap or emasculate the treaty long *before* any new deployments are initiated. It is hard to see how the U.S. SDI program, or Soviet counterpart, could proceed to the point of advanced testing and development *without* violating the treaty's provisions.

This brings us to another concern about arms-control agreements. It is quite clear that the Reagan administration's commitment to the SDI has been a factor in its reluctance to pursue negotiations to limit or halt the development and deployment of anti-satellite (ASAT) capabilities. This is troubling, because some restraint, whether covered by formal agreement or not, could well be to the advantage of both superpowers. This is so because the development of BMD technology under the SDI is likely to be very useful for ASAT purposes; this will work to the disadvantage of both superpowers, but particularly the U.S., given its dependence on military satellites.

The most worrisome aspect of the SDI may finally be its political-psychological consequences. Our concern focuses primarily on four groups: the general American public, American elites, the leadership of countries allied with the U.S., and the Soviet leadership.

The immediate response to President Reagan's March 23 speech and subsequent information derived from polls indicate a great deal of resonance with the president's vision. Americans want their children, if not themselves, to look forward to a world without the threat of nuclear weapons; they find appealing the notion of a defense so effective as to render nuclear weapons impotent and obsolete. We, however, see this vision as diversionary and illusory; it deludes the public into thinking that the solution to the dual problem of nuclear weapons and a troublesome adversary can be resolved by new weapons systems, rather than by political means.

As we mentioned earlier, most of those in the technical community, and many others who are well informed and concerned about security matters, share our belief that a perfect, or even a very good, defense against a massive Soviet attack against U.S. population and industry is just not realizable. Yet, the president and his secretary of defense have continued to espouse the goal of a perfect defense not just against ballistic missiles, but against bombers and cruise missiles as well.

We are troubled that the SDI may raise special concerns in the case of Allied governments. They may well see the SDI and U.S.-Soviet competition in strategic defense as evidence of growing isolationism in the U.S., or as additional evidence that the U.S. is trying to deal with the Soviet problem mainly through military means, as opposed to political measures. There is also the fear that if limited American and Soviet defenses are deployed, both superpowers might be more willing to take increased risks in a crisis, which in turn could increase the vulnerability of Europe to a limited nuclear war. We discount this possibility, believing that there is no realistic prospect of defenses becoming that effective, nor even of their being perceived as that effective, but we cannot discount totally the impact of such beliefs in a time when actual deployment is a distant prospect. And there are those who hold that the prospect of a Soviet defense must call into question the effectiveness of British, French, and Chinese nuclear deterrents.

Finally, there is the Soviet reaction. The SDI must have raised questions in Moscow concerning the commitment of the U.S. government to the ABM treaty and current arms-control efforts. It must have reinforced the Soviet view that, despite the recent resumption of U.S.-Soviet arms-control talks, the Reagan administration is more interested in dealing with Soviet-American problems through military and technological competition than by political means. Indeed, because of Reagan's commitment to the SDI and statements by Secretary Weinberger to the effect that the ABM treaty should not stand in the way of strategic defense, the Soviet leadership must feel that the U.S. might trade off arms-control considerations in favor of military opportunities. In addition, there are probably at least some in the Kremlin who see the SDI as militarily threatening, as raising the possibility of first-strike capability.

In summary, we see virtually no chance of developing and deploying the perfect defenses that would be required to meet President Reagan's objective of "rendering nuclear weapons impotent and obsolete." We do not even see the possibility of deploying near-perfect ones. The president's public profession of his objective, as well as his and Secretary Weinberger's optimism about its realization, seem to us a triumph of wishful thinking and fantasy over reality: an act of surrender to the promise held out by technical fixes as the preferred means of dealing with nuclear arms and a difficult adversary—two situations that must ultimately be dealt with by political means.

The Strategic Defense Initiative stands out as the most bizarre episode in the sad history of the nuclear arms race. It is extraordinary that a U.S. president should include this theoretically laudable, but technically baseless, goal of negating nuclear weapons in his address to the nation; that he should continue to solicit support for the SDI as the centerpiece of U.S. strategic policy, both at home and abroad; that the community of ABM advocates has been reassembled and its goals legitimated; that technical experts who do not accept the unrealistic final goals of the SDI program have been able to include as intermediate goals a variety of partial defenses that require very different justifications from those put forth by the president; and that the suspension of good judgment has allowed the launch of an exploratory ABM program on a scale that is irreconcilable with both state-of-the-art technology and with reasonable pursuit of nuclear stability.

ENDNOTES

[1] McGeorge Bundy, "To Cap the Volcano," *Foreign Affairs*, Oct. 1969, p. 10.
[2] For an unclassified summary of the Future Security Strategy Study, directed by Dr. Fred S. Hoffman, see "Ballistic Missile Defenses and U.S. National Security," (Washington, D.C.: Department of Defense, Oct. 1983).
[3] According to Gen. James Abrahamson, "When they [the Russians] see that we have embarked on a long-term effort to achieve an extremely effective defense, supported by a strong national will, they will give up on deployment of more offensive missiles...." *Science*, Aug. 10, 1984, p. 598.

David Holloway

The Strategic Defense Initiative and the Soviet Union

ADVOCATES OF THE STRATEGIC DEFENSE INITIATIVE have pointed to the Manhattan Project and the Apollo program as examples of the marvelous feats that American technology can perform. But the SDI is a different kind of undertaking. The scientists and engineers who built the first atomic bomb and placed the first man on the moon did not have to contend with an opponent who was trying to foil their efforts. Those who wish to build a ballistic missile defense (BMD) system, however, have to take into account the fact that the opponent whose offensive missiles they wish to destroy will do everything he can to render the defense ineffective. None of the important questions that have been raised about the technical and operational feasibility of the SDI, or about its effect on U.S. security, the risk of nuclear war, and the prospects for arms control, can be answered without some consideration of the Soviet leaders' view of President Reagan's initiative and of their likely response to it.

DETERRENCE AND DEFENSE IN SOVIET POLICY

Strategic defense has occupied a key place in Soviet thinking about nuclear war. At first, BMD was viewed as a natural extension of air defense, and a contribution to warfighting capabilities. In the late 1960s, however, Soviet attitudes changed as the implications of BMD for the strategic relationship with the United States were reassessed.

Soviet work on BMD began in the late 1940s or early 1950s as an outgrowth of an intensive program to build air defenses. When Soviet military doctrine and strategy were revised under Khrushchev to take account of the impact of ballistic missiles and nuclear weapons on the conduct of war, BMD was assigned a key role. Marshal V.D. Sokolovskii's *Military Strategy*, which was the most important study of strategy published during that period, declared that "one of the cardinal problems for Soviet military strategy is the reliable defense of the rear from nuclear strikes."

Sokolovskii, who had been chief of the general staff from 1952 to 1960, acknowledged that "in contemporary conditions the means and methods of nuclear attack unquestionably prevail over the means and methods of defense against them."[1] Yet claims made by Soviet leaders in the early 1960s suggested that the Soviet Union had found a way of carrying out the mission of strategic defense. In 1961, the Soviet minister of defense claimed that the problem of destroying enemy missiles in flight had been resolved. And in 1962, Khrushchev asserted that the Soviet Union had developed an ABM missile that could "hit a fly in space."[2] In the early 1960s, the Soviet Union began to deploy a BMD system around Leningrad, but this was soon abandoned. Later in the decade, work began on a system around Moscow, and this is the only system in the world now deployed.

Soviet BMD policy in the early 1960s was rooted in an unwillingness to regard vulnerability to nuclear attack as an acceptable basis for Soviet security. A leading military theorist of the time expressed a widespread Soviet view that the creation of an effective anti-missile system allows a country to rely primarily on its own defense efforts, rather than on mutual deterrence, which is dependent "on the good will of the other side." In 1967, the Soviet premier, Aleksei Kosygin, in language similar to that used by President Reagan sixteen years later, told a press conference in London that "I think that a defensive system which prevents attack is not a cause of the arms race. . . . Perhaps an anti-missile system is more expensive than an offensive system; but its purpose is not to kill people but to save human lives."[3]

Soviet military strategy assigned BMD an important role in the conduct of war. According to Sokolovskii, the National Air Defense Forces (which have responsibility for BMD) would have the primary role in protecting the country from nuclear strikes and in repelling the enemy's nuclear attack. BMD would minimize losses resulting from a nuclear strike, preserve the viability of the country's rear, and extend the combat capability of the armed forces. The main function of BMD in war, in short, was to enable the state and the armed forces to continue to function.

In 1967 and 1968, as the Soviet leaders were preparing for the SALT negotiations, there were signs of a change in their thinking about BMD. There seemed to be disagreement in the high command about BMD's effectiveness. The commander-in-chief of the National Air Defense Forces claimed that defenses could "reliably protect the territory of the country against ballistic missile attack." The commander-in-chief of the Strategic Rocket Forces, however, asserted that ballistic missiles had characteristics that, in practice, guaranteed their invulnerability in flight, "especially when employed in mass." More direct evidence of Soviet doubts came to light when work on the Moscow system slowed down in 1967 and 1968; only four of the eight complexes then under construction were completed.[4]

Alongside the growing doubts about the technical effectiveness of BMD came an increasing confidence in the deterrent power of the Soviet Union's offensive strategic forces, which were now approaching parity with those of the United States. In 1973, Maj. Gen. M. Cherednichenko, one of Marshal Sokolovskii's closest collaborators, wrote in the classified general staff journal *Military Thought* that the Soviet Union had now

> acquired the capability of delivering a devastating nuclear response to an aggressor in any and all circumstances, even under conditions of a sneak nuclear attack, and of inflicting on the aggressor a critical level of damage. An unusual situation developed: an aggressor who would initiate a nuclear war would irrevocably be subjected to a devastating return nuclear strike by the other side. It proved unrealistic for an aggressor to count on victory in such a war, in view of the enormous risk for the aggressor's own continued existence.[5]

Marshal N.V. Ogarkov, the former chief of the general staff, has written that in the early 1960s the United States could count "to some extent" on the possibility of a disarming strike against the Soviet Union.[6] But by the late 1960s, the United States, in the Soviet view, could no longer do so.

The Soviet leaders acknowledged that each side was vulnerable to a devastating retaliatory strike if it attacked first. The opening Soviet statement at SALT in November 1969 declared that

> mountains of weapons were growing, yet security was not improving but diminishing as a result. A situation of mutual deterrence existed. Even in the event that one of the sides was the first to be subjected to attack, it would undoubtedly retain the ability to inflict a retaliatory blow of destructive force. It would be tantamount to suicide for the ones who decided to start war.[7]

Soviet policy at SALT was based on the recognition that the existing nuclear balance could be upset by the deployment of either offensive or defensive systems.

In May 1969, Maj. Gen. V.M. Zemskov, one of the leading Soviet military theorists, wrote in *Military Thought* that the nuclear balance of power could be disrupted if either country sharply increased its offensive forces, or deployed a highly effective BMD. If the United States tilted the balance in its favor, he wrote, the danger of nuclear war would grow.[8] Given this view of the strategic balance, it seems likely that the Soviet leaders' fear of an unconstrained BMD race with the technologically superior United States played some part in convincing them of the desirability of the ABM treaty. At one point during the negotiations, the Soviet Union proposed that BMD systems be considered separately, thus suggesting particular concern about a race in defensive systems.

But this was not the only factor in the Soviet decision to sign the treaty. The Soviet leaders were aware that the deployment of BMD systems would stimulate further development of offensive forces. During the exploratory moves before SALT, the Soviet Union had asked that defensive and offensive systems be considered together. The preamble to the treaty notes that "effective measures to limit anti-ballistic missile systems would be a substantial factor in curbing the race in strategic offensive arms," and in September 1972, Marshal A.A. Grechko, the defense minister, claimed that the treaty would help prevent competition between strategic offensive and defensive weapons.[9]

Some of the arguments advanced against BMD in the Soviet Union were similar to those expressed in the contemporary American debate: BMD would not be effective against a large-scale attack by

The SDI and the Soviet Union 261

offensive missiles; BMD deployment would spur the other side into increasing its offensive forces; a BMD race might tilt the nuclear balance in the American favor, and this would increase the danger of war. Although the Soviet conception of the arms race is rooted in a different military doctrine and a different historical experience, the Soviet Union and the United States came to some of the same conclusions about BMD and its effect on the Soviet-American strategic competition.

The ABM treaty thus codified a situation in which the Soviet Union and the United States were equally vulnerable to a retaliatory strike, no matter who struck first. Soviet leaders have continued to recognize this mutual vulnerability. Marshal Ogarkov wrote in 1983 that as a result of the numbers and technical characteristics of the nuclear weapons on either side, the defending side would always retain so many nuclear weapons that in a retaliatory strike it would be able to inflict "unacceptable damage" on the aggressor.[10] And in words that echo the opening Soviet statement at SALT in November 1969, he added that "in contemporary conditions only suicides can wager on a first nuclear strike."

Although the Soviet leaders recognize mutual vulnerability to devastating retaliatory strikes as an objective condition, they do not regard nuclear war as impossible. At SALT, they showed concern about the danger of war occurring as a result of miscalculation, and about the possibility that a third nuclear power might provoke a world war. The "hot line" and the 1971 agreement to update it, the 1971 agreement on measures to reduce the risk of accidental war, and the 1973 agreement on the prevention of nuclear war were all designed to reduce the risks of war by accident or miscalculation.

The Soviet leaders have also thought it prudent to prepare for nuclear war, in case it should occur. Even within the confines of the relationship of mutual vulnerability, Soviet military strategy still focuses on how to wage nuclear war and defeat the enemy. Soviet leaders say that their military doctrine is defensive in purpose, but Soviet military strategy sees the offensive as the primary form of military operation. In a war, Soviet forces would aim to seize the initiative and to inflict defeat on the enemy by means of decisive military actions. As Marshal Ogarkov put it:

We are speaking about being able not simply to defend oneself by opposing the aggressor with the appropriate passive means and methods of defense, but of inflicting on him destructive retaliatory blows and defeating the enemy utterly in any conditions of the situation as it develops.[11]

The Soviet leaders recognize the existence of mutual vulnerability to devastating retaliatory strikes, and at the same time feel it necessary to devise a military strategy for the conduct of nuclear war in case it should occur. BMD could, in principle, play a role in waging a nuclear war, and this element in Soviet thinking has led many Western observers to regard Soviet BMD activities since 1972 with suspicion, especially since the Soviet Union has continued to modernize its air defenses, and thus has not abandoned the idea of strategic defense as a whole.

SOVIET BMD ACTIVITIES SINCE 1972

In November 1973, the commander-in-chief of the National Air Defense Forces, Marshal P.F. Batitskii, wrote that "within the framework of the agreements limiting ABM defense, such defenses will in all probability change only qualitatively, and will remain limited in capability, able only to defend the capitals of the countries against prospective means of ballistic missile attack."[12] Since 1972, there has been virtually no discussion of BMD in the Soviet military press. But the Soviet Union has maintained a large and steady R&D effort on BMD technologies.

The Soviet Union is upgrading the Moscow BMD system by replacing the Galosh interceptor missiles with SH-04 and SH-08 nuclear-armed interceptors. The SH-04 is an exoatmospheric interceptor, like Galosh, and the SH-08 is a hypersonic endoatmospheric interceptor, like the American Sprint. New phased-array radars (the Pushkino radar) are being built to perform the engagement function. These activities are permitted by the ABM treaty, which limits the number of interceptors to one hundred, and forbids deployment at any other site. The upgraded Moscow system would be ineffective against a determined American strategic strike, but it could provide some defense against theater systems such as Pershing II, and against nuclear attacks by other powers; it would also provide some defense against accidental launches.

A recent CIA report has raised the specter of a Soviet breakout from the ABM treaty through the deployment of a nationwide BMD system or of extensive defenses for its ICBM fields.[13] The Soviet Union now has the SH-04 and SH-08 interceptor missiles and the new radars in production and, with its large missile and radar industries, could deploy them on an extensive scale. There is no conclusive proof that this is the Soviet intention, however, and it would take years rather than months to carry out a strategically significant deployment. The system would not be leakproof; it could be overwhelmed, and the radars would be vulnerable to attack. The United States could respond effectively, with relatively modest effort, by improving the penetration of its offensive systems.

The Soviet Union has also been doing extensive research into the use of directed energy for military purposes, and some of this work may be intended for BMD. Most estimates of the quality of Soviet directed energy research put it on a level with that in the United States. But in other technologies that are crucial for BMD, the evidence suggests that the United States enjoys a considerable advantage. According to the U.S. Department of Defense, the United States leads in: computers, optics, automated control, electro-optical sensors, propulsion, radar, software, telecommunications, and guidance systems.[14] There is little doubt that, with its more advanced technology, the United States could make more rapid progress than the Soviet Union towards the development of a space-based BMD system. The Soviet Union, on the other hand, could move more quickly to the deployment of a conventional BMD system for mid-course and terminal defense.

The modernization of Soviet air defenses to deal with cruise missiles and improved bombers has given them the ability to handle smaller radar cross-sections and shorter reaction times, and has thus made them more capable against ballistic missile reentry vehicles (RVs). The problem is most dramatically illustrated by the SA-12 air defense missile, which is reported to have been tested not only against aerodynamic systems, but also against ballistic missile RVs. It may have been designed for use against tactical and theater ballistic missiles, as well as against aerodynamic systems, and it may also have some capability against SLBM RVs. Developments in surface-to-air missile (SAM) technology have made the problem of surface-to-air-missile upgrade, which caused difficulties in the ABM treaty negotiations, even trickier. The treaty could be modified to cope with the way in which technological advances have blurred the distinction between air defense and BMD. This problem could also be addressed unilaterally by the United States if it deployed modest decoys and used existing technology to reduce RV radar cross sections.

A more specific issue of compliance is raised by the new radar near Krasnoyarsk, which appears to be similar to the early warning radars at Pechora, Komsomol'sk-na-Amure, and Kiev. It is oriented outwards, towards the northeast, and could detect Trident SLBMs launched from the Bering Sea or the Gulf of Alaska. If this radar, which is not yet operational, is indeed an early-warning radar, it violates article VI.b of the treaty, which limits the deployment of ballistic missile early-warning radars to locations along the borders of the two countries and requires that they be oriented outwards. The significance of this radar for BMD is negligible because it is not accompanied by interceptor missiles and engagement radars of the Pushkino type, and would be vulnerable to destruction or blackout in a nuclear attack. But it does raise a serious issue of compliance with the treaty.

The Soviet BMD effort since 1972 appears to have had a number of purposes. It has sought to provide Moscow with some defense against accidental strikes or attacks by third nuclear powers. It has tried to develop defenses in those areas not covered by the treaty—against Pershing IIs, for example—and thus supplements Soviet defenses against bombers and cruise missiles. It has provided a hedge against United States breakout from the ABM treaty. It has also explored new technologies that might one day radically alter the balance between offense and defense. Whatever the reasons behind this effort, however, the arguments against BMD advanced by the Soviet Union in the late 1960s—concern about the effectiveness of BMD, the possibility of an arms race in both offensive and defensive systems from which the United States might emerge with an advantage, and the danger of war—still seem to hold sway. Soviet BMD activities since the ABM treaty do not necessarily imply a shift in attitude.

It is nonetheless prudent for the United States to monitor Soviet activities very carefully, and to question them in the Standing Consultative Commission if they are suspicious or ambiguous. It is also prudent for the United States to conduct research into penetra-

The SDI and the Soviet Union 265

tion aids and BMD technologies in case the Soviet Union should decide to break out of the treaty. But current Soviet BMD activities do not justify the claim that the Soviet Union has already broken out of the treaty, or is intending to do so in the near future.

THE SOVIET ASSESSMENT OF THE SDI

The Soviet reaction to President Reagan's "Star Wars" speech of March 23, 1983 was quick in coming. In a statement issued four days later, General Secretary Andropov said that the defensive measures Reagan spoke of would seem defensive only to "someone not conversant with these matters."[15] The United States, said Andropov, would continue to develop its strategic offensive forces with the aim of acquiring a first-strike capability.

Under these conditions the intention to secure itself the possibility of destroying with the help of ABM defenses the corresponding strategic systems of the other side, that is of rendering it incapable of dealing a retaliatory strike, is a bid to disarm the Soviet Union in the face of the United States nuclear threat.

Andropov claimed that the attempt to build a BMD system would intensify the arms race. At SALT I, he said, the Soviet Union and the United States had agreed that there was an inherent connection between strategic offensive and defensive weapons. Progress in limiting and reducing offensive weapons would be possible only if there were mutual restraint in defensive systems. Now, said Andropov, "the United States intends to sever this interconnection. Should this conception be translated into reality, it would in fact open the floodgates to a runaway race of all types of strategic arms, both offensive and defensive."

Andropov's response to Reagan's speech was hardly surprising. It was inevitable that the American goal of building an effective BMD should be seen by the Soviet leaders as, first of all, a threat to their ability to retaliate in the event of an American first strike. Both President Leonid Brezhnev and Defense Minister Dmitri Ustinov had claimed in 1982 that the Reagan administration was pursuing military superiority as part of its program to destroy socialism as a socio-economic system.[16] The MX ICBM, the Trident D-5 SLBM, the Pershing II, and the land-, sea-, and air-based cruise missile programs were all portrayed as part of a concerted effort to achieve superiority. The Soviet leaders apparently feared that even if these programs did not enable the United States to escape from the threat of retaliation, they might nonetheless give it a preemptive superiority, aimed at reducing the effectiveness of a Soviet retaliatory strike.[17]

Since this was already the official Soviet interpretation of United States policy, President Reagan's "Star Wars" speech was bound to be viewed as a new and dangerous stage in the drive for strategic superiority.

Andropov's response to Reagan's speech also expressed dismay at the implicit intention to abrogate the ABM treaty. Along with other agreements negotiated by the Soviet Union and the United States in the early 1970s, the ABM treaty embodied some measure of common understanding on how best to manage the strategic arms competition without precipitating nuclear war. Andropov chided Reagan for ignoring the link between offensive and defensive strategic weapons, and for failing to understand that BMD deployment would stimulate the competition in offensive systems. Other Soviet commentators have warned that BMD would make war more likely because it would create the illusion of invulnerability, and thus increase the temptation to strike first. It is ironic that these are precisely the arguments that Americans used in the late 1960s to persuade the Soviet Union that BMD was destabilizing.

Soviet commentary on the SDI has taken its cues from Andropov's statement. President Reagan's speech has not provoked any public reconsideration of BMD by the military, who have made the same arguments as other Soviet commentators. The most detailed analysis of the SDI to have been published in the Soviet Union has come from a group of scientists, headed by Academician R.Z. Sagdeev, director of the Academy of Sciences' Institute of Space Research, and by Dr. A.A. Kokoshin of the Academy's Institute of the USA and Canada, working under the direction of Academician E.P. Velikhov, director of the Kurchatov Institute of Atomic Energy and vice-president of the Academy. The title of the working group's report is *The Strategic and International Political Consequences of the Creation of a Space-Based Anti-Missile System Using Directed Energy Weapons*.[18]

This report examines the technological feasibility of a large-scale space-based BMD system, the potential cost of such a system, and the strategic and international political consequences of deploying it. It argues that the creation of a space-based BMD system to destroy one thousand ICBMs in their boost phase is beyond current technological capabilities, and would require significant expansion of R&D work. Even if it could be developed, such a system would be extremely vulnerable to both active and passive countermeasures.

The report's strategic conclusions follow from this technical assessment. If a space-based BMD system is vulnerable to destruction, then it cannot provide effective defense against a first strike, since the attacking side will be able to destroy the system. But such a system might give rise to the illusion that it could provide a relatively effective defense against a strategic force that had already been weakened by an attack. Its deployment would therefore be seen by the other side as a very threatening move. Under these circumstances, each side—both that which had a BMD system and that which did not—would have an incentive to strike first. The net effect of deploying a BMD system would not be to provide escape from mutual deterrence, but rather to make that relationship less stable.

This is the first time that this kind of technical and strategic assessment of new military technologies has been published in the Soviet Union. In the late 1960s, some very eminent Soviet scientists—among them Andrei Sakharov, Lev Artsimovich, and Mikhail Millionshchikov—took part in the discussion of BMD, and their advice may have influenced the change that took place in Soviet thinking at that time (although their assessments of BMD were not published). It is not clear, however, whether Velikhov's committee has any influence on policy-making, or whether its views are the same as those who are engaged in directed energy weapons R&D. Moreover, most of the technical analysis in the report is based on United States sources, and this points to a reluctance to reveal the progress of Soviet research, or to discuss its implications.

The report of the Velikhov committee should perhaps be understood as part of a larger Soviet discussion about the appropriate response to the SDI. In an interview given to the *Los Angeles Times*, Velikhov said that after the "Star Wars" speech he had organized a discussion in the Academy of Sciences. "Its result was very surprising for me," he said. "Not everybody had a real understanding of the issue because rhetorically it is quite attractive to move from offensive weapons to defensive weapons. But the real problem is it's just rhetoric."[19] Velikhov expressed the hope that the Soviet Union would not copy the United States in trying to develop such a system, but added that it would be hard to resist doing so. Even if most Soviet experts counseled against it, he said, those who argued in favor would gain the support of the political leadership if the United States went ahead with the development and deployment of such a system.

Although Soviet spokesmen have been vocal about the SDI, they have been reticent about the purposes of their own BMD activities, and this contrast, though not unusual in Soviet practice, has aroused suspicion in the West. It is true that a great deal of what they have said has been designed to influence Western public opinion. Nevertheless, the Soviet leaders' assessment of the SDI is quite consistent with the view of the strategic relationship that they have expressed since the late 1960s, when they acquired an assured retaliatory capability. Since then they have regarded this capability as a basic condition of Soviet security, and seem to fear that it may now be threatened by United States plans to develop and deploy BMD.

THE SOVIET MILITARY RESPONSE TO THE SDI

The Reagan administration's SDI envisages several years of research and development before a decision is reached, early in the 1990s, whether or not to develop and deploy a space-based BMD system. Most Soviet commentators, even those who show a clear understanding that deployment is many years off, assume that unless an arms-control agreement intervenes, this R&D work on the part of the U.S. will lead to deployment. Soviet scientists have said that they do not believe a perfect defense to be possible. Andropov made it clear that the Soviet leaders regard the SDI as part of an attempt by the United States to gain strategic superiority by augmenting its offensive forces with BMD. Seen in this light, the SDI presents the Soviet Union with the challenge of finding ways to render the defense ineffective.

Three broad options are open to the Soviet Union, either separately or in combination: it can upgrade its retaliatory forces, it can develop weapons that could destroy the space-based BMD system, or it can deploy its own BMD system. Each of these responses needs to be assessed in terms of its technical feasibility, economic cost, and contribution to Soviet military policy.

Offensive Missiles

The Soviet leaders regard their ability to inflict devastating retaliatory strikes on the United States as one of the basic conditions of Soviet security in the present circumstances. Their most obvious response to the SDI, therefore, would be to upgrade their offensive forces in order to ensure that they can penetrate, evade, or overwhelm the defense. They could increase the number of their offensive ballistic missiles, thereby complicating the task of the defense, and diversify their forces by deploying nuclear weapons in space or by making greater use of bombers and cruise missiles, which would not be vulnerable to BMD.

In order to complicate boost-phase interception, which brings the greatest gain to the defense, the Soviet Union could shield its launchers with reflective or ablative materials, thus raising the power requirement for the defense's kill mechanisms. It could also shorten the boost phase by developing fast-burn ICBMs. A payload penalty of about 10 to 30 percent might have to be paid in order to reduce the boost phase from three hundred to sixty seconds, but the gain would be an enormous increase in the difficulty of boost-phase interception.

President Reagan's science adviser, Dr. George A. Keyworth II, has written that "if the fast-burn booster is possible at all, it is probably three to five generations away from those ICBMs the Soviets already have in the works."[20] On the other hand, the Defensive Technologies Study Team, which reported to the Department of Defense in the summer of 1983, concluded that the Soviet Union could develop a fast-burn booster within fifteen years without a crash program. This seems a more realistic estimate than Keyworth's. It is true that the Soviet Union has had problems in mastering solid-fuel propulsion, but its latest ICBMs (the SS-24 and SS-25) and SLBM (the SS-N-20) are all solid-fuel missiles. Moreover, the SH-08 hypersonic endo-atmospheric BMD interceptor missile employs technology similar to that needed for a fast-burn ICBM. This indicates that the Soviet Union could develop fast-burn ICBMs before the SDI reaches the stage of deployment.

Keyworth has also argued that boost-phase defense would cause the Soviet Union "to completely change directions with a fifteen-year investment in 75 percent" of its strategic forces. But even if the Soviet Union were to deploy the new missiles at the same rate as it deployed its ICBMs between 1977 and 1982, it could replace its existing force with fast-burn boosters within ten years. According to CIA estimates, between 1967 and 1977, the Strategic Rocket Forces absorbed no more than 10 percent of Soviet military outlays. (Figures for 1977–1982 are not available, but there is no reason to suppose that the proportion exceeded 10 percent of Soviet outlays in those years either.)[21] It therefore seems that the Soviet Union could reequip its ICBM force with fast-burn boosters without devoting more than 10 percent of its military outlays to this program.

The Soviet Union could also take measures to overload or confuse the mid-course and terminal phases of the defense by deploying more RVs on its existing launchers. The Soviet ICBM force, with its large throw-weight, is particularly well-suited to countermeasures of this kind. The SS-18 ICBM, for example, which is limited to ten RVs under the provisions of SALT II, could carry up to thirty. Decoys could also be deployed, with possibly even hundreds of decoys to one RV. This would present the defense with the formidable problem of identifying the real RVs in mid-course where there is no atmospheric drag to sort them out from the decoys.

Dr. Keyworth has argued that an American BMD system would force the Soviet Union to move away from its reliance on ICBMs and to eschew any thought of a first strike. In a speech in February 1983, he claimed that the Soviet leaders, when confronted with mounting evidence that the ICBM is no longer an effective first-strike weapon, will shift their strategic resources to other weapons systems.[22] But it is not evident that the Soviet leaders will react in this way. Far from considering offensive ballistic missiles obsolete, they are likely to see them as one of the chief means for ensuring penetration and saturation of the defense, and might wish to exploit all their ballistic missile throw-weight for countermeasures.

Dr. Keyworth has also written that "the most immediate argument in favor of developing active defenses" is that "they remove the preemptive option both for the Soviet Union and the United States."[23] It is true that BMD would increase the uncertainties associated with a first strike and might lessen its effectiveness. But it would not necessarily make such a strike less likely. Soviet strategic writings have shown an interest in preemption, in striking first when it is believed that the other side is about to attack—and this is the case to which Keyworth refers in his article. The issue to be weighed in deciding on preemption is not the absolute advantage to be gained from striking first, but rather the balance of advantage between striking first and striking second. If the United States had BMD and could destroy some significant part of the Soviet offensive forces in a first strike, then the Soviet incentive to preempt in a crisis might well be increased, not diminished, by the fear of having to launch a weakened retaliatory strike against the defense.

It seems unlikely, therefore, that the SDI will make the Soviet Union abandon its ICBM force, or that it will necessarily eliminate the Soviet preemptive option. The Soviet Union is likely to want to retain, and perhaps to expand, its offensive forces in order to ensure that they can penetrate or evade the defense.

Anti-Satellite Weapons

The second response the Soviet Union could make is to develop and deploy systems to attack and destroy key elements of a U.S. space-based BMD system. Space mines, anti-satellite weapons, and ground-based lasers, for example, could attack space-based components of the BMD system, while its ground-based elements could be threatened by cruise missiles or other off-shore systems.

The Soviet Union has developed a primitive ASAT weapon, and has been working on the use of lasers for ASAT purposes. According to the latest edition of the Defense Department's *Soviet Military Power*, the Soviet Union could construct "ground-based laser anti-satellite (ASAT) facilities at operational sites" by the end of the 1980s and "may deploy operational systems of space-based lasers for anti-satellite purposes in the 1990s, if their technology developments prove successful." Velikhov has pointed to the need for the Soviet Union to develop the capability to destroy a space-based BMD system. It is possible that R&D in this area has already been stepped up. The Soviet ASAT program did not have high priority in the 1970s, but it could now be given top priority as a possible counter to space-based BMD.[24]

The acquisition of weapons that could destroy the space-based BMD system would increase the Soviet Union's offensive options, and would fit in well with dominant features of Soviet strategic thought. This would require a significant R&D effort on the Soviet Union's part, but such an effort is well in hand. And, as Ashton Carter points out in his contribution to this collection, the technical requirements for effective ASAT are much less challenging than those for effective BMD and would be less costly to develop and deploy.

Ballistic Missile Defense

The third response the Soviet Union could make to the SDI is to deploy a BMD system of its own. It was argued above that the reasons why the Soviet Union limited its BMD deployment in the 1960s still seem to be valid. But this would change if the U.S. moved to deploy strategic defenses, because once the fears that now restrain Soviet BMD activities are realized, the motive for restraint would disappear. The only concern that would remain would be cost-effectiveness.

The Soviet Union might build a BMD system in order to enhance the survivability of its offensive missiles. So far, it has tried to cope with ICBM vulnerability by developing mobile ICBMs, by diversifying its strategic forces, and perhaps by adopting a launch-under-attack policy, but it might deploy BMD if the problem became serious enough. If the United States deploys BMD, the issue of vulnerability will become more important for the Soviet Union, because it will want to ensure that as large a proportion of its offensive forces as possible will be available for retaliation if the United States should strike first. BMD deployment by the United States would thus put pressure on the Soviet Union to deploy its own system.

Apart from such strategic considerations, there are political factors that will encourage the Soviet Union to deploy BMD if the United States does so. The SDI program calls for technology demonstrations before 1990, and these will be seen as symbols of American military and technological power. The Soviet Union may well feel impelled, as it has so often in the past, to try to match the United States program. The heavy stress that the Soviet leaders lay on parity in their strategic relationship with the United States will push them in this direction.

If the Soviet leaders become convinced that the United States is going to proceed to BMD deployment, they might decide to make a "preemptive breakout" from the ABM treaty by building their own conventional BMD system on a nationwide scale. They might do this to show that the Soviet Union would not be overtaken in a BMD race, to complicate United States military plans, or in the hope of gaining some advantage in arms-control negotiations.

Soviet Priorities

The Soviet Union may already have made adjustments to its R&D programs to meet the challenge of the SDI. It is also possible that the Soviet Union, if it is convinced that the United States is going ahead with the deployment of a space-based BMD system, will gradually begin to implement its countermeasures, even before the United States begins deployment. It could start by deploying penetration aids on its existing missiles and then proceed to take further steps to negate the effectiveness of BMD.

Velikhov has recommended that the Soviet Union upgrade its retaliatory offensive forces, and develop weapons that could destroy the space-based BMD system. He, like other Soviet scientists, has expressed considerable skepticism about the feasibility of effective space-based BMD and has argued against its deployment by the Soviet Union. He has even dismissed the "Star Wars" idea as "Lysenkoism," on the grounds that Reagan, like Lysenko, is not willing to let scientific advice stand in the way of his ultimate goal.[25]

In spite of this advice, the Soviet leaders may worry that the United States will manage to develop a highly effective BMD system. They will therefore want to expand their R&D program so that they will be able to field their own system, if they should so decide. Because there is considerable overlap between ASAT and BMD technologies, an extended R&D program would contribute to ASAT as well as to BMD. It is one of the dilemmas of BMD development that it will further the development of effective ASAT, which in turn poses a serious threat to space-based BMD.

SOVIET-AMERICAN RELATIONS AND ARMS CONTROL

The Soviet Union can, in time, develop countermeasures to a possible space-based BMD system. But the Soviet leaders have also set themselves the more immediate goal of trying to stop or slow down the SDI through arms-control negotiations. They are evidently worried about the military and political aims of the United States, and fear that any superiority achieved through the deployment of BMD would not only give the United States a military advantage, but would also enable it to put political pressure on the Soviet Union. As Foreign Minister Andrei Gromyko asked rhetorically in January

1985, "Would not the ability [to launch a strike against the Soviet Union and escape retaliation] be used for pressure, for blackmail?"[26]

The Soviet leaders appear to be concerned also about the prospect of a costly technological race with the United States. Notwithstanding the advice of their scientific advisers, they may fear that the United States will develop a reasonably effective BMD system, and they will want, therefore, to forestall such a race. Moreover, they will be anxious to stop the SDI before it gathers political momentum, and Soviet commentary on the interest shown by the United States military-industrial complex in the program suggests that they are well aware of this danger.

In June 1984, the Soviet Union proposed to the United States that the two governments hold talks on preventing the "militarization of space," but then refused to take part when the United States said that it would raise the issue of offensive forces, too. The call for talks indicated that the Soviet leaders were worried by the prospect of an arms race in space, and this was doubtless one of the reasons why the Soviet Union did agree, in January 1985, to begin arms-control negotiations on intermediate-range nuclear forces, strategic offensive forces, and space.[27] In doing this, it put aside the condition it had laid down at the end of 1983, when it had quit the INF and START negotiations, that the United States would have to withdraw its GLCMs and Pershing IIs from Europe before negotiations could be resumed.

The joint Soviet-American statement issued after the January talks between Secretary of State Shultz and Foreign Minister Gromyko specifies that the three areas of negotiation "will be considered and resolved in [their] interconnection," and the Soviet leaders have insisted on the importance of this formulation.[28] The key interconnection is that between offensive and defensive systems. Gromyko has warned that the Soviet Union is not interested in a "seminar" on the SDI:

The Soviet Union is ready not only to consider the problem of strategic arms, but would even be ready to reduce them sharply—of course, while maintaining the principles of equality and equal security. And on the contrary, if there were no progress in questions of space, then it would be superfluous to speak about the possibility of reducing strategic arms.[29]

The succession of Mikhail Gorbachev as general secretary is unlikely to lead to a change in policy. In his speech to British members of Parliament in December 1984, Gorbachev said that the "non-militarization of space" and nuclear weapons "ought to be considered and resolved in [their] interconnection"—precisely the formulation that later appeared in the Shultz-Gromyko communiqué.[30]

The Soviet Union has insisted that successful reduction of offensive systems will be contingent on progress in limiting defensive systems. The Reagan administration, on the other hand, has reaffirmed its commitment to the SDI, and to the goal of moving towards a defense-dominant strategic relationship. Paul Nitze, a senior arms-control adviser to the administration, has spelled out this goal clearly:

> For the next ten years, we should seek a radical reduction in the number and power of existing and planned offensive and defensive nuclear arms, whether land-based or otherwise. We should even now be looking forward to a period of transition, beginning possibly ten years from now, to effective non-nuclear defensive forces, including defenses against offensive nuclear arms. This period of transition should lead to the eventual elimination of nuclear arms, both offensive and defensive. A nuclear-free world is an ultimate objective to which we, the Soviet Union and all other nations can agree.[31]

It is clear that the two sides have very different conceptions of the relationship between offensive and defensive systems, and this will make it difficult for them to reach agreement on arms control.

The SDI will also complicate negotiation of an ASAT agreement. The Soviet Union has pressed for such an agreement in recent years, within the broader framework of a treaty on the "militarization of space." Because ASAT systems would be one of the main Soviet responses to space-based BMD, the Soviet Union is not likely to conclude an ASAT agreement unless the United States abandons the goal of deploying space-based BMD. The Soviet Union, in its more recent statements, has dropped any reference to a separate ASAT moratorium, and is likely to press for a treaty that bans the testing and deployment of weapons that can destroy objects in space, or objects on earth from space. An ASAT agreement might put obstacles in the way of the SDI testing program, but a space treaty would have the advantage (from the Soviet point of view) of depriving the SDI of political momentum.

The Soviet Union has made it clear that it regards space weapons, and especially space-based BMD, as the most pressing issue for arms control. The United States, on the other hand, regards the reduction of offensive forces—especially the Soviet ICBM force—as a more urgent matter. Although the two sides have different goals, the possibility exists, in theory at any rate, of a trade-off between the Soviet interest in stopping the SDI and the United States interest in reducing offensive forces. A direct trade-off between the two areas is difficult to envisage, because the offensive systems are already operational, while space-based BMD is still at the research stage. Nevertheless, it is conceivable that agreements on offensive forces and defensive systems might be concluded which, although internally balanced, would be linked in the sense of being contingent on each other.

The obstacles in the way of such an agreement are considerable, however, because the two sides have different conceptions of the role of BMD in their strategic relationship. There are, besides, many difficult issues to be resolved in reaching agreement on intermediate-range and strategic nuclear forces. If the negotiations come to an impasse, the disagreements may become the focus of public accusations, of claims and counterclaims, by the two sides. The Soviet Union may make a concerted effort to encourage opposition to the SDI in the United States and Western Europe.

Unless remedial steps are taken—either unilaterally or through arms control—the net effect of the SDI on Soviet-American relations may well be to fuel the competition in offensive and defensive systems, thereby making the strategic relationship less stable; to complicate arms control, and perhaps to make it impossible; and to cast a shadow over political relations. The SDI will not provide escape from mutual deterrence, and may well make that relationship less stable and more fraught with suspicion and uncertainty.

In summary, the main appeal of the Strategic Defense Initiative is that it looks forward to a world in which the two superpowers would have defenses that would render nuclear weapons "impotent and obsolete." But as Richard DeLauer, former Under Secretary for Defense for Research and Engineering, has said, "with unconstrained proliferation [of offensive systems], no defensive system will work."[32]

If defensive systems are to contribute to a safer and more stable strategic relationship between the United States and the Soviet Union, they will have to be embedded in a strict arms-control regime that limits offensive systems. In the current political and technological circumstances, however, the attempt to build defenses may well push the other side into expanding and upgrading its offensive forces. It is thus a paradox of the present superpower rivalry that the effort to build BMD can, and very possibly will, undermine the very condition that is needed to ensure that BMD contributes to a safer world.

ENDNOTES

This article is an expanded and updated version of a discussion of the subject that appeared in *The Reagan Strategic Defense Initiative: A Technical, Political, and Arms Control Assessment,* a special report of the Center for International Security and Arms Control (Stanford: Stanford University, 1984). The author wishes to thank Sidney D. Drell, Philip J. Farley, Ted Postol, and George Smith for their help and advice.

[1] V.D. Sokolovskii, *Voennaia strategiia* (Moscow: Voenizdat, 1962), p. 231.

[2] *Pravda,* Oct. 25, 1961; and Theodore Shabad, "Krushchev Says Missile Can 'Hit a Fly' in Space," *New York Times,* July 17, 1962, p. 1.

[3] Maj. Gen. N. Talenskii, "Anti-Missile Systems and Disarmament," *International Affairs,* 1964, no. 10, p. 18; and *Pravda,* Feb. 11, 1967. For a discussion of Kosygin's statement, see Raymond L. Garthoff, "BMD and East-West Relations," in Ashton B. Carter and David N. Schwartz, eds., *Ballistic Missile Defense* (Washington D.C.: The Brookings Institution, 1984), pp. 295–296.

[4] Interview, Radio Moscow, Feb. 20, 1967. Quoted by Garthoff, loc. cit., pp. 295–296. Marshal N.I. Krylov, "Raketnye voiska strategicheskogo naznacheniia," *Voenno-istoricheckii zhurnal,* 1967, no. 7, p. 20. For an excellent analysis of Soviet BMD programs see Sayre Stevens, "The Soviet BMD Program," in Carter and Schwartz, eds., op. cit., pp. 182–220.

[5] Maj. Gen. M. Cherednichenko, "Military Strategy and Military Technology," *Voyennaya Mysl',* 1973, no. 4, FPD 0043, Nov. 12, 1973, p. 53.

[6] *Krasnaia Zvezda,* Sept. 23, 1983.

[7] As summarized by Gerard Smith, the chief American negotiator, in *Doubletalk: The Story of SALT I* (New York: Doubleday, 1980), p. 83.

[8] Maj. Gen. V.M. Zemskov, "Wars of the Modern Era," *Voyennaya Mysl,* 1969, no. 5, FPD 0117/69, p. 60.

[9] *Pravda,* Sept. 30, 1972.

[10] *Krasnaia Zvezda,* Sept. 23, 1983. See also the article on military strategy in N.V. Ogarkov, ed., *Voennyi Entsiklopedicheskii Slovar'* (Moscow: Voenizdat, 1983), p. 712.

[11] N.V. Ogarkov *Vsegda v gotovnosti k zashchite otechestva,* (Moscow: Voenizdat, 1982), p. 58.

[12] Marshal P.F. Batitskii, "The National Air Defense Troops," *Voyennaya Mysl',* 1973, no. 11 FPD 0049, August 27, 1974, p. 36.

[13] See Clarence A. Robinson, Jr., "Soviets Accelerate Missile Defense Efforts," *Aviation Week & Space Technology*, Jan. 16, 1984, pp. 14–16.

[14] On directed energy research, see Department of Defense, *Soviet Military Power*, 4th ed. (Washington, D.C.: U.S. Government Printing Office, 1985), pp. 43–5. See also the FY 1983 *Department of Defense Program for Research, Development and Acquisition*, statement by the Hon. Richard D. DeLauer, Under Secretary of Defense for Research and Engineering, to the 97th Congress (Washington D.C.: U.S. Department of Defense, 1982), p. II–21.

[15] *Pravda*, March 27, 1983.

[16] *Pravda*, July 12 and Oct. 27, 1982.

[17] *Pravda*, July 12, 1982.

[18] Committee of Soviet Scientists for Peace Against Nuclear Threat, *Strategic and International-Political Consequences of Creating a Spaced-Based Anti-Missile System Using Directed Energy Weapons*, (Moscow: Institute of Space Research, USSR Academy of Sciences, 1984).

[19] Robert Scheer, "A Soviet Scientist on the Real War Games," *Los Angeles Times*, July 24, 1983, Section IV, p. 7.

[20] Dr. George A. Keyworth, "The Case for Strategic Defense: An Option for a World Disarmed," in *Issues in Science and Technology*, Fall 1984, p. 42.

[21] Central Intelligence Agency, *Estimated Soviet Defense Spending in Rubles, 1970–1975*, SR 76-10121U, May 1976; Central Intelligence Agency, *Estimated Soviet Defense Spending: Trends and Prospects*, SR 78-10121, June 1978.

[22] Dr. George Keyworth, *Reassessing Strategic Defense*, (Washington D.C.: Council on Foreign Relations, Feb. 15, 1984), p. 18.

[23] Keyworth, loc. cit., in *Issues in Science and Technology*, p. 38.

[24] Department of Defense, *Soviet Military Power*, 4th ed. (Washington D.C.: U.S. Government Printing Office, 1985), p. 44; interview with Velikhov on Radio Moscow, May 25, 1984, *Foreign Broadcast Information Service*, June 6, 1984, USSR International Affairs, p. AA11; on the Soviet ASAT program see Stephen M. Meyer, "Soviet Military Programmes and the 'New High Ground,'" *Survival*, Sept.–Oct. 1983, pp. 204–215.

[25] Interviews, Radio Moscow, May 23 and 25, 1984, *Foreign Broadcast Information Service*, June 6, 1984, USSR International Affairs, pp. AA 9, AA 11; and the interview in note 19 above.

[26] "Beseda A.A. Gromyko s politicheskimi obozrevateliami," *Pravda*, Jan. 14, 1985, p. 4.

[27] Other reasons were that the withdrawal from arms-control talks had been politically counterproductive, and that the Reagan administration had changed its rhetoric—and perhaps its policy—towards the Soviet Union at the beginning of 1984.

[28] The joint statement was published in *Pravda*, Jan. 9, 1985, p. 4; Gromyko stressed the importance of this formulation on Soviet television on Jan. 13, 1985; see *Pravda*, Jan. 14, 1985, p. 4.

[29] Gromyko, loc. cit.

[30] *Pravda*, Dec. 19, 1984, p. 4.

[31] Hedrick Smith, "Arms Control Talks Scheduled in March, Administration Says," *New York Times*, Jan. 26, 1985, p. 5.

[32] Richard Halloran, "Higher Budget Foreseen for Advanced Missiles," *New York Times*, May 18, 1983, p. 11.

Christoph Bertram

Strategic Defense and the Western Alliance

THE CENTRAL ISSUE IN THE AMERICAN DEBATE over the Strategic Defense Initiative understandably revolves around the extent to which strategic defenses could improve the security of the continental United States. Yet since the end of World War II, that security has been sustained in a collective system that links America to its allies in Europe and the Far East. The linchpin of these security alliances, which have been so successful in assuring postwar stability, has been the ability of U.S. strategic nuclear forces to deter a possible Soviet attack.

This alliance security system has proved remarkably resilient and effective, despite changing strategic circumstances and the evolution from a U.S. nuclear monopoly in the late 1940s to the emergence of superpower strategic parity in the mid-1960s. It has also remained effective in the face of changing political circumstances, from the cold war tensions of the 1950s to the beginnings of detente in the late 1960s. The alliance security system has survived the Korean War, the Berlin crises, the Cuban Missile Crisis, the Vietnam War, and the recent bitter debate in Europe over the Pershing and cruise missiles. It has functioned from the presidency of Harry Truman to that of Ronald Reagan, keeping the peace between East and West and providing an assuring framework for allied security.

By any standard, this is a remarkable achievement. That a disparate group of sovereign states, widely separated by geography, historic traditions, and political culture should maintain an alliance for more than forty years is a situation that has few, if any, precedents. It is equally astonishing that the system has rested, not on the basis of adequate military defense, but only on that of adequate military deterrence. The raison d'être of NATO has not been that Allied territories be protected by superior war-winning capabilities, but rather by a deterrent capability—provided by the U.S.—sufficient to threaten the survival of any major attacker, even at the risk of America's own survival. Especially once the Soviet Union acquired its own long-range nuclear weapons, the "nuclear guarantee" extended by the U.S. to its European allies implied that America would no longer remain immune from massive destruction if deterrence failed.

Until now, the United States has maintained the credibility of that deterrent threat. The Strategic Defense Initiative, however, could fundamentally alter this state of affairs. With its promise to achieve an effective defense against nuclear weapons, and to alter the strategic relationship between the superpowers, the SDI concerns not only the U.S., but the security of its allies as well.

The European perspective on the SDI in particular, and weapons in space generally, is shaped by a number of factors. First, while European countries are becoming more active in space technology through their expanding civilian space programs, they are still bystanders rather than participants in the military uses of space. This will probably change over time; there are currently plans to develop European reconnaissance satellites and, at a much later date, a manned space station. Nonetheless, it will be many years before the Europeans even begin to approach the scale of military space operations that characterizes the U.S. and USSR programs.

Second, most Europeans have little more than instinctive reactions to proposals such as the Strategic Defense Inititative. The relative dependence of the Europeans on the U.S. for both information and concepts concerning new space-related technologies has hindered Europe from responding to the challenges that the SDI poses for strategic doctrine.

Third, Europeans tend to view major technological developments in political, rather than military or even strategic terms. Their experience over the centuries has led them to conclude that security is, above all, a political task. Europeans ask, more persistently than Americans, what the political consequences of a new weapon system are likely to be. They suspect that, whatever new sophisticated weapons are being introduced, the political problems will remain the same. Many Europeans instinctively regard the introduction of major new military technologies either as a threat to stability or as a futile attempt to provide hardware answers to political questions.

My own view is that the space dimension of military competition will matter profoundly to Europe, just as it will to the superpowers. Regardless of whether the promises of the advocates are fulfilled, whether adequate defense against ballistic missiles is feasible, or whether the Soviet Union can keep pace with the United States in the arms race, the effect of a major, purposeful effort to deploy defensive weapons in space will be to generate a political shift of historic proportions. It will introduce into a remarkably stable strategic relationship between East and West an unprecedented degree of uncertainty and nervousness. And it will introduce into the European-American relationship—a relationship that despite repeated strains and occasional dissent has on balance remained harmonious and well-functioning—a profound rift that could break up the Western Alliance for good.

This chapter will address the European perspective by examining the European interests that are relevant to a heightened military competition in space; by detailing specific European reactions to both the ABM debate in the late 1960s and the Strategic Defense Initiative of the early 1980s; and by considering the essentials of the European Alliance relationship with the United States and how they are likely to be affected by a major arms race in space.

BASIC ATTITUDES

In general, the European outlook on deterrence and the likely consequences of deploying weapons in space differs from the U.S. perspective in three important ways: Europeans, despite cyclical doubts, remain convinced that nuclear deterrence is essential to their security and that it can be based only on mutual assured destruction; they have a vested interest, both for international and domestic reasons, in East-West detente and arms control; and they are pursuing, through both the indigenous nuclear forces of France and Britain and the increasing European investment in the scientific and commercial aspects of space activities, interests of their own that are not always congruent with American interests.

Deterrence through Mutual Assured Destruction

Even though France and Britain have minimal nuclear forces, West Europeans fully recognize that their security depends on the nuclear guarantee of the United States and on the ability of either of the two superpowers to obliterate the other should deterrence fail. The potential for mutual assured destruction has seemed to most Europeans, with the exception of those totally opposed to all nuclear weapons, a logical and basically desirable condition. In the European historical experience, vulnerability is a natural state of affairs. With the exception of Great Britain, all European countries have in recent times experienced invasion, occupation, defeat, and victory. Deterrence based on mutual vulnerability corresponds to this experience.

Moreover, Europeans are profoundly convinced that their security rests on America's recognition of its *own* vulnerability. For Europeans, American-European solidarity is not just a matter of declared interests, but of shared fate. Herein lies the reason for the many logical contradictions evident in European arguments that in the past have often annoyed policy-makers and irritated public opinion in the United States. It also helps explain the European reluctance to rely primarily on conventional forces for their security, although this could reduce the likelihood that nuclear weapons might have to be used, as well as European opposition to such American nuclear weapons doctrines as the 1960s strategy calling for the early and massive use of tactical nuclear forces, or to the 1980s doctrine calling for selective nuclear operations, and thus to the militarily more relevant use of nuclear weapons. Finally, there is the European insistence that NATO policy not be predicated on the notion that a nuclear war could be limited to Europe.

These contradictions stem from the simple geographic fact of Western Europe's proximity to Soviet military power. West Europeans are convinced that the United States will remain vitally concerned about Europe only if its own survival is at stake. And they have no doubt that a nuclear war, however limited it may be, could well mean the end of the European states and European civilization.

This European attitude should not be dismissed merely as a parochial, regional perspective. In addition to the aim of tying the power of the United States to the fate of Europe, it reflects a more general view of the nature of strategic stability and security in the nuclear age. Its basic tenet is that war as a rational strategy has been overtaken by the dreadfulness of nuclear destruction. Accordingly, security cannot be based on the hope of either winning a nuclear war or successfully resisting a nuclear attack. Europeans are profoundly skeptical that any technological breakthrough will change this basic state of affairs; they instinctively adhere to the view that new military technologies will sooner or later be matched or countered by the other superpower. As a result, most Europeans tend to dismiss as apolitical those strategic abstractions frequently made by American thinkers concerning "intrawar deterrence," "escalation dominance," and "war-winning" or "prevailing" strategies. Even the important distinction between stabilizing and destabilizing technologies is met with skepticism. Most Europeans regard any major change in the structure of nuclear forces as either destabilizing, irrelevant, or both. They dismiss as inherently implausible the vision that new efforts in space might rid the world of the threat of nuclear devastation.

Detente and Arms Control

Americans may differ among themselves and with their European allies about the significance, prospects, and limitations of East-West cooperation. What distinguishes the European from the American debate, however, is that Europe, because of its geographical position, has no alternative to the persistent search for cooperation between East and West. America does have an alternative, one that has been reflected throughout its history in the tension between the isolationist and internationalist strains in American policy.

While Europeans do differ among themselves in the value they place on East-West relations, the experience of the past two decades has produced a consensus that transcends both regional and political differences. Nothing illuminates this more than the current political constellation in Western Europe. In France, there is a socialist government that, in contrast to its conservative predecessor, started off with deep misgivings about Soviet policies but has recently sought to intensify diplomatic contacts with Moscow. In Great Britain, the conservative government of Margaret Thatcher, outspoken in its Atlanticism and in its distrust of Soviet objectives, nonetheless emphasizes the need for constructive East-West relations. West Germany's liberal conservative government today pursues, virtually unchanged, that same *Ostpolitik* of its liberal Social Democratic predecessor it had once severely criticized.

This extraordinary European consensus stems from both international and domestic considerations. Dealing with the regional superpower, the Soviet Union, is a primary task of any European government. Even if Western Europe could ever muster the military strength and political unity to deter Soviet aggression and pressure without outside assistance, that ongoing task would remain. Reinforcing this search for a cooperative relationship with the Soviet bloc are the historical ties that bind Western and Eastern Europeans, ties that are by no means restricted to the Germans, whose divided nation straddles the line between the North Atlantic Alliance and the Warsaw Pact. What makes the risks involved in a policy of detente tolerable to West Europeans is the deep-rooted conviction that the United States is inherently stronger than the Soviet Union, and that, in the long run, the forces making for political pluralism in Eastern Europe are stronger than those of repression. Unlike many in the United States, Europeans have always been conscious of the limitations to Soviet power, both in Europe and beyond.

There are also important domestic reasons why Western European countries, at least those that profess to depend on the American nuclear guarantee, have sought an alternative to confrontation with the Soviet Union. Spurred on in part by the European peace movements, governments have realized that nuclear reliance on the United States is politically acceptable only if it is constantly accompanied by arms-control efforts. The art of compromise, which is sometimes regarded by Americans as a camouflage for defeatism, is a natural occurrence in European politics.

Arms control, then, has instinctive support in European public opinion as well as in most government circles. The specter of a new arms race in space is therefore met with grave concern. European support for arms control may at times be too eager and too uncritical, as was the case during the 1982–83 Soviet-American negotiations over intermediate-range nuclear forces (INF). Europeans are often naive in their assumption that any compromise is better than none. Nevertheless, there is in the European body politic a deeply rooted feeling in favor of arms control that shapes basic attitudes towards both existing arms-control negotiations between the superpowers and major new military programs like the Strategic Defense Initiative.

Ultimately, initiatives such as the SDI, unless accompanied by *bona fide* arms control efforts, will be viewed with skepticism and concern.

Specific European Interests in Space

The European attitude to military competition in space, including both space-based ballistic missile defense and anti-satellite capabilities, is also influenced by the effect this competition will have on the nuclear forces of Britain and France and on the growing European commercial investments in space.

Originally, the French and British nuclear forces were an expression less of military concern than of political prestige. But in recent years, the strategic rationale for these forces has become more pronounced. As the strategic and political limits of American extended deterrence have become more marked, not least as a result of the bitter INF debate in Europe, the supporters of an indigenous nuclear capability in Great Britain and France have become more determined to maintain these capabilities. Even among the non-nuclear countries of Western Europe, the French and British nuclear forces are no longer seen merely as costly prestige objects of otherwise declining powers.

At present, both Britain and France are trying to keep abreast of the technological advances in the arsenals of the superpowers. France has decided on a major program to multiply the number of missile-carried warheads from the current 98 to 594 by the early 1990s, in particular through the introduction of MIRV technology, and little opposition has been voiced against these plans by any of the French political parties. In Great Britain, the Thatcher government has firmly committed itself to the purchase of Trident D-5 SLBMs from the United States. When deployed in the 1990s, the number of British warheads could increase from the current 64 to as many as 896 (assuming a maximum deployment of 14 warheads per missile). Although the opposition Labour Party is committed to canceling the program if the party should come to power, it is worth remembering that Labour has usually supported the British nuclear forces when in government. If anything curtails the size of the Trident program, it is more likely to be cost considerations than political opposition.

Even with the realization of these modernization programs, however, a major BMD effort by the Soviet Union would weaken French and British nuclear deterrence significantly. While most experts doubt that any BMD system could protect vital targets against a massive attack, such a large-scale attack would stretch the strategic capacity that France or Great Britain could mount, even with the expanded arsenals they are planning for the 1990s. It is true that, given the high number of warheads available by then, the Soviet Union could not hope for total immunity from French or British strikes.[1] Yet if the Soviets were to embark on an ambitious defense program, the deterrent credibility of both countries would inevitably erode. For these reasons, both London and Paris have been less than enthusiastic about eventual deployments.

Another European interest is the freedom to operate satellites in space without interference from other powers. Although the European space program lags significantly behind that of the United States, the European Space Agency (ESA) is establishing a presence in space that is beginning to be commercially viable. Of the total commercial satellite launches forecast for the next three years, fully one-third are or will be European. In addition, the member states of ESA have agreed on the development of a new rocket booster, the Ariane 5, which in the early 1990s could launch a manned space station into orbit.

While the ESA programs are designed exclusively for non-military purposes, they do have two consequences for European reaction to the introduction of weapons into space and to the development of anti-satellite capabilities. First, as an actor in space, Europe has a vested interest in prohibiting military interference with its own satellites. Second, European countries might be forced into a more active, military role in space to protect their assets, should the superpowers engage in a major effort to track and destroy enemy satellites. While there is no European military space program at present, the technical capabilities for such a military role are evolving.

To sum up, America's European allies have reacted quite negatively to any major weapons efforts in space, even those ostensibly designed to provide greater security for the West. Europeans regard with skepticism, if not outright apprehension, any attempt to shift from the present structure of deterrence. They worry about the impact an unbridled military competition in space will have on arms control and East-West relations, as well as on their growing space assets.

EUROPEAN REACTIONS TO AMERICAN PROGRAMS: ABM AND SDI

The general European concerns about ballistic missile defense have been expressed specifically on two occasions: in the late 1960s during the American anti-ballistic missile (ABM) debate and recently in reaction to the Reagan administration's Strategic Defense Initiative.

In the late 1960s, when the first debate on strategic defense took place, two questions dominated European reactions: how a functioning Soviet BMD system would affect the credibility of the then much smaller British and French strategic forces, and whether such a system on both sides would enhance or weaken the American nuclear guarantee to Europe. While some British and French strategists worried about the former, the latter issue was far more important, and here European strategic opinion was almost unanimously skeptical that BMD systems would make much of a difference to the credibility of the U.S. nuclear deterrent. As Alastair Buchan wrote at the time, "The American response in a crisis is envisaged more as an act of political will than of strategic calculation... BMD would have to be shown to reduce American casualties to something near zero in the event of central war to affect the argument about the credibility of the U.S. guarantee for Europe."[2]

Moreover, it is important to remember that ABM, in the late 1960s, was never presented even by American advocates as anything other than an adjunct to the established system of deterrence. The ABM debate did not involve a fundamental reversal of strategy, but merely issues of cost-effectiveness and marginal improvements within the existing doctrine. The U.S. government itself seemed to have deep-seated doubts about the technical feasibility and the strategic relevance of the new technologies. Since the debate did not suggest a change in the fundamental nature of deterrence, European opposition to the American plans could afford to be fairly muted.[3]

Today, the situation is very different. The Reagan administration has repeatedly emphasized that it seeks not merely to improve the current system of deterrence, but to effect a total reversal of strategy: from the predominance of deterrence by threat of second-strike retaliation, to the predominance of defense by threat of destroying attacking ballistic missiles in flight. It is this radical change in attitude—regardless of whether the strategic defenses are ever deployed—that is creating the deepest worry among European governments.

These European concerns can be summarized as follows. First, it will be technically impossible to create a leak-proof defense against attacking missiles. An imperfect defense, on the other hand, will produce strategic instability, especially if both sides retain or increase their sizable offensive forces so as to overcome any projected defensive system. Moreover, the deployment of even an imperfect system could be interpreted by the adversary as an attempt to obtain a first-strike capability.

Second, Europeans suspect that the deployment of defenses will provide little additional security for Western Europe, bordering as it does on the countries of the Warsaw Pact. The result, they fear, will be a strategic decoupling of America's security from that of Western Europe, resulting in a "fortress America," with an unprotected and unprotectable *glacis* in Europe.[4]

Third, the bitter and protracted debate that took place in the early 1980s over the deployment of U.S. intermediate-range nuclear forces (INF) in Europe has left its mark. Those who favor the INF deployment believe that these systems foster an increased solidarity of risk within the Alliance, precisely because these weapons can reach targets in the Soviet Union, thus further exposing the U.S. to Soviet retaliation. Those who oppose INF, however, maintain that these systems manifest an American desire to limit a possible nuclear war to Europe. A major effort to build a BMD shield around the United States would undercut the arguments of those who have stood by the INF deployment decision.

Finally, Europeans are concerned with the impact of SDI on the resources of the Alliance as a whole. The SDI raises the specter of a new strategic arms race that could siphon off much-needed defense funds as well as scientific and industrial resources. Defense resources are likely to remain limited; even the United States will be unable to maintain its current defense spending increases in the face of mounting budget deficits. Given the enthusiasm in the White House for the SDI, Europeans fear that when the squeeze comes, the administration will not only increase pressure on them to carry more of the conventional military burden, but might also reduce its NATO-related defense spending. An American preoccupation with strategic defense could thus weaken the existing defense in Europe.[5]

The Reagan administration has sought to diminish such fears by claiming that defense against missiles targeted on Western Europe is both possible and intended. But so far, European governments, including both political elites and most defense experts, have remained unpersuaded. Europeans, in short, doubt not only that BMD can improve NATO security; they profoundly question the wisdom of a shift from a policy of deterrence to a defense that most probably will remain very imperfect. Apart from a few isolated voices in European strategic circles, the European reaction to date has been almost unanimously hostile. Especially outspoken against the proposed Strategic Defense Initiative have been a number of staunch pro-Atlanticists, including President Mitterand of France; West German Defense Minister Manfred Wörner; British Foreign Secretary Sir Geoffrey Howe; and former French Foreign Minister Claude Cheysson.[6] The French government in June 1984 went so far as to propose formally to the Geneva Disarmament Conference that a five-year ban be placed on the deployment and testing of directed energy anti-ballistic missile systems on the ground, in the atmosphere, or in space. And both France and Great Britain have expressed concern over the effect of a superpower BMD race on their strategic forces.

Yet, with the notable exception of the French government, European leaders have recently muted their original opposition, choosing instead to play down the issue. This should not be misinterpreted as a decline in European uneasiness, but rather as a growing awareness of Reagan's commitment to the project. No European government wishes to confront single-handedly the strategic option chosen by a U.S. president. The possibility of banding together, to speak with one European voice, has foundered on the familiar European inability to formulate a joint position on controversial security issues.

Instead, European governments have so far adopted a "wait and see" attitude. They do not object to the continuation of SDI research in the U.S., and have even expressed an interest in being involved in these efforts. But they have also made it clear that they regard the ABM treaty as an important contribution to arms control that they would like to see maintained. And they have assigned to the United States the burden of proof that strategic defenses will neither decouple Europe from America, nor undermine strategic stability between the superpowers.[7] In the meantime, European governments are hoping that other obstacles will impede the SDI; these could include the U.S. Congress reducing SDI funding, an inability to demonstrate the technical feasibility of various BMD systems, or an arms-control agreement to block an arms race in space.

Thus, President Reagan's vision of a world in which ballistic missiles have lost their threat is not shared by his European allies, even those who, by and large, stand firmly behind the stationing of American Pershing and cruise missiles in Europe. For, underlying the European misgivings is a deep fear that, once firmly underway, the SDI could shake the very foundation on which the Alliance and Western security have rested so far in the nuclear age.

SDI AND THE ESSENTIALS OF THE WESTERN ALLIANCE

These European fears are justified. If the research program of the SDI is followed by an attempt to attain virtual protection against nuclear missile attacks, then the basic philosophy that has maintained the cohesion of the Western Alliance since 1949 would be called into question. This is so not only because of the technical implications of the SDI, but, even more important, because of the political dynamics to which it would give rise.

What are the direct strategic implications for the Alliance? In answering this, two assumptions can and must be made. The first is that, whatever the current doubts over the technical feasibility of a leak-proof, comprehensive defensive system of both point and area defense, the analysis of its strategic and political implications must rest on the premise that such a system will eventually be deployed. It is true that some limited protection of missiles and command centers could prove strategically valuable. Yet this is not the goal of the SDI; the president of the United States has unambiguously rejected this modest and partial objective. As a result, all American efforts under his presidency, and possibly beyond, will appear as steps towards a comprehensive protective strategy. To judge the SDI on its possible intermediate applications and not on its professed goal would run counter to the objective defined by the president himself.

The second assumption is that the Soviet Union will sooner or later acquire a defense capability similar to that of the U.S.. Admittedly, the Soviets are believed to be well behind in many of the technologies—computers, optics, guidance systems—necessary for BMD, and this has led many in the U.S. to believe (and in the Soviet Union to fear) that it may be possible for America to obtain a one-sided advantage in strategic defense. However, there is nothing in the history of arms competition in the nuclear era to suggest that such an advantage will be durable. As Albert Wohlstetter has pointed out, American strategic analysts in the past have often tended to underestimate the vigor and depth of the Soviet armament effort. The USSR has repeatedly demonstrated its ability to catch up with the United States in modern weaponry, even at very high costs to its economy. Any sound analysis of the longer-term prospects must therefore base itself on the assumption that the superpowers will attain roughly equal capabilities, not only to protect themselves against missile attacks, but also to penetrate the enemy's defenses.

From the European perspective, the major issue is whether only U.S. targets would be protected against ballistic missile attacks, or those in Western Europe as well. If the former were true, then clearly the sense of shared risks between Europe and the United States would be called into question. Extended deterrence would cease to be credible, and the only credible use of Soviet nuclear forces would be against non-American allied targets (i.e., in Western Europe and the Far East). Similarly, the only credible use of U.S. nuclear forces would be against targets outside the USSR (i.e., in Eastern Europe). In short, this would conjure up the vision of a nuclear war limited to Europe, a scenario the West Europeans have been striving for thirty years to avoid.

The consequences of this would be politically unbearable for the Western Alliance. America's European allies would lose what they have always regarded as an essential component of their security: the deterrent effect of the link to the U.S. strategic nuclear arsenal. Neither American nuclear forces stationed in Europe nor indigenous European strategic forces could make up for this loss of security. Because it would be possible for the first time in the nuclear age to limit nuclear conflict to the European region, the deterrent effect of American tactical nuclear forces (TNF) in Europe would be reduced to the damage their limited use could inflict. No longer would this be backed by the uncertainty of escalation. A preemptive Soviet strike against American tactical nuclear forces in Europe would lose its risk for the attacker, just as a nuclear strike against other Western European targets would, for the first time, become militarily rational.

An increased effort to build up indigenous nuclear forces in Europe would merely multiply systems that have lost their threat. Europeans would simply be relying on deterrent forces that could be blunted by Soviet action but that could not retaliate against Soviet territory. Even if the Western Europeans should try to circumvent Soviet defenses against ballistic weapons by opting for nonballistic nuclear delivery vehicles instead, the absence of the threat of American nuclear involvement would deprive such a move of any significant deterrent effect.

In other words, Western Europe would find itself for the first time in the history of the Atlantic Alliance exposed to the very strategic situation the Alliance was created to avoid: a Soviet superiority in nuclear and conventional military strength that could not be offset by the threat of U.S. strategic involvement in a European war.

Would European reaction to the Reagan proposals change significantly if the defensive shield against ballistic missile attack were to cover *both* the United States and its European allies (assuming, on the other side, that the Soviet Union would extend its defensive shield over its East European allies)? Probably not, although Western European governments in their first reactions to the Strategic Defense Initiative have implied otherwise. Leaving aside the political problem of the presence of additional large-scale missile deployments in Europe, the major feature of a BMD world would remain: the security cord of potential escalation to the strategic nuclear level between Europe and the United States would be broken. The risk for the Soviet Union of a major attack against Western Europe would be significantly lower than it has been throughout the nuclear age. Moreover, even if a BMD shield were to cover Europe as well, the Allies would, by the mere fact of geography, remain exposed to other nuclear weapon attacks: by artillery, aircraft, or cruise missiles.

Conversely, the other side of the Atlantic would remain relatively protected by sheer geography against shorter-range or slow-flying nonballistic nuclear systems. At the same time, the U.S. would be unable to deter Soviet action by raising the specter that any major conflict might escalate to the nuclear strategic level.

But could not *some* degree of defense against missile attacks—the protection of vulnerable military targets such as missile silos and command centers, for example—enhance the deterrent capability of the Alliance? Could not a certain measure of protection for strategic forces and command centers enhance strategic stability by improving the survivability of second-strike forces, thus gaining time for controlled reaction in case of crisis and war?

If this were a realistic objective, it might be possible to reconcile American and European views. However, the SDI philosophy of the Reagan administration is quite opposed to any limitation of defensive capabilities to specific military targets. Whatever limited protection might become possible is not an end in itself, but a step towards the ultimate objective: to replace deterrence by defense. As long as this is the case, it matters little for America's allies (or, indeed, her opponents) whether the outcome will remain modest in the end. What matters is the strategic conviction behind it. The vision of a leak-proof defensive system that would remove the threat of nuclear missiles is a political objective, an expression of what the administration would like to achieve and an indication of its view of a desirable strategic future. This view is incompatible with the European understanding of, and European interests in, the Western Alliance.

Throughout the postwar period, America's central contribution to the NATO security partnership has been its willingness to regard an attack on Western Europe as an attack on itself. Such an extension of a nation's security commitment cannot be made credible by words, or even by the overseas deployment of sizeable military forces; it can only be made credible by a real solidarity of risks. For the past thirty years, the vulnerability of the United States to nuclear attack, coupled with her military presence in Europe, has manifested that solidarity of risk. Although it could never be guaranteed, and never precisely defined, it was nevertheless clear to friend and foe alike that any major attack against Western Europe contained a real possibility of U.S. strategic retaliation. Deterrence against war in Europe was and still is based on this threat. The absence since 1945 of East-West conflict in Europe, a continent filled with weapons, is attributable not least to the United States becoming, through the link of nuclear vulnerability, a direct member of the European balance.

It is true that this central link has always been difficult to define with precision. In part, it is the product of strategic interest as well as traditional relationships and the sense of common threat. Those strategists who regard deterrence as the result of specific force relations have long been uneasy with it, particularly since the emergence of U.S.-USSR strategic parity in the early 1970s. Since then, the risk for Americans of the U.S. commitment to European security has no doubt increased. Yet the very question whether America would, in the event of a war, be prepared to risk its own cities for the sake of preventing a strike on Hamburg or Paris has been besides the point, because no attacker could be quite certain about the answer. In the late 1960s, British defense secretary Denis Healey formulated a law that still holds today: it does not matter whether the West Europeans are 90 percent certain that the United States will uphold its nuclear guarantee: what matters is whether the Soviet Union is 10 percent uncertain that the United States might. Despite the growth of Soviet nuclear forces since the 1960s, this uncertainty remains, and provides the basis for extended deterrence and the U.S. nuclear guarantee to Europe.

The objectives behind the Strategic Defense Initiative of the Reagan administration threaten to undermine this basic bargain of the Atlantic Alliance. The SDI indicates that the United States wants to escape from the risk of a European conflict. In the past, deterrence has worked because the U.S. was both unable and unwilling to forgo this solidarity of shared risk. The United States may still be unable to do so, but there is no concealing the fact that the United States is now actively pursuing a course that, if implemented, will amount to restricting conflict to Europe. This policy profoundly weakens both deterrence and Alliance cohesion. The U.S. nuclear guarantee that has been so central to the Alliance becomes meaningless unless the United States is both vulnerable to nuclear attack and capable of adding its nuclear power to deter an attack on Europe. If the scenario of roughly equal superpower BMD capabilities were realized, however, the U.S. would be neither.

The president's Star Wars vision and the Strategic Defense Initiative are not just another case of American infatuation with the promises of military technology. They suggest a fundamental shift away from past strategic concepts as well as from the basic philosophy of the Alliance itself. For Europeans, the American strategic vision is one more expression of a shift in the American world outlook, away from coalition politics and towards an assertive, protected United States acting on its own.

It is true that the transatlantic strains of the past several years have generated in Europe as well a search for alternatives to the present Atlantic security arrangement. In particular, Europeans have developed an aversion to an over-reliance on nuclear deterrence. Moreover, there are growing anxieties that modern weapons, with their rapid reaction times and dependence on real-time intelligence, might make the survival of the world a hostage to computer systems. Strategic defense systems, with their reliance on automated command-and-control, could heighten these concerns and, with them, popular resentment to nuclear deterrence.

The SDI therefore risks undermining, on the U.S. side, the basic bargain of the Alliance, and on the European side, the acceptability of nuclear weapons and nuclear deterrence. Furthermore, the two trends are reinforcing each other. The more Europeans display an aversion to nuclear weapons, the more the United States will avoid committing its security to the uncertainties of European politics.

The final question is this: is it worth risking the future of the Western Alliance for the sake of uncertain and doubtful technological promises? Is it worth, on such shaky foundations, instilling in Europeans the fear that they will be left to themselves in the face of Soviet military power, and in Americans the illusion that a European war would not profoundly shake their own security? Since the birth of the Alliance, the Soviet Union has understandably sought to undermine the unity of the West. So far, it has not succeeded; Western cohesion has remained strong.

Ironically, those who believe that the world can be made safe against nuclear attack will themselves end up undermining this cohesion.

ENDNOTES

[1] See Lawrence Freedman, "The Small Nuclear Powers," in Ashton B. Carter and David N. Schwartz, *Ballistic Missile Defense* (Washington, D.C.: The Brookings Institution, 1984).

[2] Alastair Buchan, "Western European Reactions to BMD," *Hudson Institute Discussion Paper*, mimeograph, June 24, 1966, p. 4.

[3] Laurence Martin, "Europe and Ballistic Missile Defense," The Atlantic Institute, mimeograph, Oct. 1968.

[4] These doubts were first formulated by West German Defense Minister Manfred Wörner, with the apparent approval of his European colleagues, after they were

briefed on the SDI by Caspar Weinberger at a meeting of NATO's Nuclear Planning Group, held in Turkey in April 1984. See *Süddeutsche Zeitung*, April 5, 1984.

[5] This also seems to be the reason why U.S. General Bernard Rogers, the Supreme Allied Commander in Europe, has repeatedly displayed skepticism towards the SDI project.

[6] For a report of a televised interview given by President Mitterand, see *Le Monde*, Dec. 18, 1984. On March 15, 1985, Sir Geoffrey Howe gave a major policy address on the SDI at the Royal United Services Institute in London; see the *Times* (London), March 16, 1985. See *La Croix*, July 11, 1984 for an interview with Cheysson.

[7] The speech given by West German Chancellor Helmut Kohl on Feb. 9, 1985, at the international Wehrkunde meeting in West Germany was the first to summarize these points.

Jeffrey Boutwell and F.A. Long

The SDI and U.S. Security

FROM ITS INCEPTION in early 1984, the Strategic Defense Initiative (SDI) has become one of the most controversial and widely discussed new programs introduced by the Department of Defense in many years. The objective President Reagan announced for the U.S. in 1983 and that he, Secretary of Defense Weinberger, and others have reiterated is dramatically simple: to replace our current reliance on deterrence of nuclear attack by threat of retaliation with reliance on a fully effective defense to protect the U.S. and eventually its allies from all attacks by nuclear weapons. If this defense can be obtained, first against ballistic missiles and then against other weapons systems, and if the U.S. can have high confidence in the continuing effectiveness of the defense system, the threat of nuclear destruction of the U.S. will have been eliminated. This is a grand and appealing objective; is it feasible?

Before examining this question and other complex issues raised by the SDI, it may be helpful to recapitulate the principal points made by supporters and opponents of the program.

Supporters of the SDI assert that:

(1) the United States must strive to move away from deterrence based on the threat of nuclear retaliation towards protection by complete, nationwide defense;

(2) even if a complete defense against nuclear weapons is unattainable, partial defenses will be feasible and will reinforce deterrence by retaliation;

(3) the U.S. enjoys both technological and economic advantages over the Soviets and ought to exploit its strengths;

(4) at least while the SDI is in the research stage, it is essential that the United States not be constrained by arms-control agreements that might limit research into the development and utility of strategic defenses. Besides, it is difficult to arrive at joint agreements with the Soviets and they cannot be counted on to honor them;

(5) major new space-based technologies for defense appear to be possible and must be studied, if for no other reason than the Soviet Union might develop them first. Even though some of the technologies that hold promise may not work out, the research program will discover others that will. Finally, the Strategic Defense Initiative is a research program; development decisions will be made only after some years, when the effectiveness of new technologies has been established.

Opponents of the SDI maintain that:

(1) the goal of an impregnable defense is an illusory vision that is unobtainable for several reasons: the task is inherently difficult, requiring major advances in technology; the Soviets will devise countermeasures to U.S. deployments; defenses suffer from unfavorable cost effectiveness, particularly if one strives for a near-perfect system; a space-based defense system is highly vulnerable;

(2) while limited defenses, especially those of hardened military targets, are feasible, their cost-effectiveness is uncertain, there are other ways of protecting these targets (mobility, redundancy), and they do not eliminate dependence on deterrence by second-strike nuclear retaliation;

(3) the Soviets can respond to an American build-up of defenses against their nuclear weapons with a variety of relatively low-cost measures including shortened launch times for missiles; increased numbers of missiles; increased numbers of warheads, decoys, and other penetration aids to accompany the warheads; maneuverable warheads to counter terminal defenses; and deployment of ASAT systems to counter U.S. space-based systems;

(4) the cost of a multi-tiered BMD system will be enormous: $60 billion for research alone in the next eight years, and estimates

ranging from several hundred billion to $1000 billion for final deployment of the full system. The need for defenses against other modes of delivery will aggravate the problem and increase the costs;

(5) a major U.S. program to develop strategic defenses will only succeed if offensive forces are greatly reduced and constrained, yet the SDI, if pursued unilaterally, could foreclose the possibility of negotiating joint reductions in nuclear arms;

(6) the projected U.S. defense program is an attempt to apply military-technical solutions to what is basically a political problem between the U.S. and the Soviet Union. The SDI must be a well-integrated component of an overall political strategy aimed at reducing tensions and confrontation between the two nations, enhancing mutual security, and decreasing the probability of nuclear war.

This simplified list by no means exhausts the issues raised by the SDI. Supporters and critics of the program also argue over the value of strategic defenses for U.S. allies in Europe and the Far East and express concern over the effects escalating SDI costs might have on funding for other military programs, especially in light of increased costs for military weapons in general. However, the most controversial issues of the SDI remain the feasibility of President Reagan's central objective and the impact the SDI and follow-on programs will have on U.S. national security—in particular on the character of the relationship between the U.S. and the Soviet Union.

The overwhelming response of informed U.S. experts, including scientists, diplomats, and retired military officers, is that in today's world of opposing superpowers, the goal of perfect defense is unattainable. The reason is clear. If the U.S. moves toward a perfect defense, the Soviet Union will be forced to respond, since a fully effective U.S. defense would eliminate Soviet deterrence capability. Furthermore, the Soviets have the wherewithal to deploy a variety of effective and comparatively inexpensive countermeasures: they can build more nuclear weapons, deploy more warheads per missile, shorten greatly the boost time for missile launch, deploy decoys for missiles on the ground and for warheads in flight, and attack and destroy vital components of the U.S. defensive system. The development of these countermeasures will certainly impose costs and burdens on the Soviet defense establishment, yet the time scale for researching, developing, and testing any of these countermeasures is well within the period predicted for the development and deployment of U.S. defenses.

Paul Nitze, a senior advisor to the administration, recently stated that a U.S. defense system must be survivable and cost-effective, and "if the new technologies cannot meet these standards, we are not about to deploy them."[1] But obvious Soviet countermeasures can insure that a fully effective U.S. defense will be neither cost-effective nor fully survivable.

Cannot new technologies overcome these obstacles? Since the U.S. does not yet have even the initial technologies necessary for its projected defense, the question would seem to be unanswerable. However, to attain the goal of perfect defense—one that will free Americans from the threat of nuclear destruction—in the face of Soviet opposition seems indefinitely remote, regardless of new technologies we may succeed in developing. After all, both nations can and will play the new technology game, each for its own objectives.

Consideration of defense against a delivery mode whose technologies are already quite well understood will illustrate the point. The U.S. will eventually require fully effective defenses against the Soviet delivery of nuclear weapons by intercontinental bombers. Can we expect to develop an impenetrable defense? For thirty years, the Soviets have been developing costly, elaborate, and steadily improving air defenses against U.S. bombers; nevertheless, the U.S. Air Force remains confident that its bombers, equipped with penetration weapons such as nuclear-armed cruise missiles, can still attack and destroy targets inside the Soviet Union. To underscore its conviction, the Air Force is developing two new intercontinental bombers, the B-1 and the Stealth. In contrast, can the U.S. realistically remain confident in its ability to develop and deploy air defenses capable of preventing any penetration by next-generation Soviet bombers and their missiles? It does not seem conceivable.

Arguments against the feasibility of an impregnable defense do not apply to the feasibility of partial defenses that the U.S. might deploy in order to protect valuable military targets such as command-and-control centers and missile silos. The problem of maintaining cost-effectiveness against Soviet countermeasures would persist, but would not involve the exceedingly unfavorable cost ratios that characterize the final stages of developing an impregnable overall defense. In addition, these partial defenses need not always strive for full protection of the targets they defend. For example, if partial defenses could limit destruction from a major Soviet attack to three-fourths of the Minuteman and MX missiles in silos, the remaining 250 missiles, in conjunction with surviving U.S. bombers and SLBMs, would still constitute a formidable retaliatory force.

Limited and focused defenses of this sort would not eliminate the threat of destruction of cities and industry; the U.S. would have to continue to rely on deterrence by the threat of second-strike nuclear retaliation. The operative question for the United States would then be whether the benefits of deploying such limited defenses outweigh their military and political costs—especially as their deployment would necessitate the abrogation or re-negotiation of the ABM treaty and could lead to an increase in Soviet offensive nuclear weapons.

THE POLITICAL APPEAL OF STRATEGIC DEFENSES

Although many administration spokesmen and supporters of the SDI have argued in favor of limited objectives, the president has continued to stress the goal of achieving a total defense of the United States and its allies. There are obvious political reasons for doing so. Apart from the intrinsic appeal of building perfect defenses against nuclear weapons, there is the danger that presenting the SDI as only a means for bolstering retaliatory deterrence might not generate the domestic support needed to win Congressional approval of a greatly expanded SDI program. George Keyworth, the president's science advisor, has directly confronted those in the administration and elsewhere who advocate limited defenses, saying that "if these arguments continue to be used as the basis to achieve congressional and allied support ... the opportunity for strategic change—and the president's objective—is lost."[2]

The political appeal of the president's goal of "saving lives rather than avenging them" makes it easier to understand why Reagan took the unusual step in March 1983 of announcing the goal of a perfect defense after only negligible preliminary study and consultation.[3] There is little doubt that President Reagan genuinely believes that a nationwide defense against nuclear weapons is possible. Yet the president and his advisors might also have reasoned that his call for a revolution in nuclear strategy would deflect much of the American and European criticism at that time of his administration's nuclear force buildup and arms-control policies.[4] It was certainly more than coincidence that Reagan's call for perfect defenses came at the end of a speech aimed at generating public support for increased defense spending, and during a period when there was considerable Congressional opposition to the MX missile program.

The administration has also sought to take the political high road concerning the SDI and the March 1985 resumption of superpower arms-control negotiations. In comparison to 1983, when there was widespread criticism of new U.S. missile deployments in Europe, the administration's support of the SDI is far more politically tenable. By emphasizing the exploratory nature, non-nuclear character, and defensive goals of the SDI, the administration can more easily shift the burden of proof onto the Soviet Union for lack of progress in the Geneva arms-control talks. The U.S. refusal to agree to limitations on the SDI program is plausible in the sense that restraints on technology research programs are difficult to negotiate, much less monitor. Nonetheless, to claim that restraints should not be placed on an effort that holds out the promise of ending reliance on offensive retaliation could be counterproductive if the SDI effort proves to be a major stumbling block in reaching agreement on reducing offensive forces.

Finally, SDI supporters can also reasonably claim that the program provided essential leverage in getting the Soviets back to the arms-control table in March 1985. It is also true, however, that the Soviet walkout from the 1983 arms-control talks was a demonstrable failure. Having failed to stop the deployment of U.S. Pershing and cruise missiles by their propaganda campaign and refusal to negotiate, the Soviets were gaining nothing by staying away from Geneva. Only by resuming negotiations could the Soviets hope to constrain the continuing U.S. offensive nuclear build-up (MX, Trident D-5 SLBM, the B-1 bomber and hundreds of sea-launched cruise missiles), which by the late 1980s will increasingly threaten Soviet strategic forces, especially their ICBMs.

Yet, whatever short-term benefits the SDI may have provided in the way of domestic politics and arms-control negotiations, a greatly accelerated strategic defense program also poses numerous short-term risks. It will be many years before conclusive technical and military evaluations can be made of the president's goal of total defenses. Within a few years, however, the SDI could have severe political ramifications that would guarantee that the U.S. will never reach the goal of a comprehensive defense—regardless of the ultimate feasibility of the technologies now being explored.

THE POLITICAL RISKS OF THE SDI

It is generally recognized by both supporters and critics of the SDI that in order to provide effective nationwide protection, strategic defenses must be accompanied by substantial reductions in offensive forces. Administration spokesmen claim that the SDI program, and Soviet apprehensions over U.S. technological know-how, could prove beneficial in negotiating successful arms reductions. There could be some truth in this. While the Soviets have the capability for building very effective countermeasures to U.S. BMD systems, Soviet leaders probably do not relish the prospect of a heightened competition in both offensive and defensive weapons systems. The Soviet Union has devoted many years to reaching strategic parity in offensive systems with the U.S., and there is little chance that a new competition in strategic defenses would improve the overall Soviet military position.[5] There is also, however, little chance that such a competition would benefit the U.S., given that the Soviets will devote the resources necessary for countering an American BMD system.

Any attempt to move from a world of offensive deterrence to one of protection by defense must therefore be a mutual one. Again quoting Nitze, the transition period to a defense-dominant world must be "a cooperative endeavor with the Soviets," in which "arms control would play a critical role."[6] Within the administration, there are currently fundamental differences of opinion on the utility of arms control that must be resolved. There are also serious contradictions between the administration's rapidly expanding SDI program, its offensive nuclear weapons buildup, and its arms-control efforts.

To take one example: the U.S. has begun deployment of hundreds of sea-launched cruise missiles (SLCMs), and the Soviets are following suit. If unconstrained by arms control, these Soviet SLCMs will provide an excellent counter to a U.S. BMD system, no matter how effective such a system is against Soviet ballistic missiles.[7] Indeed, the Soviets have a distinct geographical advantage in deploying SLCMs, because of our country's long coastlines and the fact that the majority of our population lives along these coasts. A BMD system and other types of strategic defenses cannot hope to begin reducing the nuclear threat unless constraints are placed on the production and deployment of SLCMs and other types of cruise missiles, despite the verification problems such constraints will pose for arms control.

An even more immediate problem concerns the continued viability of the 1972 ABM treaty and the consequences that a breakdown of the treaty would pose for arms control in general. Administration officials continue to maintain that SDI research efforts are fully consistent with the ABM treaty,[8] but in the period since the SDI was established in 1984, projected tests of BMD components that could violate the treaty have been increasingly moved up. Space-testing of BMD pointing and tracking equipment is now set for 1987, while other tests are planned in the near future for SDI lasers, kinetic energy projectiles, and surveillance equipment. The Defense Department has maintained that these tests will not violate ABM treaty provisions, but this rationale is based on the fact that the tests will be conducted against anti-satellite interceptors, not ballistic missiles, and that the power requirements used will not be sufficient for an actual BMD system.[9] Whether or not such arguments represent "creative lawyering," as some critics maintain, the fact remains that encroaching on the ABM treaty is hardly likely to be conducive to negotiating reductions in offensive nuclear forces.[10]

Until such time as the U.S. and Soviet Union have jointly agreed on how to achieve deterrence based on defenses, it makes little sense to call into question the viability of existing arms-control agreements. Senior administration officials, however, have openly questioned both the existing SALT II limits and the ABM treaty.[11] In contrast, President Reagan's Commission on Strategic Forces (the Scowcroft Commission) has stressed the importance of the ABM treaty, noting in its 1984 report that "the strategic implications of ballistic missile defense and the criticality of the ABM treaty to further arms control agreements dictate extreme caution in proceeding to engineering development in this sensitive area...."[12]

There are a number of other political risks in expanding the SDI program as quickly as the administration proposes. The claim is made that SDI research is needed as a hedge against possible Soviet technological breakthroughs, and to provide future presidents with greater options in the field of strategic defenses. The first argument is true only up to a point. Research in weapons technologies, sensing and optical systems, and computer processing is needed both to determine their feasibility for a variety of military applications—not just BMD—and to keep abreast of possible Soviet discoveries. A rapidly expanding SDI program, however, the cost for which is projected to total over $60 billion through 1993, is neither a rational nor a cost-effective method of conducting this research. It is even questionable whether the program can effectively utilize the large increases in funding requested by the administration. Former Defense Secretary James Schlesinger has testified before Congress that annual increases of 35 percent for military R&D programs stretch the limits of what can usefully be spent,[13] yet funds for the SDI are projected to increase by 160 percent between fiscal years 1985 and 1986.

The rapid acceleration of the SDI program, and its focus on systems that will not be deployable for another ten to fifteen years at least, also call into question the president's stated rationale of providing options for future presidents. If increases in SDI funding and organizational momentum continue, the accelerating pace of the program could well foreclose options for the next administration, both politically and militarily. For example, many administration spokesmen see the SDI as necessary insurance against a possible Soviet breakout of the ABM treaty. Yet one of the differences between the SDI program and U.S. BMD efforts prior to 1984 is a reduced emphasis on currently available technologies for terminal defense being developed by the U.S. Army. It is these systems, and not space-based BMD systems that are years away, that would prove beneficial should the Soviet Union seek to gain a strategic advantage by rapidly deploying a nationwide defense system.

Supporters of the SDI maintain that, given the greater economic strength of the U.S., we will be in a better position to compete with the USSR on what will assuredly be very expensive weapons systems. The notion of exerting economic leverage on the Soviets in the military sphere is appealing, but the argument cuts both ways. While the American economy is much stronger than that of the Soviet Union, the Soviets do not face comparable domestic constraints in allocating resources. The combination of mounting federal deficits, increased defense budgets, and reductions in social services could well lead to domestic and Congressional resistance to spending large sums on strategic defenses of unproven effectiveness. While the Soviet leadership may be faced in the years ahead with economic constraints regarding its military programs, there is little doubt that the USSR will invest whatever is necessary to ensure the deterrent capabilities of its nuclear forces.

In addition to complicating U.S.-Soviet relations and arms-control efforts, the administration's handling of the SDI has also created uncertainty and apprehensions among America's allies in Europe and the Pacific. In his 1983 speech, President Reagan emphasized the importance of U.S. commitments to our allies, noting that their safety and ours are one. Despite such assurances, it was inevitable that a significant new policy direction by the U.S. would raise questions among our allies, particularly as this policy involves major changes in the military-strategic area. It is not surprising that the long-term goal of deploying defenses has aroused both apprehensions and support. Britain and France are concerned about the utility of their own nuclear arsenals, should the Soviet Union respond to the U.S. initiative by developing and deploying defenses of its own. Some Europeans have expressed concern about a possible U.S. return to a "fortress America" concept, should defenses prove more valuable for the U.S. than for Europe. Still others have welcomed an initiative whose stated objective is to free the world from the threat of nuclear war.

The SDI program itself is probably not yet fully understood abroad, but there appears to be general approval among our allies (as among Americans) that research on new technologies for defense should be carried out, especially since the Soviets are similarly conducting BMD research. Few of our allies, though, have agreed to take the next step of actively participating in SDI research, even though they could benefit economically and technologically from such participation.[14]

The major apprehension that is voiced by many of our allies is that the new U.S. initiatives will not lead to arms reductions and a mutual U.S.-Soviet exploration of strategic defenses, but rather to still further deterioration in East-West relations. In this view, serious U.S.-Soviet negotiations for major reductions in nuclear armaments will become difficult—perhaps impossible—if the U.S. is unilaterally working to develop a strong defense against Soviet nuclear weapons with an explicit goal of achieving a total defense against them. Skeptics of the SDI envision a difficult transition period leading to a more polarized and less stable world, and to diminished security for the U.S., the Soviets, and the allies. Both supporters and skeptics maintain that

The SDI and U.S. Security   307

real progress toward enhanced stability and security will not be attained through unilateral efforts, but must involve joint negotiations and agreements, whether they be for arms reduction, mutual and parallel defense buildup, or both.

STRATEGIC DEFENSES AND MUTUAL SECURITY

That the problem of avoiding nuclear war is one that the U.S. and Soviet Union share is recognized by the Reagan administration and its advisors. The Secretary of Defense, in an article on the SDI, refers to "our new approach, stressing defensive measures for mutual survival...."[15] Yet during the first round of the Geneva talks between the U.S. and Soviet Union in the spring of 1985, the U.S. firmly resisted any notion of limiting the SDI program. It remains to be seen whether the administration can, or even intends to, use the SDI as bargaining leverage to gain substantial results in Geneva. At some point, there will have to be agreement with the Soviets on how strategic defenses might enhance the security of both countries. The prospect of an increased level of arms competition involving both offensive and defensive weapons cannot be acceptable to either nation.

Who would have predicted in the early 1950s, when both nations had nuclear weapons and relations between them were chilly indeed, that in only a few years the two countries would be seriously involved in arms-control negotiations, and that these negotiations would lead to some twenty formal agreements and treaties, almost all designed to decrease the risk of war? The basic reason for this surprising level of joint effort was cogently described recently by Dean Rusk, secretary of state to Presidents Kennedy and Johnson, when he remarked that "there is a special feature about our relations with the Soviet Union. We, and they, are the only two nations in the world who, if locked in mortal combat, can raise serious questions about whether this planet can any longer sustain the human race. We, and they, share a massive, common responsibility. That is the prevention of all-out nuclear war. They know that; we know that."[16]

With this common recognition of their joint responsibilities and concerns—concerns validated by the many years of arms-control negotiations—it is impossible to believe that the United States and Soviet Union will not eventually include the possible role and possible control of defensive weapons in their discussions of arms limitations and reductions. How much more preferable to confront this issue promptly, before strong momentum toward deployment of defensive weapons diminishes the scope of any possible agreement, or even makes agreement impossible.

While it may be too soon to negotiate direct controls on the new BMD technologies, there are important related issues that have been raised in previous U.S.-Soviet discussions and negotiations. Prompt negotiation in some of these areas could renew the sense of joint responsibility and pave the way to broader and more substantive arms reductions and limitations. What is desperately needed at this juncture is a more profound recognition by both nations that if they are to escape the terrible dilemma that nuclear weapons have engendered, it must be by joint action. Recognition of this fact could lead to both mutual agreements and the realization that there are significant measures for promoting stability that each nation can take by itself. The following suggestions point to several mutually useful and significant arms-control measures for which there is much precedent and where agreements on specific measures might be rapidly obtainable.

One way the U.S. by itself can improve opportunities for mutual agreements is to exercise restraint on SDI funding. The current FY 1985 budget for SDI is $1.4 billion. For FY 1986, Congress has proposed cutting the administration's request of $3.7 billion to $2.75 billion. Yet even this amount is too rapid an acceleration of the program, especially as no limits exist on the evolution of SDI projects from basic research to engineering development. As an example of how the SDI program could be restructured, Senators Dale Bumpers, John Chafee, Charles Mathias, and William Proxmire introduced an alternative SDI budget of $1.86 billion for FY 1986. This budget would increase basic BMD research by 30 percent, but would limit funding for those SDI demonstration projects that could impinge on the ABM treaty.[17] This type of BMD program would allow the U.S. to explore the military feasibility of new technologies and keep abreast of Soviet BMD work, while avoiding the political complications of a greatly accelerated SDI effort.

In considering joint negotiations, there are at least two relevant areas where negotiations between the superpowers could be entered into promptly. One is updating and improving the 1972 ABM treaty.

The other is negotiating a treaty that bans anti-satellite (ASAT) weapons, or at least prohibits further testing of them; this would effectively limit development and deployment of useful ASAT systems.

For more than a decade, the ABM treaty has been an effective agreement in promoting stability between the U.S. and Soviet Union. At present, there are serious allegations that the Soviets are violating the agreement with the construction of the Krasnoyarsk phased-array radar in Siberia. In addition, the Soviets have accused the U.S. of certain treaty violations, and there are a number of gray area issues that need to be resolved if the treaty is to remain viable.

In the end, however, none of these issues is so militarily significant that the security of the U.S. would be improved by withdrawing from the treaty. Instead, both countries should work to clear up those areas of ambiguity that have arisen over the years. ABM treaty provisions and SCC consultation procedures could be strengthened with respect to: the building of new phased-array radars, anti-tactical ballistic missile (ATBM) systems, the dividing line between BMD research and development, stricter definitions of what constitutes a BMD component, and the testing of BMD components under an ASAT rubric.[18] Until the prospects of a survivable and cost-effective defense against nuclear weapons become much clearer, it should be in the interests of both nations to maintain as effective an ABM treaty as possible.

It should also be in the interests of both nations to negotiate a treaty that would inhibit the development of ASAT systems by either party. This possibility was the subject of U.S.-Soviet arms-control talks in 1978–79, and drafts of possible treaties have been presented by the Soviet Union. While the ASAT issue has since been complicated by the attention recently given new BMD technologies, the fact is that an unrestrained ASAT competition between the U.S and Soviet Union would preclude any hope of strategic defenses increasing our security.

If both countries are going to commit themselves to eliminating the threat of nuclear weapons, they will need to engage in a long-term cooperative effort involving substantial arms reductions as well as joint consideration of defensive systems. Even with such unprecedented cooperation, it might well take two decades or more before the threat of nuclear destruction is minimal or absent. The buildup to the present immense stockpiles of nuclear weapons has taken over three decades, and the process of reversing this trend will not be achieved quickly. But the essential first step is to begin the reversal.

If the U.S. persists unilaterally in working towards a reliance on defensive forces against nuclear weapons, it will face the prospect of determined Soviet opposition. The transition period of moving from offensive deterrence to a defense-dominant posture will be long and will be filled with political and technical uncertainties. The objective of even very good defenses will surely remain unattainable. Even if some of the projected new technologies become feasible and prove to be militarily attractive, their deployment will contribute little to strategic stability if they are introduced in an environment where arms-control limitations are absent, both countries are expanding their offensive nuclear forces, and political relations are dominated by mutual suspicion that the other country is seeking to gain a unilateral strategic advantage.

ENDNOTES

[1] Paul Nitze, "On the Road to a More Stable Peace," speech delivered to the Philadelphia World Affairs Council, Feb. 20, 1985, p. 4.

[2] *Aviation Week & Space Technology,* April 29, 1985, p. 225.

[3] For an account of events leading up to the March 23 speech, see "Reagan's 'Star Wars' Bid: Many Ideas Converging," *New York Times,* March 4, 1985, p. 1.

[4] Deputy National Security Advisor Robert McFarlane was reportedly especially influential in convincing President Reagan that a strategic defense program might attract bipartisan support in the U.S. and be a useful lever in arms-control negotiations. See "Reagan Seized Idea Shelved in '80 Race," *Washington Post,* March 3, 1985, p. 1.

[5] See Raymond Garthoff, "BMD and East-West Relations," in Ashton B. Carter and David N. Schwartz, eds., *Ballistic Missile Defense* (Washington, D.C.: The Brookings Institution, 1984).

[6] Nitze, op. cit., p. 5.

[7] See "Banquet Address: The Honorable James R. Schlesinger," given at the *National Security Issues Symposium 1984: Space National Security, and C^3I,* Oct. 25, 1984 (Bedford, Mass.: The MITRE Corporation), p. 57.

[8] In January 1985, Secretary of State George Schultz maintained that "no decisions to go beyond (SDI) research have been made nor could they be made for several years." See "Excerpts from Schultz's News Conference About Arms Talks," *New York Times,* Jan. 9, 1985, p. A10.

[9] See the appendix entitled "Compliance of the Strategic Defense Initiative with the ABM Treaty" that was attached to the Defense Department "Report to the Congress on the Strategic Defense Initiative" (Washington, D.C.: Department of Defense, 1985).

[10] Paul Warnke, chief U.S. negotiator during the SALT II talks, called the Defense Department's justification of the SDI tests "a total fraud," adding that "if this were a Soviet report, we would decide there was no reason to continue negotiating with them...." See "U.S. Changing 'Star Wars' Test Plan," *Boston Globe,* April 21, 1985, p. 18.

[11] In May 1985, Richard Perle, assistant secretary of defense, said that the U.S. should break out of the SALT II limits on offensive forces in September 1985, when a new Trident submarine goes to sea. See "Arms aide: U.S. should let SALT II lapse," *Boston Globe,* May 8, 1985, p. 8. Secretary of Defense Weinberger has openly questioned the utility of the ABM treaty—for example, on the NBC television program, "The Real 'Star Wars': Defense in Space," aired Sept. 8, 1984; text reprinted in *Current News: Special Edition,* Oct. 25, 1984 (Washington, D.C.: Department of Defense), pp. 26–30.

[12] *Final Report, President's Commission on Strategic Forces,* March 21, 1984 (Washington, D.C.: The White House), pp. 7–8.

[13] See "SDI Director Defends Rise in Research Funds," *Washington Post,* March 20, 1985, p. A12.

[14] At the 1985 Economic Summit in Bonn, European resistance to the SDI was led by President Mitterand of France, who proposed instead that the Europeans should cooperate in a civilian space science program called Eureka. See "Mitterand's Goals: 'Preserving Skills,' 'Protecting Europe' " *New York Times,* May 6, 1985, p. A13.

[15] Caspar W. Weinberger, "The Strategic Defense Initiative," *Harvard International Review,* Jan.-Feb. 1985, p. 7.

[16] "Perspectives on Power: An Interview with Dean Rusk," *East-West Outlook,* March 1985 (Washington, D.C.: American Committee on East-West Accord), p. 4.

[17] "Senators Propose Alternative SDI Budget," *Military Space,* April 15, 1985, p. 1.

[18] See Thomas K. Longstreth, John E. Pike, and John B. Rhinelander, *The Impact of U.S. and Soviet Ballistic Missile Defense Programs on the ABM Treaty,* a report for the National Campaign to Save the ABM Treaty (Washington, D.C.: 1985).

Hans A. Bethe and Richard L. Garwin

Appendix A: New BMD Technologies

A VARIETY OF TECHNICAL MEANS are available or under consideration for defense against nuclear-armed ballistic missiles in flight. Some of the technologies for attacking the missiles as they approach the U.S. and head for their targets have been under study since the 1960s and are fairly mature. Other technologies, especially those for attacking ICBMs and SLBMs in the early stages of flight, are only now undergoing research and planning, and their eventual utility is quite uncertain. A layman's guide to these technologies can be found on pages 53 through 71.

There are important advantages to destroying ICBMs and SLBMs in their initial launch phase before they have released their "payload" of reentry vehicles (RVs) that carry the nuclear weapons. It is almost mandatory that attack in this initial phase be from space-orbiting satellites that overfly the launch sites. This requirement is the main driving force for the new types of weapons that would operate from space. Our emphasis on the technology of such weapons should not obscure the crucial problem of ensuring the survival of the defense itself.

The offensive nuclear weapons systems of the two superpowers are formidable (roughly ten thousand deliverable nuclear warheads on each side), and it can be assumed that either nation will move to maintain an effective offensive capability if the other deploys defense systems. The offense can take countermeasures designed to weaken or defeat the various components of any defensive system. In many cases, these may be less expensive and more readily implemented than the defensive system itself. Given the diversity of offensive and defensive measures and responses, it becomes important to have at least an approximate notion of the comparative costs of defense systems and the possible countermeasures the offense may undertake. It is apparent that the cost of frustrating perfect or near-perfect defense is less than that of denying the goals of a less ambitious defense of similar technology. Fast-burn ICBMs are particularly useful in augmenting the offensive force as their boosters burn out at an altitude of 80 kilometers, in 50 seconds, totally negating several types of weapons for boost-phase intercept and increasing the performance required by those systems that still have some effectiveness.

This appendix confines itself to analyses of defense against ballistic missiles and offensive countermeasures, but there are other ways of delivering nuclear weapons—long-range cruise missiles and bombers being the most obvious. A complete defensive system will, of course, also need to deploy defenses against these. The large number of Soviet ICBMs, their capability for simultaneous launch, and the possible clustering together of future ICBM deployments provide a generally more difficult problem than that posed by long-range SLBMs. For that reason, most of our discussion deals with ICBMs, although short-range SLBMs—particularly those launched on "depressed trajectory"—pose special problems.

The major portion of this appendix focuses on an analysis of a comparatively new class of weapons, directed energy weapons (DEWs), for which the kill mechanism is a beam of light or of particles that travel at nearly the speed of light. This emphasis is perhaps misleading, since these weapons are for the most part only in the very early research stage of evolution, and one cannot speak with confidence even of their feasibility, let alone their characteristics. They are emphasized because they are a new mode of defense with exceedingly short response times, and the use of such defensive weapons seems essential if substantial numbers of offensive ballistic missiles are to be destroyed in their initial launch phases.

DIRECTED ENERGY WEAPONS

Types of Weapons Considered

More than a decade ago, it was proposed that a defense against ICBMs and SLBMs might best be based on the use of directed energy

Appendix A: New BMD Technologies

beams or particles. Traveling at or near the velocity of light, such beams or particles would traverse outer space essentially unhindered. Offensive missiles could travel only a short distance before being attacked by weapons moving at such great speed. This would be a great improvement over previous ABM systems, whose chemically propelled missiles did not move as fast as their targets.

Five different types of directed energy weapons have been proposed. Four would employ lasers; the fifth, particle beams. Hydrogen and fluorine could be combined to lase with the emission of light in the infrared range, with a wavelength of about 2.7 micrometers. Excimer lasers produce light in the ultraviolet range, with a much shorter wavelength. The beam of light from either of these lasers, if focused on the skin of an ICBM booster by one or more large mirrors deployed in space, could heat a spot about one meter in diameter, causing it to melt or evaporate and leading to gross failure of the booster.

In a free-electron laser (FEL), beams of arbitrary wavelength, including that of visible light, would be extracted from a powerful beam of high-energy electrons. This technique is very much in its infancy, so that neither its performance nor its cost can be estimated. While high efficiency may be achievable, it is not yet known how multi-megawatt optical powers can be handled on the small mirrors that FELs will apparently use. Because of the preliminary state of the technology, and because most of the systems questions will be illustrated in our discussion of the excimer laser, we shall not discuss the FEL further here.

An X-ray laser would be powered by a nuclear explosion that would strongly illuminate a material surrounding the explosive, leading to the propagation of a laser beam along "wires" embedded in the material. X-ray lasers could deliver a heavy jolt to an ICBM booster, damaging its internal mechanisms.

A beam of atomic particles accelerated to high velocity could damage the electronics of an ICBM booster by penetrating deeply. Neutral hydrogen atoms could be used for this purpose to avoid deflection by the magnetic field of the earth, which would affect beams of charged particles.

Possible Missions for DEWs

Space would be used in many ways in ballistic missile defense (BMD). First, the early-warning satellites would detect possible ICBM or

SLBM launches by the enemy. Second, one or more battle-management stations would track and assess each of the launched missiles and assign it to a specific BMD device. These battle-management stations would have to be able to keep in view both the ICBM silos and the BMD devices; it would therefore be best for them to be deployed in geosynchronous orbit (GEO). They would also have to be equipped with computers far beyond the present state of the art. Third, there would be the BMD weapons themselves: they could be either permanently stationed in orbit, or launched ("popped up") only when needed. The pop-up weapons, assumed invulnerable before launch, raise starkly the problem of use of such systems to augment a first-strike offensive force. It is hoped that with their tremendous speed these weapons would be able to reach and kill offensive missiles in the first phase of their flight, the so-called boost phase that occurs between launch and the end of booster acceleration. Such capability would greatly enhance the leverage of the defense. It would be far easier, for instance, to track an ICBM in boost phase than in later parts of its trajectory, because the booster flame is discernible at great distances, especially in the infrared. Also, the booster skin is relatively thin and weak and hence more vulnerable than the reentry vehicles the booster will deploy. Finally, there are fewer boosters than reentry vehicles, and they are more difficult and costly to simulate or "decoy."

All the directed energy weapons mentioned here could theoretically be effective in boost phase against the current generation of liquid-fueled ICBMs, whose boost phases last 200 to 400 seconds and reach altitudes of 400 kilometers. However, the duration of flight and final altitude of ICBM boosters are likely to decrease as new weapons are developed in response to the threat of boost-phase intercept. Already the MX, under development in the U.S., has reduced the boost phase to some 180 seconds duration and 200 kilometers altitude. Contractor studies for the Fletcher panel indicate that further reductions to as little as 40 seconds and 80 kilometers can readily be achieved. In that case, only ultraviolet (UV) and infrared (IR) beams will be useful for boost-phase defense. X-rays and particle beams cannot easily penetrate the earth's atmosphere. They would therefore be useless against boosters that burn out at low altitudes (under 110 kilometers for X-rays; under 140 kilometers for particle beams).

Appendix A: New BMD Technologies

Both X-rays and particle beams do, however, have potential for use in the post-boost phase, after the acceleration stages have been discarded but before the booster "bus" has released all its reentry vehicles (RVs) and decoys. This phase now lasts approximately 300 seconds and takes the bus to altitudes of about 600 kilometers—well above the atmosphere.

The next phase—the mid-course—of an ICBM's flight will occupy some 15 to 20 minutes, during which the warheads, accompanied by decoys, travel through space toward their targets. Despite the comparative length of this period, mid-course interception presents formidable problems for the defense because of the enormous number of objects deployed and the weakness of any RV infrared signals that might aid in discrimination and tracking. Mid-course behavior will not be changed by fast-burn boosters or by the absence of a bus.

Directed energy weapons have little potential for use in the terminal phase when warheads reenter the atmosphere and approach their targets.

Possible Basing Modes

The principal attraction of directed energy weapons purportedly comes from their ability to "reach back" further in missile flight than was possible with the older technologies, attacking enemy missiles before they have released their warheads and when they are most vulnerable. The curvature of the earth and the fact that the ICBMs would rise no more than 400 kilometers above the Soviet Union during boost phase would allow a clear shot only from space.

This section reviews alternative ways of placing the directed energy weapons in space. These include stationing the weapons permanently in space aboard satellites in either low earth or geosynchronous orbits, launching them from the ground ("pop-up") when an attack is observed, or directing their beams from the ground to mirrors in space.

It is relatively easy to deploy a satellite into low earth orbit (LEO). Since 1957, this has been done by rockets similar to or identical with those that launch ICBM warheads. It can also be done by a space shuttle. A shuttle flight could carry about 30 tons into an eastward near-equatorial orbit, at a few-hundred-kilometer altitude, and about half that weight into a polar LEO. The cost would be about

$3 million per ton for the eastward LEO and about $6 million per ton for low polar orbit.

The basic drawback of deployment in LEO is the so-called "absentee problem." A satellite in LEO orbits the earth every 1.5 to 2 hours. Since the earth rotates, the satellite will be over a different part of the earth on each successive orbit. Any given satellite would therefore be near the intended target only a small fraction of the time. A large number would have to be in orbit to ensure that there is always one near the target. We can estimate the number of satellites that will be absent (not near the target at any time) as follows: at present, the Soviet ICBMs are distributed in a band about $L = 5000$ km in length along the Trans-Siberian railroad at latitude $\simeq 57°$, with some extension to the northwest at the western end. Let R be the "kill radius" of a laser satellite, i.e., the distance at which it could destroy an ICBM booster, and assume the satellites are in circular polar orbits at an altitude A from which a laser can see the earth's surface at a distance R (the horizon). Then the fraction of satellites that are within a distance R from the nearest point on the latitude circle of $57°$ would be $f_1 = 2R/\pi R_o$, where $R_o = 6370$ km, the radius of the earth. The circle of constant latitude λ would have a length of $2\pi R_o \cos\lambda$. Lasers at latitude λ could attack ICBM boosters if they were located within an arc of length $L + 2R$. Lasers south or north of that latitude would have a somewhat smaller effective arc, but this would be made up partly by some lasers above the northwest extension of the band of ICBMs. Thus we expect that the fraction of orbiting lasers that could be effective would be $f = (2R/\pi R_o)(L+2R)/(2\pi R_o \cos\lambda)$. For every laser that could be effective, we would have to put $N = 1/f$ laser satellites into orbit. Setting $R = 3000$ km, for which $A = 670$ km, we find $N = 6.6$. This result seems to agree with more detailed calculations. Of the N satellites, N-1 would be "absentees" at any one time. We call N the absentee ratio. Of course, it does very little good to deploy lasers at high altitude A in order to obtain a low absentee ratio N, if the lasers cannot destroy the many boosters they can now see.

It would be desirable to make the absentee ratio small, but to do that by increasing laser *power* and altitude would have the offsetting disadvantage that more fuel would have to be carried into orbit for each satellite in order to attack at maximum range. In fact, to attack ICBMs spread over a region, the possibility of assigning a booster to the closest laser would lead one to lasers of lesser capability, lower altitude, and *higher* absentee ratio—although if mirror size were cost-free, one would always prefer the largest mirror. The absentee problem of low earth orbit can be overcome by deploying the lasers in geosynchronous orbit, i.e., in such a way that they take precisely 24 hours to circle the earth. They would be above the equator and would see the same region of the earth at all times. However, even though the laser in GEO might have the Soviet silos always in view, it would be roughly 40,000 kilometers from the target.[1] To obtain a beam small enough in diameter from such an enormous distance would require optical instruments of unprecedented size and performance. It would require about four times as much energy to lift a given weight into GEO as into LEO.

One suggestion on how to overcome the weight problem is to place the very heavy excimer lasers on the ground and beam their rays up through the atmosphere to a large mirror in GEO, which in turn might be assisted by tracking satellites and auxiliary relay mirrors in LEO. As we will discuss later in this appendix, however, such an arrangement would itself encounter formidable problems, e.g., vulnerability of the mirrors, electric power requirements, and disturbance by cloud cover and other atmospheric phenomena.

An additional possibility is to "pop up" lasers when an enemy attack is detected. This would reduce the vulnerability and much of the cost involved in deploying satellites permanently in orbit. Because of weight considerations, pop-up deployment is not a basing option for IR or UV lasers; it could be considered only with very lightweight weapons such as X-ray lasers.

A prime difficulty with the pop-up scheme is that of gaining a line of sight to the Soviet ICBM while it is still in its boost phase. To accomplish this, the lasers would have to be launched from a point nearer the Soviet ICBM fields than the American continent. This would require launching from submarines in the Indian Ocean or some Arctic seas. Even from these points, ability to sight the target in time would depend on the burn-out speed and altitude of Soviet boosters. If the Soviets make significant improvements in these regards, it is doubtful that the pop-up scheme would be viable for attacking their ICBMs. It might still be usable, however, against SLBMs.

Performance Requirements

The Fletcher panel concluded that ICBM boosters would be likely to suffer severe damage if exposed to an energy flux of 200 MJ/m^2 (megajoules per square meter) from a chemical (IR or UV) laser. For a 1 m^2 spot, this would be the energy equivalent of about 100 pounds of TNT, which on earth would cost about $100. In space, and delivered by a laser, the cost of this amount of energy would be immensely greater.

For X-ray lasers, the energy flux required would be somewhat smaller, perhaps 100 MJ/m^2. For particle beams, the requirement might be considerably less under special conditions, perhaps 0.1 MJ/m^2 for destroying electronic components.

To gain some sense of how difficult it may be to achieve the damage level contemplated by the Fletcher panel, we should consider what would be involved in focusing a laser beam on a booster so as to incapacitate it. To begin with, the beam would have to be concentrated on a spot on the ICBM booster with a diameter of about one meter. Moreover, the laser would have to dwell on the target for several seconds, during which time the booster would move many kilometers. To keep such a small spot stationary on the booster would pose a tremendous challenge for the pointing and tracking equipment, especially as the booster would be about 10 meters ahead of the flame that it generates. The configuration of plume and booster would have to be very well known beforehand to permit finding a given spot on the booster to one meter accuracy. In addition, the tracking would have to be extremely accurate, since in the time required for the light to pass from laser to ICBM—about 0.01 seconds for an infrared laser in LEO—the booster would have moved several tens of meters.

We would also need to face the fact that the needed beam diameter (and hence energy density) can be obtained only when the laser is within a certain maximum distance from the target. Because of the wave nature of light, every laser beam has a minimum angular spread of $\alpha = (4\lambda)/\pi d$, where λ is the wavelength of the light, and d the diameter of the mirror. A beam with spread α is called "diffraction limited"; practical laser beams may have a wider spread. A precise derivation of the maximum power density at any point in a diffraction-limited beam gives the result that the maximum power

Appendix A: New BMD Technologies

density within a focused laser spot is $(\pi/4)P(d^2/\lambda^2R^2)$. This *limit* on power density is the laser power divided by $(\pi/4)S^2$, giving a precise meaning to the concept of spot diameter S. In conventional terms, the power density at range R cannot exceed B/R^2, with B the "brightness" of the laser, which we have here calculated as $B = PA/\lambda^2$, with A the (projected) area of the (uniformly illuminated) mirror. Therefore, if we wish the beam to form a spot of diameter $S = \alpha R = (4/\pi)(\lambda/d)R$, the laser can at most be at a distance $R = S/\alpha = (\pi d/4\lambda)S = 0.8\, Sd/\lambda$.

In the above example, we assumed that the mirror can have a diameter of 10 meters. We can appreciate the difficulty of achieving such a size if we consider that the largest operating astronomical mirror, that of the Palomar Observatory, has a diameter of only 5 meters; that it took years to build; and that many corrections were needed before it had the required quality. A larger astronomical mirror with a diameter of perhaps 10 meters is planned for the future but will probably take at least four years to build. The Hubble Space Telescope planned for space operation in 1986 will have a diameter of only 2.4 meters and is expected to cost some $1.2 billion.

Note also that we have postulated that the beam spot on the target would be much smaller than the mirror. Thus it would not be enough to focus the laser to infinity; it would have to be focused to the approximate distance of the target.

From the foregoing, one can conclude that the performance requirements we have postulated for directed energy weapons are formidable indeed. There is no assurance at this time that their achievement is technically feasible. Nevertheless, the uncertainty in technical feasibility is dwarfed by the difficulty in ensuring the survival of the defense and its economic competitiveness with augmentation of the offense—the twin criteria set by Paul H. Nitze that the defense be survivable, and that it be "cost-effective at the margin."

ALTERNATIVE DEWS AND THEIR BASING MODES

Infrared Chemical Lasers

The preferred infrared chemical laser is hydrogen fluoride (HF), which emits radiation with a wavelength of about 2.7 micrometers. Such lasers could be placed in satellites circling the earth at an altitude of perhaps 1000 kilometers. If there were a Soviet ICBM launch, it would be observed by the early-warning satellite, which would transmit this information to the battle management satellite. The latter would then select one of the chemical lasers and command it to direct its beam against a Soviet booster. The laser beam would heat a spot perhaps 0.3–1.0 meter in diameter on the booster skin, and make it melt or evaporate; this would lead to gross failure of the booster.

IR lasers can only work satisfactorily from low earth orbit. With the HF laser of wave length $\lambda = 2.7$ μm, and taking a mirror $d = 10$ m and requiring a spot size $s = 1$ m, we find that the maximum distance from laser to target is $R \simeq 3000$ km—the "kill radius." (The distance to target from a satellite in GEO, as calculated earlier, would be nearly 40,000 kilometers; clearly, an IR laser could not usefully be deployed in GEO.) For a target at this distance to be visible from the satellite, the latter must be at an altitude of at least 670 kilometers. As we have indicated, for every satellite viewing the target, there would be several absentees. For our kill radius the formula gives the absentee ratio as 6.6.

Aviation Week & Space Technology (*AW&ST*) has published an abstract of the conclusions of the Fletcher panel. It includes an artist's conception of an HF laser of 25 megawatts; this is far beyond the present state of the art. A more detailed fantasy of this same laser is to be found in the October 1984 *Scientific American*.[2] A spot of one meter diameter, which we assumed above, has an area of not quite a square meter, so that we would need not quite 200 megajoules (MJ) to destroy the target. Because of the unavoidable imperfections of optical focusing, and the fact that our derivation obtained this power density only at a single point, we shall assume a requirement of 250 MJ. A laser of 25 MW power would then need to dwell on the target for 10 seconds to destroy it.

Some of the laser satellites might be closer to the target than 3000 kilometers—"the kill radius"—so that under ideal conditions the laser could focus a smaller spot S on the target, and need less time to destroy it. Under ideal conditions, the average dwell time of this laser on its target might be only 5 seconds. This, of course, assumes that the battle management computer is near-perfect in assigning lasers to targets.

Appendix A: New BMD Technologies 323

The report in *AW&T* also postulates (arbitrarily) that the laser should have fuel for 100 seconds. It would then be able to destroy 20 ICBMs if it was in sight of the Soviet ICBM field. Using the absentee ratio we have calculated (N = 6.6), we find that to target each Soviet ICBM we would need to deploy 0.33 laser satellites. For the 1400 Soviet ICBMs now existing, this would mean a total of 460 laser satellites in orbits about 1000 kilometers high.

The number of laser satellites needed might be slightly diminished by aiming for a laser spot on the booster of the smaller diameter S. The price for this reduction in the number of satellites, however, would be an increase in the difficulty of aiming and pointing the laser at the target. However, Gregory Canavan of Los Alamos and Christopher Cunningham of Livermore have proposed moving in this direction.[3] Cunningham calculates that lasers deployed at 300-km altitude can destroy more than 2 boosters per second. But once the dwell time of the laser on this target gets this short (42 milliseconds for some of Cunningham's engagements with boosters at 200-km altitude), it is essential to consider the time required for the laser mirror to slew from one target to another. For guidance, we note that the space telescope, with a mirror of 2.4-meter diameter, takes 3 seconds to move by 0.02 arc-seconds and then stabilize. Of course, that telescope is not designed for fast motion, but it seems excessively optimistic to believe that the 10-meter mirrors assumed for the BMD lasers could be stabilized in as little as one second, after moving through several degrees of arc.

Clearly, many factors need to be considered in estimating the number of satellites required.[4]

We will now sketch the result of an analytical derivation of the number of satellites needed for targeting Soviet ICBMs (See figure 1, next page). For M boosters clustered in a region of diameter h or less (h the altitude of the satellite constellation), the required satellite *density* s (in satellites per square megameter) is given by

$$M = (\pi s B T_0/J) \ln[1 + E(BT_s/J)^{-1/2}]$$

where B is satellite brightness (units of 10^{18} watts per steradian), T_0 the boost time of the boosters (engagement time), J the booster hardness (megajoules per square meter), E the radius of the earth, and T_S the retarget time. The result holds for the optimum satellite altitude $h = (BT_S/J)^{1/2}$. B cannot exceed PA/λ^2, with P the laser power, A the area of the laser mirror, λ the wavelength, on the assumption

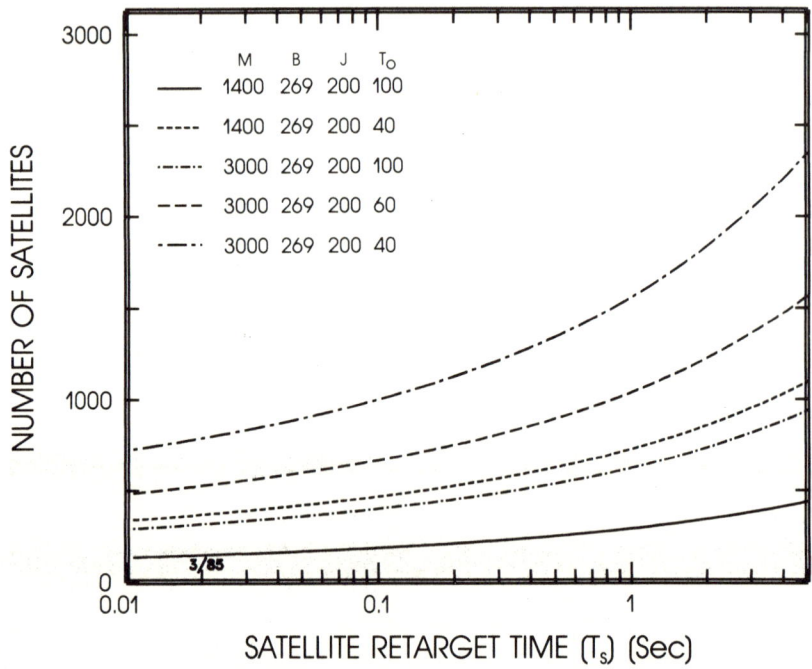

fig. 1 Dependence of satellite numbers on booster numbers (M), hardness (J), boost-time (T_o), and satellite brightness (B)

The number of our standard laser satellites is plotted against assumed laser retarget time for two assumed booster numbers and three assumed engagement times during boost phase. Against the clustered launch of 3000 boosters of 40-sec engagement time, 1344 satellites of 0.5-sec retarget time in optimum orbit are required, working perfectly reliably, to dispatch the boosters.

that the laser is focused at the distance of the target. For P = 25 megawatts, and for a mirror diameter of 10 meters and a wavelength 2.7 micrometers, B = 2.69 x 10^{20} W/steradian.

The *number* of satellites needed is minimized by choosing an orbit of inclination i equal to the latitude of the northernmost silo, assuming the silos uniformly distributed between latitudes λ_1 and λ_2. The required number of satellites is N = $4\pi sE^2/y$, with the concentration factor y = 3.1 for silos distributed between 50° and 60° north. This approach does not use the vague "absentee factor" N, but for clustered launch gives N = $(4\pi E^2)/(\pi y H^2)$, with H the distance from laser to horizon. For T_S = 0.5 sec, h = 820 km and H≡$[(E + h)^2 - E^2]^{1/2}$ = 3.3 Mm. Thus, N = 4.72. For T_S = 0.1 sec, h = 370 km and H = 2.20 Mm and N = 10.8. For 3000 boosters of T_0 = 40 sec and hardness J = 200 MJ/m², clustered in a region of diameter less than

Appendix A: New BMD Technologies 325

h, the number of laser battle stations in optimum orbit required to overcome them is 2044 if a retarget time $T_S = 3$ sec can be achieved; 1003 with the SDI goal of 0.1 sec.

Although the above "absentee factor" was computed for Soviet ICBM launch sites, IR chemical lasers would also be able to take care of Soviet SLBMs. Since these would be in locations different from the ICBMs, different lasers would be engaged. Figure 1 shows the dependence of satellite numbers on booster numbers, hardness, burn time, and on satellite retarget time.

It is difficult to estimate the cost of infrared satellites. An absolute *minimum* might be deduced from their weight (although the cost of a pair of binoculars or of a fine mechanical watch bears no relation to its weight). Some 500 joules of energy per gram of fuel may be achieved in an HF laser. If the satellites are to operate for 100 seconds with an output of 25 megawatts of laser light, the fuel must be sufficient to produce 2500 megajoules. Accordingly, the weight of the fuel would be 5 tons. The satellite must weigh at least twice as much as the fuel, or 10 tons. This would be a bare minimum. *AW&ST* estimates that its more elaborate conceptual satellite would weigh 100 tons. The cost of simply bringing the weight into space would be about $600 million for each *AW&ST* satellite, $60 million for our minimum weight satellite. These figures do not include the cost of manufacturing the delicate devices which would go on the satellites.

Another way to estimate costs would be to start with the actual costs of present satellites, which vary from $100 to $200 million, including their launch into space. The laser satellite would be bigger and more complicated. Written testimony by a strong supporter of the SDI states, "Every laser-equipped satellite will cost about as much as a Trident submarine—several billion dollars."[5]

Still another approach has been taken by George Field of the Harvard Astrophysics Center. He based his estimates of the cost of laser satellites on the known cost of optical mirrors in space. Allowing that mass production would reduce the cost by two-thirds, and neglecting the cost of the lasers, he found that it would cost between $400 million and $2 billion to destroy a single Soviet missile. This would make the cost of the laser satellite system alone on the order of a trillion dollars.

Excimer (Ultraviolet) Lasers

If we wish to produce a spot of one meter diameter on an ICBM booster at a range of 40 megameters, we would need an angular spread better than $\alpha = 1/40$ μr. If this were to be done with infrared lasers, the mirror would need to have a diameter of 140 meters, which is clearly out of the question. Therefore, it has been proposed that lasers in GEO should use ultraviolet light produced by an excimer laser. Such lasers have been demonstrated in the laboratory at much lower outputs. Their wavelength is roughly 0.3 micrometers, so that the required mirror diameter would be only 15 meters. The UV lasers would perform similarly to the infrared chemical ones, by melting a hole in the booster skin. The ten-fold more stringent pointing and tracking of UV lasers in GEO would, however, be even more difficult than for infrared lasers in LEO.

The lasing gas in excimer lasers would be a compound of a noble gas and a halogen, e.g., KrF. These molecules are essentially unstable in their ground state, but in an excited electronic state they are chemically stable. Methods have been developed to form them in excited states from the elementary gases. They will then emit ultraviolet light when the molecules return from the excited to the ground state. The ground state of the molecule will, in turn, immediately dissociate into two atoms, thereby allowing a population inversion, i.e., an occupied excited state but an empty ground state. These excimer lasers could have 6 percent efficiency, i.e., the energy in the laser beam could be 6 percent of the electric energy supplied to the gas mixture.

A problem with excimer lasers would be their weight. If it is possible to concentrate the laser beam on one square meter of the ICBM booster, we would again need 250 MJ of laser light to destroy one booster. Because the excimer would be powered by an electron beam rather than by a chemical reaction, the weight of fuel required to produce the necessary electricity is likely to be at least 3 tons, compared with half a ton for an IR laser. Worse than this, the lasing gas KrF would get hot during the laser action and would have to be replaced. Its weight may be estimated as $\simeq 25$ tons per booster destroyed. Thus, we would be confronted with the need for fifty times the weight of an IR laser. This would far more than offset the extra weight necessitated by the 6.6 absentee ratio for lasers in LEO,

Appendix A: New BMD Technologies   327

especially since it takes about four times as much energy to lift a given weight into GEO as compared to LEO. All this indicates that it would cost much more to deploy excimers in GEO than to deploy IR lasers in LEO.

An alternative to space deployment would be placing the excimer lasers on the ground and then beaming their rays through the atmosphere at a large mirror in GEO, which would redirect the beam to destroy one booster after another. This would considerably ease the weight problem, because now only the mirror and some pointing equipment would have to be carried up. Many problems would remain, however, chief among them the vulnerability of the mirror-carrying satellites. Still other difficulties that would have to be addressed include the following:

• The ICBM booster would be detected by the emission of IR from its plume, not by UV. But if an IR detector were put on the mirror in GEO, it would not have the resolution necessary to form an image of the booster adequate for pointing the UV laser with the required accuracy. If the booster plume were a tiny point like a star, fixed in position with respect to the booster itself, "beam splitting" could be used with a smaller mirror—but such is not the case. It has therefore been proposed that an auxiliary satellite be placed in LEO to observe the IR plume, do the pointing and tracking, and then transmit electronically the exact position and velocity of the booster to the mirror station in GEO. This would be an extremely complicated system, with many possibilities of malfunction.[6]

• A large amount of electric power would be needed for the excimer laser, even on the ground. Assuming 6 percent efficiency from power plant to laser, we would require about 4000 megajoules of electric power for destruction of each booster. If 1000 boosters had to be dealt with in a hundred seconds, 40,000 megawatts of electric power would have to be available. Because any given location of the ground-based lasers, even in the desert, might be covered by clouds, we should probably double this requirement. This would mean we would need the output of 80 full-size electric power stations. Since these stations would not have to give steady power, they could be of a very simple type, perhaps with oil-fired gas turbines and without cooling towers, and could probably be built for about $300 per kilowatt (instead of the $1,000 to $2,000 required for a modern coal- or nuclear-power station). Even so, the cost of the electric power stations alone would be $25 billion. We have not taken into account a factor-2 or -3 absorption in the atmosphere of the ultraviolet light, which would double or triple our requirements.

• Cloud cover would prevent the excimer laser's UV beam from penetrating through the atmosphere. Even with a clear sky, the normal turbulence in the atmosphere would disturb the beam, broadening it from an earth-based 5-meter diameter telescope to 1000 times that area at GEO. Furthermore, the high-intensity laser would itself cause a strong disturbance of the air, one that would change with time as the beam was operating. Schemes such as adaptive optics or phase conjugation of a pilot pulse from the GEO mirror might reliably provide a clean beam near the diffraction limit for weak lasers, but it is not clear that this could be done for the power levels needed in weapons.

Aviation Week & Space Technology, reporting on the Fletcher panel's findings, concluded that by 1989 an excimer laser could be developed that would have a power of 2 MW and a beam spread of 0.05 μr. Such a power output would clearly be insufficient in relation to the need, which we have assumed to be 25 MW.

X-Ray Lasers

As described above, the X-ray laser would be powered by a nuclear explosion that would strongly illuminate nearby material with the bomb X-rays. The actual lasing action may take place in a bundle of long, thin wires. By the nature of the process, the quantum energy in the laser beam would be considerably less than that of the X-rays in the surrounding plasma. An estimate of 1 keV for the energy of the laser quanta seems reasonable.

The X-rays would damage an ICBM booster by evaporating a very thin layer at its surface. The depth of penetration would be determined by the absorption coefficient of X-rays in the booster skin material; it would be only a micrometer or so. The surface material would be given a very high outward velocity by the X-rays. One can calculate that, for an incident X-ray flux of 100 MJ/m^2, the impulse given to the evaporated material would be P ≃ 5000 taps. (One tap is 1 g/cm² cm/sec—the momentum carried by one gram of mass per square centimeter of surface, moving at a velocity of 1 cm/s.) The material below the blow-off layer would be pushed inward with a momentum equal to P. This would have two consequences. One is that the booster as a whole would recoil with a velocity $\Delta v = PA/M$, where M is the mass of the booster and A its lengthwise area. A reasonable estimate is $M/A = 100$ g/cm², and the corresponding recoil is $\Delta v \simeq 50$ cm/s. Such a recoil, if uncorrected, would cause the reentry vehicles carried by the booster to impact about 1 kilometer away from the intended target. If that target was a hardened missile silo, this amount of deflection would save the silo. However, if the target was soft, and especially if it was in or near a city, the deflection would make hardly any difference. Moreover, modern inertial guidance devices should be able to correct for such a recoil incurred in powered flight.

The other effect of the X-ray impulses would be elastic, and possibly plastic, deformation of the booster skin. The velocity given to the booster skin would be $V = P/m$, where m is the mass of the skin per cm². We estimate m = 1 to 3 g/cm². The kinetic energy per gram, $1/2\ V^2$, would have to be taken up by the elastic deformation energy or by special mitigation means. Reasonable assumptions on the strength of the booster indicate that an impulse of $P = 5000$ taps would be just enough to damage a booster skin of reasonable strength and thickness. We therefore estimate that the X-ray flux to the booster would have to be at least 100 MJ/m² in order to do substantial damage.

Other "kill mechanisms" are conceivable. In particular, certain booster weaknesses might reduce the required X-ray flux by as much as a factor of 10. However, we must assume that the Soviets, faced with the existence of an X-ray laser, would be very careful to design the booster skin in such a way that it would not have such weaknesses.

A further advantage of the nuclear-powered X-ray laser is that it would give its pulse in a very short time, perhaps 10–100 ns (nanoseconds). During such a short time, the booster would not move appreciably, in contrast to the optical lasers which would have to dwell on the booster for several seconds. But the nuclear explosion scheme has the obvious disadvantage that it would be just a single shot.

An X-ray laser beam could not be focused by a mirror because there are no materials that back-reflect intense soft X-rays. Instead, the beam would be defined by the geometry of the wires in which the lasing action takes place, and would be rather broad.[7] This would put high demands on the total laser energy, which would have to fill with lethal energy a spot much *larger* than the booster target; although it would relax the requirement for pointing accuracy.[8] An attractive feature of the X-ray laser is that the weight of the device could be moderate since the energy source would be a nuclear explosion. This makes it possible to consider deploying the X-ray laser in a pop-up mode, a technique that would not be available for the much heavier and larger IR or UV lasers. In pop-up, the device would be launched only when an enemy ICBM attack was detected. In this way, the vulnerability of weapons deployed in space would be avoided, but not the vulnerability of sensors in space. Pop-up also would avoid violation of the 1967 treaty prohibiting the stationing of nuclear weapons in space, but that treaty is in any case unlikely to survive a space test of the X-ray laser, which would also violate the 1963 Limited Test Ban.

The principal difficulty that the pop-up scheme has to overcome is that the laser must have the ICBM in its line of sight while the ICBM is still in its boost phase. To achieve this, the X-ray laser must be raised to a very high altitude that will increase as the distance from the point of the pop-up launch to that of the ICBM launch site increases. No point on the territory of the U.S. is closer than about 6000 kilometers from most of the Soviet ICBM launch sites, and 9000 kilometers is more typical. Thus, it has been proposed to launch X-ray lasers from submarines located either in the Indian Ocean south of Pakistan, at a latitude 22–24°, or in some Arctic seas. The latter alternative would put the laser launch within a great-circle distance of about 4000 kilometers from some of the ICBM silos.

These relationships are demonstrated in figure 2 and in the following calculations:

The earliest possibility for the Soviet ICBM to come into view is "beyond the horizon," as is shown in figure 2. The X-ray grazes the atmosphere at a certain height H_O, the target is at a height $H_T > H_O$, the laser rocket at a height H_H. The horizon is determined by the

Appendix A: New BMD Technologies 331

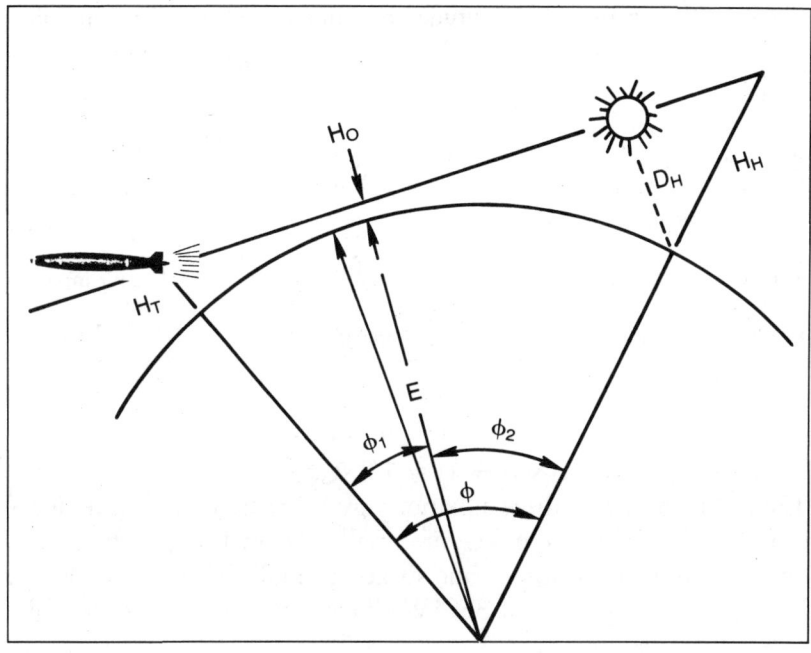

Fig. 2 Geometry of Pop-up Intercept. For a target at height H_T and a line of sight that grazes the earth at altitude H_O, the interceptor ("hunter") must be at altitude exceeding H_H, as shown. Also indicated is the "interceptor flight" distance D_H to the line of sight.

condition that the X-rays passing through the atmosphere at H_o should not be too much attenuated. To determine H_o, we must calculate the amount of matter traversed along the horizontal line in figure 2. The density of air decreases approximately exponentially with altitude $\rho = \rho_o e^{-H/L}$, where L is the scale height. At the relevant altitudes, L is $\simeq 18$ km. By integration, the amount of matter along the horizontal line is then $\sigma = (2\pi L R_o)^{1/2} \rho(H_o)$, where H_o is the minimum ray altitude and E or R_o the radius of the earth. An X-ray beam of 1 keV will penetrate $\sigma = 3.0 \times 10^{-4}$ g/cm^2 of air, at which point its intensity has decreased to $1/e = 37$ percent. Inserting σ, L, and R_o, we find $\rho(H_o) = 3.5 \times 10^{-12}$ g/cm^3, corresponding to $H_o = 155$ km.

Let now H_T and H_H be the altitudes of target and "hunter." Then the geometry of figure 1 shows that

Table 1 Pop-Up Intercept Altitude and Time from 4000-km Standoff

	SS-18	MX	Medium-fast burn	Fast-burn
Burn-out altitude (km)	400	200	160	90
Burn-out time (sec)	300	180	130	50
Interceptor flight (km)	571	995	1355	—
Time to intercept (sec) given $M_o/M = 30$	85	127	162	can't intercept
Time to intercept (sec) given $M_o/M = 15$	93	146	190	can't intercept
Intercept possible?	yes	marginal	no	no

$$H_T - H_o = R_o(1/\cos\phi_1 - 1) \simeq 1/2\, R_o\phi_1^2$$
$$H_H - H_o = R_o(1/\cos\phi_2 - 1) \simeq 1/2\, R_o\phi_2^2$$

The last members of these equations are calculated with the assumption that the angles ϕ_1 and ϕ_2 are small. This will only be true if the angular distance of hunter and target is small. With our choice of launch points, ϕ is about 35°. We have, of course, $\phi_2 = \phi - \phi_1$. Evaluating, we find

$$H_H = [\phi(R_o/2)^{1/2} - (H_T - H_o)^{1/2}]^2 + H_o$$

With our choice of ϕ we have $\phi(R_o/2)^{1/2} = 35$ (km)$^{1/2}$

The altitude to which the X-ray laser must rise in order to have a line of sight to the booster therefore depends sensitively on the booster's burn-out altitude. If the booster burns slowly, its burn-out altitude H_T would be great (to achieve ICBM speed of 7 km/s) and the hunter would not need to go very high. In addition, the hunter would have much more time to gain the required altitude and would not have to reach such a high speed; thus a much smaller fraction of its initial mass would need to be rocket fuel. On the other hand, a fast-burn booster would make necessary a very large interceptor to deliver a small X-ray laser payload very rapidly to a high altitude. Table 1, above, gives the necessary interceptor flight distance for several booster burnout altitudes, as well as the time required to fly to those points (calculated exactly, not with the assumption that ϕ_1 and ϕ_2 are small). The table is calculated for a great-circle distance of 4000 kilometers, a minimum X-ray beam grazing altitude of 160 kilometers, and intercept just at booster burnout. Both interceptor times are calculated for an interceptor that has the same acceleration

Appendix A: New BMD Technologies

as a fast-burn booster—i.e., one acquiring 7 km/s of speed in 40 seconds (constant 17.5 g acceleration)—and with two alternative values of the overall mass, either 30 tons or 15 tons per ton of payload. For a rocket propellant with a specific impulse of 300 seconds and perfect staging, this corresponds to interceptor burnout speeds of 10.2 km/s or 8.1 km/s respectively, and the times as shown in the table. A one-ton X-ray laser payload would thus require a 30-ton or 15-ton submarine-based interceptor missile. (The Poseidon SLBM weighs about 30 tons.)

In this table "interceptor flight" is the shortest distance from an interceptor launch point to a line drawn from booster burnout to graze-altitude, calculated exactly. Times and mass ratios ignore gravity. Interceptor speed is obtained from rocket exhaust velocity $gI_{sp} = V_e$; and $M_o/M = \exp(V/V_e)$. Interceptor burn time is then $T_o = V/g'$ (with $g' = 0.175$ km/s^2), and interceptor flight time to distance D is $T = T_o/2 + D/V$.

We have to add to the numbers in table 1 the time needed before the interceptor rocket can start. This includes the time for detection of the ICBM launch by early-warning satellites, for communication of this warning to the rocket-launching submarine, for the actual launching of the rocket until it breaks out of the water, and finally for aiming of the X-ray laser. At present, the first of these alone takes about two minutes. Allowing for improvements, surely 30 seconds is a low estimate for all these activities. Thus, if the booster burned out at 200 kilometers, the interceptor rocket would have to reach its station and be ready to fire its X-rays about 150 seconds after it was launched.

Most of the boosters presently in the arsenals of both sides have a boost phase that lasts longer than 150 seconds. The boost phase of the MX (presently being developed by the U.S.) is about 180 seconds, allowing barely enough time for attack by a pop-up X-ray laser. The Soviets could certainly equal this performance if the existence of a U.S. X-ray laser gave them the motivation to do so. Of course, a greater interceptor speed would cut the interceptor boost time; but to attain even 12 km/s interceptor speed would require an interceptor with a gross weight of 55 tons for a one-ton payload. Considering the relative ease of shortening the boost phase of ICBMs compared with the difficulty of reducing the intercept time of submarine-launched interceptors armed with X-ray lasers, we believe that the pop-up scheme of attacking Soviet ICBMs can never be viable.

As we mentioned earlier, the laser would be launched about 4000 kilometers from the Soviet ICBM sites. Assuming then a beam width of 50 μr and requiring a flux of 100 MJ/m² on the booster, the X-ray laser would have to emit a total energy of 1 kiloton (kt). But this laser energy would be only a fraction of the total energy of the nuclear explosion. The exact fraction is not known, but it is unlikely to be as high as 1 percent. Therefore, a successful X-ray laser would have to start with an explosion of at least 100 kilotons.

But 100 kilotons is unlikely to be sufficient. The reason has to do with the fact that X-ray laser beams cannot penetrate far into the atmosphere. At 1 keV, the beam penetrates $\simeq 3.0 \times 10^{-4}$ g/cm² of air, at which point its intensity has decreased to 37 percent. At vertical incidence, this penetration would bring the beam to $\simeq 105$ km altitude. If beams of very high intensity could be attained, let us say 100 MJ/m² on average, they would remove some of the electrons from the N and O atoms and thus "burn through" some of the air; this might increase the penetration to $\simeq 7 \times 10^{-4}$ g/cm². To accomplish this, initial laser energy would have to be raised from 1 to about 3 kilotons (the explosive energy from 100 to at least 300 kilotons, a very high yield) if targets are to be attacked even at vertical incidence to an atmospheric depth of 7×10^{-4} g/cm².

While popped-up X-ray lasers thus seem unlikely to be effective against ICBMs, there remains a possibility that they could be effective against SLBMs launched at close range against targets near the U.S. coast. In this case, the rockets could be popped up from the American continent, and the angular distance ϕ between the rocket and the SLBM would be small. The X-ray laser could therefore be in place before the SLBM burns out. An operational countermeasure by the offense would be to launch the SLBM on a strongly depressed trajectory so that the booster would burn out at such a low altitude that the X-rays could not penetrate, i.e., below $\simeq 110$ km, with no fast-burn booster needed.

Particle Beams

It has been suggested that a beam of very fast particles could be used to attack an ICBM or SLBM. For instance, protons of 500 MeV (three-fourths the velocity of light) would penetrate all the way into

Appendix A: New BMD Technologies 335

an ICBM or reentry vehicle, severely damaging its delicate electronics, and thus disturbing the guidance system of the missile or the firing circuit of the nuclear weapon.

Protons (hydrogen nuclei), however, cannot be used directly in a particle beam because they would be strongly deflected by the magnetic field of the earth, and therefore be difficult to aim. To make matters worse, the magnetic field fluctuates. This problem can be avoided by the use of a beam of neutral hydrogen atoms, available at lower intensities in various laboratories. To accelerate such atoms to high velocity, one starts with negative hydrogen ions, H^-, i.e., hydrogen that has two electrons rather than one orbiting the nucleus. It is not easy to produce large currents of H^-, but it has been done in some particle accelerators, notably TRIUMF in Vancouver and LAMPF at Los Alamos. In these facilities, H^- beams are produced and accelerated to energies of 500 and 800 MeV, respectively. Such energies would be useful for beam weapons. After being accelerated and aimed precisely at its target, the H^- would be passed through a very thin foil or a gas cell. The extra electron would be stripped off, the gas pressure having been carefully chosen so that, insofar as possible, only one electron is stripped. The result would be a beam of neutral hydrogen.

An important property of any beam is its angular spread. Even with a perfectly parallel beam from the accelerator, the neutral H beam will have a certain minimum spread because the extra electron removed from the H^- has a certain momentum relative to the nucleus of the atom. The distribution of this momentum can be calculated from quantum mechanics, but neither the actual magnitude nor its direction can be predicted or controlled for a particular atom.

When the extra electron is removed by collision with an atom, the remaining neutral H will have a momentum equal and opposite to that of the electron. In principle, neutralization might be accomplished by laser removal of the second electron, and if the laser wavelength is barely short enough to eject the electron, the divergence so introduced into the beam could be less than that from neutralization by collision. But it does not seem practical to laser-strip a beam of large area and near-continuous current. Stripping by collision would give a minimum angular spread of the neutral H beam of $\simeq 1$ μr (microradian), no matter how well the H^- beam was collimated. This means that at 3000 kilometers from the accelerator, the beam would be spread over 3 meters. The designer must try to obtain an H⁻ beam which has a divergence of < 1 microradian. At the same time, the intensity of the beam must be maximized.[9]

To penetrate deeply enough into the booster and then into the RVs, in order to affect their semiconductor components, the energy of the particles in the beam must be about 300 MeV or greater. To destroy silicon semiconductor circuitry, one may need $\simeq$ 10,000 rad.[10] Taking into account that the energy deposited by a proton of 100 to 300 MeV is about 3 MeV per g/cm², the required incident energy would be about 0.2 megajoule/m², much less than for those lasers previously discussed. To fill a beam spot of 3-meter diameter from 3000 kilometers, we would thus need 1 MJ of particle beam. At an energy of 300 MeV, this would mean 3 mA-sec (milliampere-seconds). An H⁻ beam of 1 mA would need to dwell on the target for 3 seconds. We could consider having an accelerator work for 100 seconds, so that one accelerator could destroy 30 boosters. At a beam current of 100 mA, one accelerator could destroy the electronics of 3000 boosters in 100 seconds, providing it could itself survive and target accurately and instantly one booster after another.

The weight of the accelerator is likely to be rather large, while that of the power source required for the particle beam probably smaller. An attractive feature of the particle beam is that it may be steered to a certain extent by electromagnetic means before it is neutralized; this would make it possible to attack several targets in succession without swinging the entire accelerator. Because the magnetic steering operates on a beam of large diameter at the output of the beam expansion telescope (perhaps 40 centimeters in diameter), if large steering angles are required the magnetic steering system can become very massive—perhaps weighing hundreds of tons.

Particle beams have a serious limitation: the neutral atoms will not remain neutral once they enter the atmosphere; collisions with the air atoms will strip off the electron. We estimate that this will happen at an air density of $\simeq 10^{-11}$ g/cm³, which corresponds to an altitude of about 140 kilometers. Once the electron has been stripped from the H-atom, the remaining proton is subject to deflection by the earth's magnetic field (as discussed above). Since the stripping occurs at different altitudes, the beam will change into a fan whose width will be $\simeq$ 300 m at $\simeq$ 130 km altitude. This means that the density of particles in the beam would be decreased by a factor of 100, so that

Appendix A: New BMD Technologies

a hundred-fold increase in the beam intensity would be necessary to compensate. This is beyond the foreseeable beam intensity. We have previously mentioned that it may be possible to develop boosters that would acquire their full velocity at $\simeq 90$ km altitude. A particle beam would be worthless against such a booster.

Another possibility is that particle beams might be used against the "bus" that remains after the various acceleration stages of the booster have been discarded. The bus will go to a higher altitude than the booster. It will not have as conspicuous a flame as the booster and will therefore be more difficult to detect and target. It would have to be hit, moreover, before it dispenses its RVs and decoys. The bus's electronics are essential only for providing guidance in the boost and post-boost phase. If the particle beam were to be used in mid-course, it would probably have to destroy structural components or the high explosive, which would require much greater beam power. Single-warhead missiles and even MIRVed missiles can be built without a bus. In mid-course, there would normally be no air to obstruct the particle beam from reaching its target, but a shield of air might be raised by large nuclear explosions near the top of the atmosphere specifically as a countermeasure to particle-beam intercept.

It has been suggested that a particle beam could be produced by a nuclear explosion, rather than by a particle accelerator. No method is known however, for forming a highly collimated beam of protons, let alone neutral hydrogen atoms, from a nuclear explosion. In part, this is because all the energy must be used in the small fraction of a microsecond before the system is destroyed. This would also be a one-shot affair, and could only with difficulty be directed against more than one booster or bus.

This view of the effectiveness of a neutral particle beam weapon is based on a 10,000 rad semiconductor kill dose. But MESFET circuits made of gallium arsenide (GaAs) may withstand gamma radiation doses far beyond this—perhaps 10 megarads (100 joules/g)[11]—which is in the dose range required to overheat or melt structural materials, destroy explosives, and the like. Such a dose corresponds to $\simeq 160$ MJ/m^2 and could be achieved in 10 seconds only with a 100 megawatt accelerator. Defect generation by the protons from the beam (or by secondary neutrons) would set a comparable dose limit.

The SDI Program Office has mentioned approvingly a suggestion to use anti-matter beams (just 5 micrograms of anti-matter could produce 1000 MJ of annihilation energy). To produce 5 milligrams of anti-matter to feed the accelerators so that they could deliver 1000 MJ against each of 1000 boosters, 5000 particle accelerators as powerful as the most powerful existing one, that at Batavia, Illinois, would have to work for a full year. Even then, the problem would remain of preserving the anti-matter and accelerating it (as anti-hydrogen atoms) against the Soviet boosters. Calculation shows that anti-matter would in any case have little advantage over normal hydrogen.

HOMING KILL (KINETIC ENERGY) VEHICLES

Use for Boost-Phase Intercept

Kinetic energy kill of ICBMs in boost phase is entirely feasible, especially considering the fragility of the booster and its very substantial area. The difficulty is not in killing the booster, but in *reaching* and striking it while it is still boosting.

In ground warfare, the requisite velocities are obtained either by a gun—in which the velocity is given impulsively to the projectile while it is within the tube of the launcher—or by a rocket, for which gas expelled at high velocity from a nozzle at the back of the projectile continually increases the momentum of the projectile while reducing its mass. Many decades of effort have gone into the development of anti-tank guns that produce velocities of 2 km/s; such a velocity from a gun in space would carry the projectile only 400 kilometers during a nominal booster burn time of 200 seconds—not a useful distance. The limitation is that the gases from an explosion in the stationary gun barrel move only at rocket-propellant velocities, and can no longer push on the projectile when it is going faster. Thus, approaches to higher velocity guns include the use of hot-hydrogen propellant, with higher sound speed and hence larger maximum velocity of expansion.

Current interest in the use of guns for ICBM boost-phase intercept centers on space-based electromagnetic "railguns," in which the projectile would complete the circuit between two conducting bars or rails. A modest voltage applied to these rails would produce very large currents, the magnetic field behind the projectile expelling and continuously accelerating the projectile. Sub-kilogram projectiles

Appendix A: New BMD Technologies

have been accelerated in this fashion to speeds of 10 km/s, and somewhat more is feasible. Nevertheless, railgun-launched projectiles would be useful for boost-phase intercept only if they are *guided*, in order to cope with booster maneuver and uncertainties in predicted booster position.

Another approach to intercepting the ICBM before the end of its boost phase would be to use space trucks (so named by the High Frontier project). Each of these would carry enough rocket-propelled kill vehicles to destroy all ICBMs that might be launched within its range. There would be a clear trade-off between the spacing of such space trucks (and the number of kill vehicles required) and the velocity given to each kill vehicle (beyond the 8 km/s orbital velocity of the space truck). In its 1982 publication, High Frontier considered 1 km/s velocity, which in a nominal 100-second boost phase could move a kill vehicle only 100 kilometers. We will shortly consider a projectile velocity of 8 km/s, which could be imparted to a 5-kg homing warhead that had an initial total rocket mass of some 50–100 kg.

Railguns would of course impart velocity in a *very* small fraction of the distance to the ICBM, and we assume here that rocket-propelled projectiles would achieve their burnout velocity in only a few seconds. Either type of projectile would coast the rest of the way. Both would miss the booster if they did not keep it under constant observation and adjust their trajectories as necessary. As they near their quarry, these homing kill vehicles (HKVs) would observe the flame of the rocket booster and readjust their course accordingly. In the case of the HKV associated with the U.S. F-15-launched ASAT, these adjustments are made by firing one or more of fifty-six small rockets clustered around the spinning HKV and directed by an infrared telescope. A similar technique was used by the HKV tested June 10, 1984 on a Minuteman-I booster. For boost-phase intercept, the infrared sensors need not be cooled, since the rocket flame would emit short-wave IR, rather than the thermal IR that is used for homing on satellites or ICBM RVs. Nevertheless, the rocket flame is *not* the vulnerable part of the booster, and the HKV has to exercise a great deal more intelligence than is required simply to direct itself toward the booster flame. This autonomous intelligence necessary in HKVs is also a source of their vulnerability.

One can hardly object to modest investigations of railgun propulsion, but old-fashioned rockets seem to have the advantage in firing rate and in the efficiency with which they convert thermal to kinetic energy. For a perfectly staged rocket, final velocities of 4, 5, 6, and 7 times the exhaust velocity (3 km/s for normal rocket propellents) would provide payload kinetic energies of 29, 17, 9, and 4.5 percent of the initial fuel energy. The corresponding rocket masses would be 55, 148, 403, and 1000 times the homing payload of the rocket. To match the 9% energy transfer from propellant to HKV, a railgun obtaining its electrical energy from propellant at 30% efficiency would have to convert this energy to kinetic energy of the HKV with a further 30% efficiency. One must add to these advantages the very high peak-firing capability of a rocket cluster (because of the independent acceleration of each of the rockets) and the absence of investment in large masses and power supplies. It would therefore seem that railguns with a velocity below 20 kms/sec are not competitive for launching HKVs against boosters. Nor are there any indications that railguns would work well at higher speeds.

Boost-phase intercept by HKV has assets and liabilities. With a perfect telescope 10 centimeters in diameter operating at a near-IR wavelength of 3 micron (suitable for boost-phase intercept), the diffraction limit of an HKV would be 30 microradian. At 100 kilometers, this would correspond to a 3-meter diameter resolution spot. The best the HKV could do from such a distance would be to determine the center of the booster flame and guide itself to a preset "lead" in order to have the minimum final correction.

The imperfections of the HKV system could easily be exploited by the use of various offensive countermeasures. For example, boosters at high altitude could improve their survival by modest irregular accelerations. While this would reduce the rocket's maximum range slightly, it would impose unattainable maneuvering requirements on the HKV. Large foil screens carried by the booster could obscure the flame, thereby preventing homing altogether. Also, the explosion of nuclear weapons at high altitudes could provide background heat that would make more difficult the tracking of the boosters by near-IR detectors. Finally, if directed against fast-burn boosters, near-IR homing would be impeded at altitudes up to 100 kilometers by the glow of air heated by the passage of the HKV itself. Special

Appendix A: New BMD Technologies 341

filters could, however, help the HKV continue to see the booster in the presence of this hot-air background.

The principal problem an HKV would have in intercepting fast-burn boosters would be *reaching* the booster in the 40-seconds available and at altitudes of 80-kilometers or lower, where burnout would occur. For instance, in 40 seconds an 8 km/s added-speed HKV could move no further from its moving space truck in orbit at 300 kilometers (in order to have a reasonable lifetime against orbital drag) than to a 230-kilometer radius circle (at 80-kilometer altitude). Such a space truck might have to handle up to 200 ICBMs; about 400 HKVs would have to be provided by that single space truck or others at the same point. Since something like 1000 space trucks would have to be in optimum orbit in order to have an average of *one* within range of a given ICBM (for 230-kilometer range), some 400,000 HKVs would have to be in orbit in order to counter 200 fast-burn boosters launched from each of 10 launch fields (if there were no other problems). At 100 kilograms per HKV, this would amount to 40,000 tons in low near-polar orbit at a launch-cost alone of \$240 billion.

Use for Mid-Course Intercept

Once an RV is identified in mid-course and an HKV assigned, as much as 10 minutes would be available for the intercept. In this time, HKVs moving at 4 km/s could be marshaled from distances as great as 2400 kilometers. For defense of the U.S., the intercepts would take place north of 70 degrees latitude over Canada. A satellite in polar orbit has $2\times(90\text{-}70)/360 = 0.11$ probability of being above 70° north, and $2\times 2400/40{,}000 = 0.12$ of being within an additional 2400 km of 70° north. That part of Canada below 70° north occupies about 22 degrees of latitude and 60 degrees of longitude. The added probability of a polar orbit satellite being within 2400 km (22° of arc) of Canada $\simeq (60+(2\times 22)/\cos\lambda)/360 \times (22/360) = 0.027$, for a total fraction of satellites $(0.11+0.12+0.03) = 0.26$ of space trucks able to contribute their HKVs—an overestimate.

Based on the above calculation, we arrive at an absentee ratio of 4, i.e., for every space truck within 2400 kilometers of the battle area and thus able to contribute an HKV, three others could do nothing regardless of whether they were above the horizon and able to see. Hence, against 10,000 RVs (with no decoys), about 40,000 HKVs in orbit would be required, or perhaps 80,000 to provide greater reliability. If the targets were more concentrated, even fewer space trucks could participate.

These calculations assume that there would be perfect discrimination, so that no HKVs would be launched against decoys. Once discriminated, RVs would still not necessarily be vulnerable to HKVs. An RV need emit a decoy tethered at a distance of 10 meters or more only once in a while to totally confuse an HKV. If the HKV had enough intelligence to recognize that there were two or more objects in its field of view as it approached the RV, it would probably still be unable to choose the correct one. (The "centroid problem," familiar from air defense, arises in space as well.) One can imagine a command link to each HKV so that the SATKA (surveillance, acquisition, tracking, kill assessment) system could help it discriminate as it approached an RV, but there is no reason to believe that SATKA itself would be able to discriminate (or that it would survive to do so).

For mid-course intercept, HKVs would need cooled IR detectors and would have to pursue, not 1000 to 2000 boosters, but 10,000 RVs and 100,000 to one million decoys. Alternatively, a cooperative SATKA could occasionally illuminate each of the mid-course targets in turn with weak laser light to provide a short-wave (say one micron) signal for the HKV homer; this target-designating "semi-active homing" scheme is less demanding than a radar or laser homer in each HKV. With the absentee ratio of 4, however, we could need some 400,000 to 4 million HKVs in orbit to attack the decoys alone.

Use as Anti-Satellite Weapons

Kinetic energy projectiles can be extremely effective as anti-satellite weapons. Because satellite orbital velocity ranges from 3.6 km/s in GEO to 8 km/s in LEO, and because satellites are for the most part light and very vulnerable, even a 1-gram projectile in the path of a satellite could cause its destruction. As is discussed in the chapter by Kurt Gottfried and Richard Ned Lebow, the direct-ascent ASAT interceptor would work in exactly this way, interposing in the path of a satellite a non-explosive few-kg "warhead," a similar warhead equipped with a steel mesh or net to increase its effective area, or a pellet cloud.

Appendix A: New BMD Technologies 343

Against satellites in GEO, direct ascent would provide an impact velocity of 3.6 km/s or more—greater than that of the fastest anti-tank weapon available. Without defensive satellite measures, a counter-rotating satellite in GEO would provide some 7.2 km/s relative impact velocity, which would allow use of a warhead smaller by about a factor of 4 or more. Against shields, pellets could be made more lethal by proper shaping. Shaped as a sphere, a projectile might penetrate a shield 10 times as thick as its own diameter. A projectile in the form of a dart aligned with its impact velocity could penetrate 10 times the length of the dart. One could arrange to orient a payload of darts appropriately, even those mounted on an umbrella or net. We may conclude that kinetic energy weapons of the types described might play a crucial role in countering satellite-based BMD.

Use for Terminal Defense

At the end of its flight, the RV will reenter the atmosphere and move towards its target. For reasons that will become evident, HKVs would be the defensive weapons of choice at this stage.

The HKV used in a high-endoatmospheric defense would have the apparent advantage of being non-nuclear; it thus would not interfere with the defense against other RVs. It also might have a low enough cost to make intercept of non-discriminable decoys cost-effective. Neither of these advantages is likely to be sufficient, however. The offense can always "salvage fuse" to provide a megaton detonation (instead of the few kilotons typical of nuclear-armed interceptors). In addition, it would seem difficult for an HKV to match the low delivery cost of a high-altitude decoy.

The general basing requirements for terminal BMD would be much the same as those used for other phases of missile defense. Basing must insure the survivability of the defensive system itself and must make it possible for the system to acquire and destroy the ballistic missile threat effectively.

The demands on city defense would obviously be much greater than those on hard target defense. It would be very difficult to defend a city because, relative to silos, for example, buildings in the city are very soft, and people are vulnerable to fires as well as to collapsing buildings. Therefore, any enemy missile, even if exploded quite high above a city, would cause enormous damage. Modern missile technology has aggravated this problem by the introduction of maneuvering RVs: they may appear during mid-course to be aimed at a point perhaps 100 kilometers away from a city, but after entering the atmosphere, they may pursue a horizontal flight for some distance and then descend towards the city.[12] There could be many variations on this procedure. Furthermore, the use of non-nuclear armed interceptors within the atmosphere would not preclude the explosion of the RV warhead at the moment of intercept. Aside from complicating greatly the task of other sensors and interceptors in their attack on other RVs, such an explosion could devastate the urban target with an intense thermal pulse. But low-yield nuclear-armed interceptors (which do no damage to the defended cities) are themselves no guarantee against powerful enemy detonations at altitudes somewhat above those at which intercept can be accomplished—so-called "ladder-down" tactics.

Another important difference between city defense and missile silo defense is that, in the case of the silos, the strategic planner would be satisfied if only a certain fraction, say 20 percent, survived, whereas it would be totally unacceptable if only 20 percent of the population of a city were to survive. In addition, there are far more silos than cities, with each silo a more difficult and less valuable target.

Use of HKVs in Terminal Defense of Silos

Interception in the terminal phase is easier than in mid-course, provided one is satisfied to defend small, numerous, preferably hardened targets such as missile silos.

The classic "hard point" ABM system has a radar on the ground that locates and tracks incoming reentry vehicles, and then launches an interceptor against them. If one allows the incoming RV to descend to an altitude of about 10 kilometers before launching the interceptors, the incoming swarm of decoys will automatically have been sorted out. Plastic balloons will have burned up long before, and heavier decoys will have been greatly decelerated. RVs, on the other hand, will still have a velocity not far below their original velocity in space. This ability to discriminate RVs from decoys at low altitude makes it possible to defend against an ICBM attack on silos with a limited number of defensive missiles, perhaps two interceptors per RV to be destroyed. If the task is to preserve a small fraction of the silos for a short time, "preferential defense" will provide the defense with

Appendix A: New BMD Technologies 345

an additional cost leverage so long as it can conceal which silos are being defended.

Ground radars that detect and track incoming missiles will be less hardened than the ICBM silos they defend. Proposals have been made to use many very small radars for each field of missile silos, and to consider them expendable. These small radars would be sufficient if we wished to detect incoming RVs at only a short distance, say 10 kilometers.

If a radar and BMD system need only enforce a keepout range of 500 meters from the single silo it defends, and if it need do that against only a few missiles, totally different BMD approaches become feasible. One of these involves the use of silo-based mini-radars. They have come to be known as "Kleenex radars," because after one is used in combat it is discarded and another takes its place. (Others have used this name because there is only "one blow per radar.") Such single-silo defense radars could be cheap if they were located a few kilometers north of the silo, looking up, and could see the incoming RVs side-on, rather than having to see them nose-on at a distance of hundreds of kilometers. A corresponding weapon for killing the RV could be a clustered launcher for small (few-kg) unguided rockets, of which perhaps 10,000 would be launched against the point at which an RV is predicted to appear at a range of 500 meters or 1 kilometer from the silo target. Such a system has been dubbed Swarmjet, and there is little doubt that it could increase the price of a silo to the attacker from 2 RVs to 3 or 4—more if the attack were not concentrated in time. The individual Swarmjet rockets would be blown off course if fired into the high winds of the preceding nuclear explosion, but in an attack extending over hours, many RVs could be defeated by successive waves of Swarmjet interceptors. If the warheads (large pellets) are carried in a single stabilized fast-burn interceptor rocket to the intercept point at 500 or 1000 meters from the silo, immunity to high winds could be achieved and the non-nuclear silo defense extended to RVs arriving in a brief attack.

Still another approach would use the same silo-based mini-radars to identify individual approaching RVs and to detonate a nuclear explosive buried in the ground 1 kilometer north of the target silo. The column of earth so produced would bar access by RVs for a short time. A number of such explosions would produce a local pall in the upper troposphere that could bar access by accurate high-velocity RVs for 30 minutes or more. A combination of Swarmjet rockets, mentioned above, with these buried nuclear explosives could provide an effective, though temporary, defense of each missile silo being defended. Since defense of silos is presumably feasible and a far-from-perfect defense suffices, not much is gained by adding to it other layers of defense, e.g., against the boost phase or in mid-course. If money is to be spent on defending silos, it is better spent entirely on defensive installations in their immediate vicinity.

Such defenses would not be effective against attacks on cities. They also could not effectively defend a small number of valuable targets, such as command posts, against a large force, even if the targets were very hard. Over a considerable period, very many reentry vehicles could be launched against each of these valuable targets and would eventually overwhelm them.

SATKA SYSTEMS

For almost two decades, the U.S. has operated short-wave infrared telescopes on satellites in synchronous orbit with the mission of detecting the launch of ICBMs and SLBMs, whether for a peacetime test or a wartime attack. These early-warning satellites report via radio to U.S. command headquarters and would play an important role both in the decision to launch U.S. retaliatory strategic forces and in the choice of targets.

Similar satellites would play a key role in a system of boost-phase intercept or in a layered defense against ballistic missiles. For BPI, one would need a survivable, near-IR satellite system able to deal with up to 2000 simultaneous launches without missing more than a very small fraction. This system would also be required to discriminate real launches from laser excitation of the satellite sensors designed to simulate launch, and from fake dummy boosters that might be launched to attract costly and scarce BPI weapons. The satellite system must furthermore operate in the presence of laser excitation intended to desensitize or overload the warning and tracking system. In addition, the communication links must be reliable and invulnerable, perhaps even closing the loop in space so as to avoid the vulnerability of ground-based segments and any delay in human intervention. For BPI, the job of the warning satellites would largely

Appendix A: New BMD Technologies

be accomplished when one or more weapons had been assigned to their targets. With homing kill vehicles, there would be no time for re-attack, although there might be with speed-of-light directed energy weapons. In the latter case, the kill-assessment role of the system would come into play.

In BPI, each of the assigned weapons would have to have its own navigation and reference system, so that when a target is identified at given geocentric coordinates, the weapon could immediately set its long-range tracker on the booster and quickly identify and refine the required line of sight. Such identification and tracking would initially be accomplished by means of an IR telescope on the weapon itself. This telescope need have only enough diameter to receive a signal adequate for tracking and for pointing the fine sensor. However, if the directed energy of the weapon is to be concentrated on a spot of one-meter diameter, the booster itself must be *imaged* to such an accuracy. Because the offense can easily arrange for the brightness of the booster to vary in space and time, this cannot be done by "beam splitting," as is customary with high-precision anti-aircraft or anti-missile radars, but must be done by real imaging of the booster. Since the booster does not glow in the near-IR, the weapon would probably have to use a laser of moderately short wavelength and a mirror of appropriate size in order to flash-illuminate the missile often enough and with a bright enough light to provide an adequate image. Stipulating that 10 photons must enter the detecting aperture for every 10-cm resolution element on the booster, and assuming that the laser has to illuminate 200 sq m in order to have a *field* in which to locate the 1-m diameter spot, we find that at 3000-km range and for a 1-sq m receiving telescope, some 200 J of light would be required per pulse. This assumes 10 percent reflectivity of the booster. If a full 10-m diameter mirror were used in the visible or ultraviolet for this imaging task, only $\simeq$ 2 J per pulse would be required. This would be a very modest laser requirement, but a very big mirror requirement.

From synchronous orbit, and with the same mirror system as assumed for LEO, 10,000 times the imaging power would be required. Another problem is that it takes light 0.25 seconds to travel round-trip from synchronous orbit, which with 2 g random modulation of booster thrust corresponds to $\simeq$ 0.6 m jitter, even assuming otherwise perfect tracking.

Mid-course intercept must fail unless the deployment of decoys can be observed reliably, and only the real RVs held in track. Otherwise, the sensors would have to scan all of space to pick up all objects that might suddenly appear, and would continuously be expending energy and computing power tracking a million decoys. The countermeasure of using an umbrella to shield the deployment process from prying eyes seems attractive; it would be far more difficult to build the required observation capability or to defeat the umbrella that would shield against both light and radar observation. This is a classic example of an element of defense which is not "cost-effective at the margin," because the countermeasures are easier to achieve.

This section has only given a glimpse of the problems connected with integrating a BMD system. The chapter by Charles A. Zraket discusses these matters more fully.

SUMMARY EVALUATION

We find space-based BMD very unpromising as a defense against a strong offensive threat such as that posed by the Soviet Union. Even the most credible BMD, based on chemical lasers, would require staggering numbers of satellites if the Soviet Union made the modest expenditures to supplement its offensive force with 3000 fast-burn boosters.

About 1300 satellites would be needed if equipped with lasers of a power well beyond the state of the art, and with mirrors and optical systems of similar extremely high technologies, capable of 0.5 seconds retarget time. X-ray lasers, even if they could be developed, would be of doubtful effectiveness, and particle beams would probably be useless against ICBMs in boost phase.

If the defensive weapons were to be deployed in space, they would be exposed to many possible countermeasures by the potential enemy. If they were to be popped-up at the time of need, they would probably not reach the required altitude in time to be useful.

Moreover, what would happen if a layer failed catastrophically— if the anticipated attrition is not attained at all? If this occurred with the *last* layer, it is perfectly clear that the arrival of an average of 10 RVs (which *should* have been reduced to zero or one by the nominal performance of the previous layer) would result in ten RVs arriving at

Appendix A: New BMD Technologies 349

their targets. But if the *first* (BPI) layer failed, the situation would be much worse. If one is designing on the basis of the expected performance of each system, layers two, three, and four will be sized to do their job against (respectively) 150 post-boost vehicles carrying an average of (say) 5 RVs; 75 RVs and their associated decoys; and, in the terminal area, 8 to 10 RVs left over by the expected operation of the mid-course system. Should the post-boost system fail, the mid-course system would not have the firepower to achieve a continuing 90 percent attrition from the ten-fold increase in surviving warheads and decoys. And if the post-boost system has not been sized to handle total failure of the BPI, it will not work either, and the terminal defenses will have to handle 10,000 RVs instead of ten.

Furthermore, post-boost and mid-course systems are absolutely dependent on birth-to-death tracking, and *that* in turn depends on space assets that can be mined, deceived, destroyed by ASATs, and the like.

If, on the other hand, one is looking towards a BMD system that might exact a modest attrition (say 50 percent—a number that has been used by Dr. Fred Hoffman in his advocacy of the utility of a modest BMD system), then it makes no sense to provide a layered defense. In that case, one should choose the candidate layer in which the defense has the greatest (if any) advantage over the offense or in which attrition can be exacted at minimum cost, and concentrate one's modest defense efforts in that layer. For defense of ICBM silos, this would clearly be an endoatmospheric terminal defense, but for the defense of cities, soft military facilities, unique hard targets, and the like, 50 percent effectiveness does not seem useful.

Finally, the natural reaction of the Soviet Union to the U.S. development of a BMD system would likely be a further increase in its offensive capability, this being an easy way to overcome any ballistic missile defense we might put up. We therefore believe that a BMD system would intensify, rather than mitigate, the arms race, and that it would make the United States less, rather than more, secure.

ENDNOTES

[1] The radius of a GEO is 42,000 kilometers, as can easily be calculated from the gravitational attraction of the earth. We are concerned with the distance from the laser in GEO to the Soviet ICBM field, which is at a latitude of about 57°. This

distance, by simple geometry, would be 39,300 kilometers. In attacking a booster at 55° north latitude, the beam from such a GEO satellite would enter the earth's atmosphere at an angle of about 27° with the horizontal.

[2] Hans A. Bethe, Richard L. Garwin, Kurt Gottfried, Henry W. Kendall, "Space-based Ballistic Missile Defense," *Scientific American*, Oct. 1984, p. 46.

[3] C. T. Cunningham, "Critique of Systems Analysis in the OTA Study 'Directed Energy Missile Defense in Space'," DDV-84-0007, Lawrence Livermore National Laboratory, (Aug. 30, 1984).

[4] Richard L. Garwin, "How Many Orbiting Lasers for Boost-phase Intercept?" *Nature*, May 23, 1985, p. 286.

[5] Robert Jastrow, typescript for presentation to the House Republican Study Committee, Aug. 9, 1984.

[6] A more powerful approach is to use a LEO auxiliary mirror to receive the excimer laser beam reflected from the GEO satellite, and *direct* (and focus) it on the booster. A 5-m diameter mirror at 3000 km range from its target could focus 0.3 µm light to a circle 0.23 m diam, but if that same mirror were used for pointing with respect to the varying infrared image, its resolution would be 2 meters. Accordingly, we have taken a 1 m diam spot as the attainable average.

[7] E. Walbridge, in "Angle Constraint for Nuclear-pumped X-ray Laser Weapons," *Nature*, July 19, 1984, and previous authors have noted that the super-radiant beam has a contribution to angular half-width by diffraction $\alpha_d = 1.22\lambda/D$, and by geometry (from skew rays in the wire) $\alpha_g = D/L$, with λ the X-ray wavelength (about 1.2 nm for 1-keV x ray), D wire diameter, and L the wire length. For L = 2 m, the total angular beam spread is minimum at $\alpha = 1.5(\lambda/L)^{1/2}$, or about 40 microradians, at a wire diameter $D = (1.22\lambda L)^{1/2}$, or some 55 µm.

[8] Schemes for focusing X-ray lasers or obtaining greater coherence and smaller beams may be suitable for anti-satellite use, but remain to be explored.

[9] The beam spread from the original ion source or from the acceleration process itself can be reduced by a magnetic-lens telescope ("conservation of phase space" requires that the diameter be increased by the same factor that the beam divergence is reduced).

[10] A "rad" is the radiation unit commonly used to describe biological effects, namely an energy depositon of 100 erg per gram.

[11] Private communication from Sven Roosild, Nov. 19, 1984.

[12] Richard L. Garwin and Hans A. Bethe, "Anti-Ballistic-Missile Systems," *Scientific American*, March 1968, pp. 21–31.

APPENDIX B: RELEVANT DOCUMENTS

From the Address to the Nation by President Ronald Reagan March 23, 1983 "Peace and National Security"

THE SUBJECT I WANT TO DISCUSS WITH YOU, peace and national security, is both timely and important. Timely, because I've reached a decision which offers a new hope for our children in the 21st century....[1]

.... Thus far tonight I've shared with you my thoughts on the problems of national security we must face together. My predecessors in the Oval Office have appeared before you on other occasions to describe the threat posed by Soviet power and have proposed steps to address that threat. But since the advent of nuclear weapons, those steps have been increasingly directed toward deterrence of aggression through the promise of retaliation.

This approach to stability through offensive threat has worked. We and our allies have succeeded in preventing nuclear war for more than three decades. In recent months, however, my advisers, including in particular the Joint Chiefs of Staff, have underscored the necessity to break out of a future that relies solely on offensive retaliation for our security.

Over the course of these discussions, I've become more and more deeply convinced that the human spirit must be capable of rising above dealing with other nations and human beings by threatening their existence. Feeling this way, I believe we must thoroughly

---

U.S. Department of State, *Department of State Bulletin*, vol. 83, no. 2073 (Washington, D.C.: USGPO, 1983), p. 8; pp. 13–14.

examine every opportunity for reducing tensions and for introducing greater stability into the strategic calculus on both sides.

One of the most important contributions we can make is, of course, to lower the level of all arms, and particularly nuclear arms. We're engaged right now in several negotiations with the Soviet Union to bring about a mutual reduction of weapons. I will report to you a week from tomorrow my thoughts on that score. But let me just say, I'm totally committed to this course.

If the Soviet Union will join with us in our effort to achieve major arms reduction, we will have succeeded in stabilizing the nuclear balance. Nevertheless, it will still be necessary to rely on the specter of retaliation, on mutual threat. And that's a sad commentary on the human condition. Wouldn't it be better to save lives than to avenge them? Are we not capable of demonstrating our peaceful intentions by applying all our abilities and our ingenuity to achieving a truly lasting stability? I think we are. Indeed, we must.

After careful consultation with my advisers, including the Joint Chiefs of Staff, I believe there is a way. Let me share with you a vision of the future which offers hope. It is that we embark on a program to counter the awesome Soviet missile threat with measures that are defensive. Let us turn to the very strengths in technology that spawned our great industrial base and that have given us the quality of life we enjoy today.

What if free people could live secure in the knowledge that their security did not rest upon the threat of instant U.S. retaliation to deter a Soviet attack, that we could intercept and destroy strategic ballistic missiles before they reached our own soil or that of our allies?

I know this is a formidable, technical task, one that may not be accomplished before the end of this century. Yet, current technology has attained a level of sophistication where it's reasonable for us to begin this effort. It will take years, probably decades of effort on many fronts. There will be failures and setbacks, just as there will be successes and breakthroughs. And as we proceed, we must remain constant in preserving the nuclear deterrent and maintaining a solid capability for flexible response. But isn't it worth every investment necessary to free the world from the threat of nuclear war? We know it is.

In the meantime, we will continue to pursue real reductions in nuclear arms, negotiating from a position of strength that can be

Appendix B: Relevant Documents   353

ensured only by modernizing our strategic forces. At the same time, we must take steps to reduce the risk of a conventional military conflict escalating to nuclear war by improving our non-nuclear capabilities.

America does possess—now—the technologies to attain very significant improvements in the effectiveness of our conventional, non-nuclear forces. Proceeding boldly with these new technologies, we can significantly reduce any incentive that the Soviet Union may have to threaten attack against the United States or its allies.

As we pursue our goal of defensive technologies, we recognize that our allies rely upon our strategic offensive power to deter attacks against them. Their vital interests and ours are inextricably linked. Their safety and ours are one. And no change in technology can or will alter that reality. We must and shall continue to honor our commitments.

I clearly recognize that defensive systems have limitations and raise certain problems and ambiguities. If paired with offensive systems, they can be viewed as fostering an aggressive policy, and no one wants that. But with these considerations firmly in mind, I call upon the scientific community in our country, those who gave us nuclear weapons, to turn their great talents now to the cause of mankind and world peace, to give us the means of rendering these nuclear weapons impotent and obsolete.

Tonight, consistent with our obligations of the ABM treaty and recognizing the need for closer consultation with our allies, I'm taking an important first step. I am directing a comprehensive and intensive effort to define a long-term research and development program to begin to achieve our ultimate goal of eliminating the threat posed by strategic nuclear missiles. This could pave the way for arms control measures to eliminate the weapons themselves. We seek neither military superiority nor political advantage. Our only purpose—one all people share—is to search for ways to reduce the danger of nuclear war.

My fellow Americans, tonight we're launchng an effort which holds the promise of changing the course of human history. There will be risks, and results take time. But I believe we can do it. As we cross this threshold, I ask for your prayers and your support.

Thank you, good night, and God bless you.

Treaty Between the United States of America and the Union of Soviet Socialist Republics on the Limitation of Anti-Ballistic Missile Systems

Signed at Moscow May 26, 1972
Ratification advised by U.S. Senate August 3, 1972
Ratified by U.S. President September 30, 1972
Proclaimed by U.S. President October 3, 1972
Instruments of ratification exchanged October 3, 1972
Entered into force October 3, 1972

THE UNITED STATES OF AMERICA and the Union of Soviet Socialist Republics, hereinafter referred to as the Parties,

Proceeding from the premise that nuclear war would have devastating consequences for all mankind,

Considering that effective measures to limit anti-ballistic missile systems would be a substantial factor in curbing the race in strategic offensive arms and would lead to a decrease in the risk of outbreak of war involving nuclear weapons,

Proceeding from the premise that the limitation of anti-ballistic missile systems, as well as certain agreed measures with respect to the limitation of strategic offensive arms, would contribute to the creation of more favorable conditions for further negotiations on limiting strategic arms,

---

U.S. Arms Control and Disarmament Agency, *Arms Control and Disarmament Agreements: Texts and Histories of Negotiations* (Washington D.C.: USGPO, 1980), pp. 139–147; pp. 161–163.

Mindful of their obligations under Article VI of the Treaty on the Non-Proliferation of Nuclear Weapons,

Declaring their intention to achieve at the earliest possible date the cessation of the nuclear arms race and to take effective measures toward reductions in strategic arms, nuclear disarmament, and general and complete disarmament,

Desiring to contribute to the relaxation of international tension and the strengthening of trust between States,

Have agreed as follows:

Article I

1. Each Party undertakes to limit anti-ballistic missile (ABM) systems and to adopt other measures in accordance with the provisions of this Treaty.

2. Each Party undertakes not to deploy ABM systems for a defense of the territory of its country and not to provide a base for such a defense, and not to deploy ABM systems for defense of an individual region except as provided for in Article III of this Treaty.

Article II

1. For the purpose of this Treaty an ABM system is a system to counter strategic ballistic missiles or their elements in flight trajectory, currently consisting of:

(a) ABM interceptor missiles, which are interceptor missiles constructed and deployed for an ABM role, or of a type tested in an ABM mode;

(b) ABM launchers, which are launchers constructed and deployed for launching ABM interceptor missiles; and

(c) ABM radars, which are radars constructed and deployed for an ABM role, or of a type tested in an ABM mode.

2. The ABM system components listed in paragraph 1 of this Article include those which are:

(a) operational;
(b) under construction;
(c) undergoing testing;
(d) undergoing overhaul, repair or conversion; or
(e) mothballed.

Article III

Each Party undertakes not to deploy ABM systems or their components except that:

(a) within one ABM system deployment area having a radius of one hundred and fifty kilometers and centered on the Party's national capital, a Party may deploy: (1) no more than one hundred ABM launchers and no more than one hundred ABM interceptor missiles at launch sites, and (2) ABM radars within no more than six ABM radar complexes, the area of each complex being circular and having a diameter of no more than three kilometers; and

(b) within one ABM system deployment area having a radius of one hundred and fifty kilometers and containing ICBM silo launchers, a Party may deploy: (1) no more than one hundred ABM launchers and no more than one hundred ABM interceptor missiles at launch sites, (2) two large phased-array ABM radars comparable in potential to corresponding ABM radars operational or under construction on the date of signature of the Treaty in an ABM system deployment area containing ICBM silo launchers, and (3) no more than eighteen ABM radars each having a potential less than the potential of the smaller of the above-mentioned two large phased-array ABM radars.

Article IV

The limitations provided for in Article III shall not apply to ABM systems or their components used for development or testing, and located within current or additionally agreed test ranges. Each Party may have no more than a total of fifteen ABM launchers at test ranges.

Article V

1. Each Party undertakes not to develop, test, or deploy ABM systems or components which are sea-based, air-based, space-based, or mobile land-based.

2. Each Party undertakes not to develop, test, or deploy ABM launchers for launching more than one ABM interceptor missile at a time from each launcher, not to modify deployed launchers to provide them with such a capability, not to develop, test, or deploy automatic or semi-automatic or other similar systems for rapid reload of ABM launchers.

Article VI

To enhance assurance of the effectiveness of the limitations on ABM systems and their components provided by the Treaty, each Party undertakes:

(a) not to give missiles, launchers, or radars, other than ABM interceptor missiles, ABM launchers, or ABM radars, capabilities to counter strategic ballistic missiles or their elements in flight trajectory, and not to test them in an ABM mode; and

(b) not to deploy in the future radars for early warning of strategic ballistic missile attack except at locations along the periphery of its national territory and oriented outward.

Article VII

Subject to the provisions of this Treaty, modernization and replacement of ABM systems or their components may be carried out.

Article VIII

ABM systems or their components in excess of the numbers or outside the areas specified in this Treaty, as well as ABM systems or their components prohibited by this Treaty, shall be destroyed or dismantled under agreed procedures within the shortest possible agreed period of time.

Article IX

To assure the viability and effectiveness of this Treaty, each Party undertakes not to transfer to other States, and not to deploy outside its national territory, ABM systems or their components limited by this Treaty.

Article X

Each Party undertakes not to assume any international obligations which would conflict with this Treaty.

Article XI

The Parties undertake to continue active negotiations for limitations on strategic offensive arms.

Article XII

1. For the purpose of providing assurance of compliance with the provisions of this Treaty, each Party shall use national technical means of verification at its disposal in a manner consistent with generally recognized principles of international law.

2. Each Party undertakes not to interfere with the national technical means of verification of the other Party operating in accordance with paragraph 1 of this Article.

3. Each Party undertakes not to use deliberate concealment measures which impede verification by national technical means of compliance with the provisions of this Treaty. This obligation shall not require changes in current construction, assembly, conversion, or overhaul practices.

Article XIII

1. To promote the objectives and implementation of the provisions of this Treaty, the Parties shall establish promptly a Standing Consultative Commission, within the framework of which they will:

(a) consider questions concerning compliance with the obligations assumed and related situations which may be considered ambiguous;

(b) provide on a voluntary basis such information as either Party considers necessary to assure confidence in compliance with the obligations assumed;

(c) consider questions involving unintended interference with national technical means of verification;

(d) consider possible changes in the strategic situation which have a bearing on the provisions of this Treaty;

(e) agree upon procedures and dates for destruction or dismantling of ABM systems or their components in cases provided for by the provisions of this Treaty;

(f) consider, as appropriate, possible proposals for further increasing the viability of this Treaty; including proposals for amendments in accordance with the provisions of this Treaty;

(g) consider, as appropriate, proposals for further measures aimed at limiting strategic arms.

2. The Parties through consultation shall establish, and may amend as appropriate, Regulations for the Standing Consultative Commission governing procedures, composition and other relevant matters.

Article XIV

1. Each Party may propose amendments to this Treaty. Agreed amendments shall enter into force in accordance with the procedures governing the entry into force of this Treaty.

2. Five years after entry into force of this Treaty, and at five-year intervals thereafter, the Parties shall together conduct a review of this Treaty.

Article XV

1. This Treaty shall be of unlimited duration.

2. Each Party shall, in exercising its national sovereignty, have the right to withdraw from this Treaty if it decides that extraordinary events related to the subject matter of this Treaty have jeopardized its supreme interests. It shall give notice of its decision to the other Party six months prior to withdrawal from the Treaty. Such notice shall include a statement of the extraordinary events the notifying Party regards as having jeopardized its supreme interests.

Article XVI

1. This Treaty shall be subject to ratification in accordance with the constitutional procedures of each Party. The Treaty shall enter into force on the day of the exchange of instruments of ratification.

2. This Treaty shall be registered pursuant to Article 102 of the Charter of the United Nations.

DONE at Moscow on May 26, 1972, in two copies, each in the English and Russian languages, both texts being equally authentic.

FOR THE UNITED STATES OF AMERICA

FOR THE UNION OF SOVIET SOCIALIST REPUBLICS

President of the United States of America

General Secretary of the Central Committee of the CPSU

*   *   *

Agreed Statements, Common Understandings, and Unilateral Statements Regarding the Treaty Between the United States of America and the Union of Soviet Socialist Republics on the Limitation of Anti-Ballistic Missiles

1. Agreed Statements

The document set forth below was agreed upon and initialed by the Heads of the Delegations on May 26, 1972 (letter designations added);

AGREED STATEMENTS REGARDING THE TREATY BETWEEN THE UNITED STATES OF AMERICA AND THE UNION OF SOVIET SOCIALIST REPUBLICS ON THE LIMITATION OF ANTI-BALLISTIC MISSILE SYSTEMS

[A]

The Parties understand that, in addition to the ABM radars which may be deployed in accordance with subparagraph (a) of Article III of the Treaty, those non-phased-array ABM radars operational on the date of signature of the Treaty within the ABM system deployment area for defense of the national capital may be retained.

[B]

The Parties understand that the potential (the product of mean emitted power in watts and antenna area in square meters) of the smaller of the two large phased-array ABM radars referred to in subparagraph (b) of Article III of the Treaty is considered for purposes of the Treaty to be three million.

[C]

The Parties understand that the center of the ABM system deployment area centered on the national capital and the center of the ABM system deployment area containing ICBM silo launchers for each Party shall be separated by no less than thirteen hundred kilometers.

[D]

In order to insure fulfillment of the obligation not to deploy ABM systems and their components except as provided in Article III of the Treaty, the Parties agree that in the event ABM systems based on other physical principles and including components capable of substituting for ABM interceptor missiles, ABM launchers, or ABM radars are created in the future, specific limitations on such systems and their components would be subject to discussion in accordance with Article XIII and agreement in accordance with Article XIV of the Treaty.

[E]

The Parties understand that Article V of the Treaty includes obligations not to develop, test or deploy ABM interceptor missiles for the delivery by each ABM interceptor missile of more than one independently guided warhead.

[F]

The Parties agree not to deploy phased-array radars having a potential (the product of mean emitted power in watts and antenna area in square meters) exceeding three million, except as provided for in Articles III, IV, and VI of the Treaty, or except for the purposes of tracking objects in outer space or for use as national technical means of verification.

[G]

The parties understand that Article IX of the Treaty includes the obligation of the U.S. and the USSR not to provide to other States technical descriptions or blue prints specially worked out for the construction of ABM systems and their components limited by the Treaty.

2. Common Understandings

Common understanding of the Parties on the following matters was reached during the negotiations:

A. *Location of ICBM Defenses*

The U.S. Delegation made the following statement on May 26, 1972:

Article III of the ABM Treaty provides for each side one ABM system deployment area centered on its national capital and one ABM system deployment area containing ICBM silo launchers. The two sides have registered agreement on the following statement: "The Parties understand that the center of the ABM system deployment area centered on the national capital and the center of the ABM system deployment area containing ICBM silo launchers for each Party shall be separated by no less than thirteen hundred kilometers." In this connection, the U.S. side notes that its ABM system deployment area for defense of ICBM silo launchers, located west of the Mississippi River, will be centered in the Grand Forks ICBM silo launcher deployment area. (See Agreed Statement [C].)

B. *ABM Test Ranges*

The U.S. Delegation made the following statement on April 26, 1972:

Article IV of the ABM Treaty provides that "the limitations provided for in Article III shall not apply to ABM systems or their components used for development or testing, and located within current or additionally agreed test ranges." We believe it would be useful to assure that there is no misunderstanding as to current ABM test ranges. It is our understanding that ABM test ranges encompass the area within which ABM components are located for test purposes. The current U.S. ABM test ranges are at White Sands, New Mexico, and at Kwajalein Atoll, and the current Soviet ABM test range is near Sary Shagan in Kazakhstan. We consider that non-phased array radars of types used for range safety or instrumentation purposes may be located outside of ABM test ranges. We interpret the reference in Article IV to "additionally agreed test ranges" to mean that ABM components will not be located at any other test ranges without prior agreement between our Governments that there will be such additional ABM test ranges.

On May 5, 1972, the Soviet Delegation stated that there was a common understanding on what ABM test ranges were, that the use of the types of non-ABM radars for range safety or instrumentation was not limited under the Treaty, that the reference in Article IV to "additionally agreed" test ranges was sufficiently clear, and that national means permitted identifying current test ranges.

C. *Mobile ABM Systems*

On January 29, 1972, the U.S. Delegation made the following statement:

Article V(1) of the Joint Draft Text of the ABM Treaty includes an undertaking not to develop, test, or deploy mobile land-based ABM systems and their components. On May 5, 1971, the U.S. side indicated that, in its view, a prohibition on deployment of mobile ABM systems and components would rule out the deployment of ABM launchers and radars which were not permanent fixed types. At that time, we asked for the Soviet view of this interpretation. Does the Soviet side agree with the U.S. side's interpretation put forward on May 5, 1971?

On April 13, 1972, the Soviet Delegation said there is a general common understanding on this matter.

D. *Standing Consultative Commission*

Ambassador Smith made the following statement on May 22, 1972:

The United States proposes that the sides agree that, with regard to initial implementation of the ABM Treaty's Article XIII on the Standing Consultative Commission (SCC) and of the consultation Articles to the Interim Agreement on offensive arms and the Accidents Agreement,* agreement establishing the SCC will be worked out early in the follow-on SALT negotiations; until that is completed, the following arrangements will prevail: when SALT is in session, any consultation desired by either side under these Articles can be carried out by the two SALT Delegations; when SALT is not in session, *ad hoc* arrangements for any desired consultations under these Articles may be made through diplomatic channels.

Minister Semenov replied that, on an *ad referendum* basis, he could agree that the U.S. statement corresponded to the Soviet understanding.

E. *Standstill*

On May 6, 1972, Minister Semenov made the following statement:

In an effort to accommodate the wishes of the U.S. side, the Soviet Delegation is prepared to proceed on the basis that the two sides will in fact

---

*See Article 7 of Agreement to Reduce the Risk of Outbreak of Nuclear War Between the United States of America and the Union of Soviet Socialist Republics, signed Sept. 30, 1971.

observe the obligations of both the Interim Agreement and the ABM Treaty beginning from the date of signature of these two documents.

In reply, the U.S. Delegation made the following statement on May 20, 1972:

The U.S. agrees in principle with the Soviet statement made on May 6 concerning observance of obligations beginning from date of signature but we would like to make clear our understanding that this means that, pending ratification and acceptance, neither side would take any action prohibited by the agreements after they had entered into force. This understanding would continue to apply in the absence of notification by either signatory of its intention not to proceed with ratification or approval.

The Soviet Delegation indicated agreement with the U.S. statement.

3. Unilateral Statements

The following noteworthy unilateral statements were made during the negotiations by the United States Delegation:

A. *Withdrawal from the ABM Treaty*

On May 9, 1972, Ambassador Smith made the following statement:

The U.S. Delegation has stressed the importance the U.S. Government attaches to achieving agreement on more complete limitations on strategic offensive arms, following agreement on an ABM Treaty and on an Interim Agreement on certain measures with respect to the limitation of strategic offensive arms. The U.S. Delegation believes that an objective of the follow-on negotiations should be to constrain and reduce on a long-term basis threats to the survivability of our respective strategic retaliatory forces. The USSR Delegation has also indicated that the objectives of SALT would remain unfulfilled without the achievement of an agreement providing for more complete limitations on strategic offensive arms. Both sides recognize that the initial agreements would be steps toward the achievement of more complete limitations on strategic arms. If an agreement providing for more complete strategic offensive arms limitations were not achieved within five years, U.S. supreme interests could be jeopardized. Should that occur, it would constitute a basis for withdrawal from the ABM Treaty. The U.S. does not wish to see such a situation occur, nor do we believe that the USSR does. It is because we wish to prevent such a situation that we emphasize the importance the U.S. Government attaches to achievement of more complete limitations on strategic offensive arms. The U.S. Executive will inform the

Congress, in connection with Congressional consideration of the ABM Treaty and the Interim Agreement, of this statement of the U.S. position.

B. *Tested in ABM Mode*

On April 7, 1972, the U.S. Delegation made the following statement:

Article II of the Joint Text Draft uses the term "tested in an ABM mode," in defining ABM components, and Article VI includes certain obligations concerning such testing. We believe that the sides should have a common understanding of this phrase. First, we would note that the testing provisions of the ABM Treaty are intended to apply to testing which occurs after the date of signature of the Treaty, and not to any testing which may have occurred in the past. Next, we would amplify the remarks we have made on this subject during the previous Helsinki phrase by setting forth the objectives which govern the U.S. view on the subject, namely, while prohibiting testing of non-ABM components for ABM purposes: not to prevent testing of ABM components, and not to prevent testing of non-ABM components for non-ABM purposes. To clarify our interpretation of "tested in an ABM mode," we note that we would consider a launcher, missile or radar to be "tested in an ABM mode" if, for example, any of the following events occur: (1) a launcher is used to launch an ABM interceptor missile, (2) an interceptor missile is flight tested against a target vehicle which has a flight trajectory with characteristics of a strategic ballistic missile flight trajectory, or is flight tested in conjunction with the test of an ABM interceptor missile or an ABM radar at the same test range, or is flight tested to an altitude inconsistent with interception of targets against which air defenses are deployed, (3) a radar makes measurements on a cooperative target vehicle of the kind referred to in item (2) above during the reentry portion of its trajectory or makes measurements in conjunction with the test of an ABM interceptor missile or an ABM radar at the same test range. Radars used for purposes such as range safety or instrumentation would be exempt from application of these criteria.

C. *No-Transfer Article of ABM Treaty*

On April 18, 1972, the U.S. Delegation made the following statement:

In regard to this Article [IX], I have a brief and I believe self-explanatory statement to make. The U.S. side wishes to make clear that the provisions of this Article do not set a precedent for whatever provision may be considered for a Treaty on Limiting Strategic Offensive Arms. The question of transfer of strategic offensive arms is a far more complex issue, which may require a different solution.

D. *No Increase in Defense of Early-Warning Radars*

On July 28, 1970, the U.S. Delegation made the following statement:

Since Hen House radars [Soviet ballistic missile early-warning radars] can detect and track ballistic missile warheads at great distances, they have a significant ABM potential. Accordingly, the U.S. would regard any increase in the defenses of such radars by surface-to-air missiles as inconsistent with an agreement.

*   *   *

Protocol to the Treaty Between the United States of America and the Union of Soviet Socialist Republics on the Limitation of Anti-Ballistic Missile Systems

Signed at Moscow July 3, 1974
Ratification advised by U.S. Senate November 10, 1975
Ratified by U.S. President March 19, 1976
Instruments of ratification exchanged May 24, 1976
Proclaimed by U.S. President July 6, 1976
Entered into force May 24, 1976

THE UNITED STATES OF AMERICA and the Union of Soviet Socialist Republics, hereinafter referred to as the Parties,

Proceeding from the Basic Principles of Relations between the United States of America and the Union of Soviet Socialist Republics signed on May 29, 1972,

Desiring to further the objectives of the Treaty between the United States of America and the Union of Soviet Socialist Republics on the Limitation of Anti-Ballistic Missile Systems signed on May 26, 1972, hereinafter referred to as the Treaty,

Reaffirming their conviction that the adoption of further measures for the limitation of strategic arms would contribute to strengthening international peace and security,

Proceeding from the premise that further limitation of anti-ballistic missile systems will create more favorable conditions for the completion of work on a permanent agreement on more complete measures for the limitation of strategic offensive arms,

Have agreed as follows:

Article I

1. Each Party shall be limited at any one time to a single area out of the two provided in Article III of the Treaty for deployment of anti-ballistic missile (ABM) systems or their components and accordingly shall not exercise its right to deploy an ABM system or its components in the second of the two ABM system deployment areas permitted by Article III of the Treaty, except as an exchange of one permitted area for the other in accordance with Article II of this Protocol.

2. Accordingly, except as permitted by Article II of this Protocol: the United States of America shall not deploy an ABM system or its components in the area centered on its capital, as permitted by Article III(a) of the Treaty, and the Soviet Union shall not deploy an ABM system or its components in the deployment area of intercontinental ballistic missile (ICBM) silo launchers as permitted by Article III(b) of the Treaty.

Article II

1. Each Party shall have the right to dismantle or destroy its ABM system and the components thereof in the area where they are presently deployed and to deploy an ABM system or its components in the alternative area permitted by Article III of the Treaty, provided that prior to initiation of construction, notification is given in accord with the procedure agreed to in the Standing Consultative Commission, during the year beginning October 3, 1977 and ending October 2, 1978, or during any year which commences at five year intervals thereafter, those being the years for periodic review of the Treaty, as provided in Article XIV of the Treaty. This right may be exercised only once.

2. Accordingly, in the event of such notice, the United States would have the right to dismantle or destroy the ABM system and its components in the deployment area of ICBM silo launchers and to deploy an ABM system or its components in an area centered on its capital, as permitted by Article III(a) of the Treaty, and the Soviet Union would have the right to dismantle or destroy the ABM system and its components in the area centered on its capital and to deploy an ABM system or its components in an area con-

taining ICBM silo launchers, as permitted by Article III(b) of the Treaty.

3. Dismantling or destruction and deployment of ABM systems or their components and the notification thereof shall be carried out in accordance with Article VIII of the ABM Treaty and procedures agreed to in the Standing Consultative Commission.

Article III

The rights and obligations established by the Treaty remain in force and shall be complied with by the Parties except to the extent modified by this Protocol. In particular, the deployment of an ABM system or its components within the area selected shall remain limited by the levels and other requirements established by the Treaty.

Article IV

This Protocol shall be subject to ratification in accordance with the constitutional procedures of each Party. It shall enter into force on the day of the exchange of instruments of ratification and shall thereafter be considered an integral part of the Treaty.

DONE at Moscow on July 3, 1974, in duplicate, in the English and Russian languages, both texts being equally authentic.

For the United States of America:
RICHARD NIXON
President of the United States of America

For the Union of Soviet Socialist Republics:
L.I. BREZHNEV
General Secretary of the Central Committee of the CPSU

Contributors

Christoph Bertram, born in 1937 in Kiel, Germany, is the chief political editor for the German weekly *Die Zeit*. He was director of the International Institute for Strategic Studies in London from 1974 to 1982. His recent articles include "Political Implications of the Theater Nuclear Balance," in B. Blechman, ed., *Rethinking the U.S. Strategic Posture* (1982) and "Europe and America in 1983," in *Foreign Affairs* magazine (1984).

Hans A. Bethe, born in 1906 in Strasbourg, France, is Professor Emeritus of physics at Cornell University. A Nobel laureate in physics, Dr. Bethe was director of theoretical physics for the Manhattan Project during World War II. He has written widely on defense and technology issues and is a consultant to the Los Alamos and Lawrence Livermore Laboratories.

Jeffrey Boutwell, born in 1950 in Boston, Massachusetts, is a staff associate at the Committee for International Security Studies of the American Academy of Arts and Sciences, and adjunct research fellow at the Center for Science and International Affairs, Harvard University. He is the author of *Nuclear Weapons and the German Dilemma* (forthcoming from Cornell University Press) and co-editor of *The Nuclear Confrontation in Europe* (Croom Helm, 1985).

Ashton B. Carter, born in 1954 in Philadelphia, Pennsylvania, is assistant professor of public policy at the John F. Kennedy School of Government, Harvard University, and assistant director of the Kennedy School's Center for Science and International Affairs. Dr. Carter is co-editor of *Ballistic Missile Defense* (1984), and the author of "MX Missile Basing" and "Directed Energy Missile Defense in Space" for the Office of Technology Assessment, U.S. Congress.

Abram Chayes, born in 1922 in Chicago, Illinois, is Felix Frankfurter Professor of Law, Harvard University. Professor Chayes served as legal advisor to the U.S. Department of State (1961–1964). Among his publications are *ABM: An Evaluation of the Decision to Deploy an Anti-Ballistic Missile System* (1969) and *International Arrangements for Nuclear Fuel Reprocessing* (1977).

Antonia Handler Chayes, born in 1929 in New York City, is a partner of the Boston law firm Csaplar & Bok and a fellow of Harvard University's Center for International Affairs. Dr. Chayes has also served as under secretary of the U.S. Air Force (1979–1981). Prior to that, she was dean of Jackson College and an associate professor of political science (1968–1970). The author of numerous articles, her most recent book is *The Corporate Legal Function* (1985).

Alexander Flax, born in 1921 in New York City, is President Emeritus of the Institute for Defense Analyses in Arlington, Virginia. Dr. Flax formerly served as the Air Force's chief scientist and assistant secretary for research and development.

Richard L. Garwin, born in 1928 in Cleveland, Ohio, is an IBM fellow at the Thomas J. Watson Research Center and holds teaching and research positions at Columbia, Cornell, and Harvard Universities. Dr. Garwin served on the president's Science Advisory Committee from 1962 to 1965 and from 1969 to 1972. Formerly a member of the Defense Science Board (1966–1969), he has been an active consultant to the U.S. government and its agencies since the 1950s.

Kurt Gottfried, born in 1929 in Vienna, Austria, is professor of physics at Cornell University. A member of the board of directors of the Union of Concerned Scientists, he has served on the High Energy Physics Advisory Panel of the Department of Energy and the National Science Foundation.

Donald L. Hafner, born in 1944 in New York City, is associate professor of political science at Boston College and has written extensively on arms control and anti-satellite weapons. As a member of the U.S. Arms Control and Disarmament Agency in 1977–78, he served as an adviser with the U.S. SALT Delegation in Geneva and as an analyst with the National Security Council's SALT and ASAT Working Groups.

David Holloway, born in 1943 in Dublin, Ireland, is a senior research associate for international security and arms control at Stanford University, and a reader in politics at the University of Edinburgh. Recent publications include *The Soviet Union and The Arms Race* (1983) and *The Reagan Strategic Defense Initiative: A Technical, Political, and Arms Control Assessment* (1985), which he co-authored with Sidney D. Drell and Philip J. Farley.

Richard Ned Lebow, born in 1942 in Budapest, Hungary, is professor of government and director of the Peace Studies Program at Cornell University. Dr. Lebow was professor of strategy at the U.S. National War College, and formerly scholar-in-residence at the Central Intelligence

Agency. He is the author of many books and articles on international politics, most recently, *Between War and Peace: The Nature of International Crisis* (1984).

F. A. Long, born in 1910 in Great Falls, Montana, is Professor Emeritus of Chemistry and of Science and Society at Cornell University's Program on Science, Technology, and Society. He was the assistant director for science and technology of the U.S. Arms Control and Disarmament Agency and participated in the 1963 negotiations in Moscow that resulted in the 1963 Limited Test Ban Treaty. He is co-editor of *Arms, Defense Policy, and Arms Control* (1975), *The Genesis of New Weapons: Decision Making for Military R&D* (1980), and *Appropriate Technology and Social Values: A Critical Appraisal* (1980).

George Rathjens, born in 1925 in Fairbanks, Alaska, is a professor of political science at the Massachusetts Institute of Technology. He has served as chief scientist and deputy director of the Defense Department's Advanced Research Projects Agency, as special assistant to the director of the U.S. Arms Control and Disarmament Agency, and as director of the Institute for Defense Analyses' Systems Evaluation Division. He is the author of numerous articles and co-edited *Arms, Defense Policy, and Arms Control* (1975).

Jack Ruina, born in 1923 in Rypin, Poland, is professor of electrical engineering at the Massachusetts Institute of Technology. The former director of the Defense Department's Advanced Research Projects Agency, Dr. Ruina has written extensively on issues of defense technology.

Eliot Spitzer, born in 1959 in New York City, recently graduated from Harvard Law School. He is currently a law clerk to the Honorable Robert W. Sweet, federal district judge, southern district of New York.

Paul Stares, born in 1955 in London, England, is a research associate at the Brookings Institution. Formerly a Rockefeller International Relations and NATO Fellow, he is the author of several books including the upcoming *The Militarization of Space* (1985; Cornell University Press). Dr. Stares is co-editor of *The Exploitation of Space* (1985; Butterworths, Guildford).

John C. Toomay, born in 1922 in Ontario, California, is a retired Major General of the U.S. Air Force. A member of the U.S. Air Force Scientific Advisory Board, he has worked extensively in research and development for U.S. strategic systems. He was a member of the Defensive Technologies Study Team established to advise President Reagan on strategic defense initiatives and is the author of numerous articles on defense and strategic technology.

Gerold Yonas, born in 1939 in Cleveland, Ohio, was director of pulsed power sciences at Sandia National Laboratories, Albuquerque, New Mexico, and a member of the Defensive Technologies Study Team established to advise President Reagan on strategic defense initiatives. Shortly after writing this chapter, Dr. Yonas was appointed chief scientist of the Strategic Defense Initiative Organization at the Department of Defense.

Herbert F. York, born in 1921 in Rochester, New York, is director of the Institute on Global Conflict and Cooperation at the University of California, San Diego, where he is also a professor of physics. A former member of the president's Science Advisory Committee and former Defense Department official, he served as the chief negotiator on the 1963 Limited Test Ban Treaty as well as participating in the delegation on anti-satellite arms control with the Soviet Union, 1978–1979.

Charles A. Zraket, born in 1924 in Lawrence, Massachusetts, is executive vice president and chief operating officer of MITRE Corporation, in Bedford, Massachusetts. He is also a trustee of the Hudson Institute, a member of the board of overseers for the Center for Naval Analyses, and consultant to the Department of Defense. His most recent publication is "Strategic Command, Control, Communications, and Intelligence," *Science* magazine (June 1984).

Index

ABM-1 system, 38–39, 41
ABM (anti-ballistic missile) systems, see BMD
ABM treaty (1972), 39, 134, 172, 186–88, 193–218, 355–70
   ASATs and future of, 163, 167, 186–87
   "breakout" from, 47, 48, 74, 181, 185, 209, 213, 240, 251, 263–64, 272, 301, 304, 305
   compliance body of, 201–14, 264, 309
   "development" as understood in, 201–3, 215
   dual-purpose technologies and, 205–10
   European attitude toward, 43, 289–90
   exclusion of exotic technologies from, 198–201
   issues in interpretation of, 201–10
   legal context of, 193–218
   principle rules relevant to SDI in, 214–15
   "research" as understood in, 48–50, 134, 201–3, 215, 298
   SDI and, 172, 194–195, 198–201, 214–16, 252–53, 308
   "testing" as understood in, 201–2, 205
   text of, 355–70
   violations and, 75, 194–95, 198, 304
Abrahamson, James, 69, 86
"absentee problem," in BMD deployment, 63, 318–19, 324–25, 341
accelerators, particle, 61–62, 335–37
accidental missile launches, 39, 232, 240, 251, 252, 264
Accident Measures Agreement (1971), 210
Acquisition Tracking and Pointing System (ATP), 204–5
aerosol clouds, 57, 58, 66

airborne warning and control system (AWACS), 141
aircraft, 20, 34, 38, 139, 152, 158, 179–80
   see also bombers, manned
air defense, 37–38
   "centroid problem" and, 342
   cost-exchange ratios and, see cost-exchange ratios
   early system of, 119–20
   BMD as natural extension of, 39–40, 257
   SAGE, 119–20
   Soviet, 37, 40, 43, 180, 221–22, 248, 263, 300
Air Force, U.S., 19, 138, 152
   ASAT program of, 151–52
   early BMD responsibilities of, 34
   Sentinel system and, 42
Airland bridge, 157
air-launched miniature vehicles, 135, 151–52, 174–75
ALCMs (air-launched cruise missiles), 21–22, 265, 300
Allen, Lew, 151
ALPHA laser, 204
Andropov, Yuri V., 265–66, 268
angular spread of particle beam, 335
anti-matter beams, 337–38
anti-satellites, see ASAT arms control; ASATs
"Anti-Satellite Weapons: Weighing the Risks" (Gottfried and Lebow), 147–70
anti-simulation technique, 45–46
anti-submarine warfare (ASW), 129, 141–42
anti-tactical ballistic missiles (ATBMs), 75, 97, 206, 208–10, 309
Ariane 5 rocket booster, 286

Index

arms-control agreements:
  bilateral, 193–94
  NTM of verification of, 197, 203, 206, 210–11
  reconnaissance in verification of, 21, 163–64
  see also specific agreements and treaties
Arms Control and Disarmament Agency (ACDA), 200–201, 203
arms-control negotiations:
  ABM treaty as foundation for, 193–95, 308–9
  ASATs in, 24–25, 135–36, 163–68, 188, 275, 308–9
  BMD race and, 272, 276, 277, 308
  directed energy ABMs and, 289
  hard-point defenses and, 225
  1978–79 talks on ASATs, 163, 309
  SDI and, 101–3, 185
  SDI as leverage in, 273–74, 276–77, 302–3, 307
  space militarization and, 274
  U.S. vs. Soviet objectives in, 163, 274–76
  whole-country defense and, 233–34
  see also ASAT arms control; deterrence; Reagan SDI policy, *specific negotiations and treaties*
Army, U.S.:
  early BMD responsibilities of, 34
  Homing Overlay Experiment of, 49, 58, 152, 165, 174–75, 207
  Project Orbiter of, 19, 131
  terminal defense developed by, 305
artificial intelligence, 120–21, 339
Artsimovich, Lev, 267
ASAT arms control, 24, 163–68, 173, 206, 309–10
  deterrence and, 113, 147, 164
  directed energy weapons and, 166
  SDI and, 135, 172, 175–76, 185, 253, 275–76
ASATs (anti-satellite weapons), 22–25, 75, 127, 144–45, 147–70, 171–89
  ABM treaty and future of, 163, 186–87, 206
  air-launched miniature vehicle, 135, 151–52, 174–75
  in arms-control negotiations, 24–25, 135–36, 163–68, 188, 275, 308–9
  BMD overlap with, 115, 166–67, 174–79, 273
  in conventional war, 155–60, 172
  crisis stability and, 147–48, 162–65, 168
  current, 85, 150–52, 174–76, 178
  dedicated vs. residual, 143, 164–68, 179
  deterrence and, 113, 147, 164
  directed energy weapons as, 142, 152–53, 160, 162, 173

early-warning capabilities and, 147, 160–61, 163, 168
  hypothetical examples in upgrading of, 182–84
  IR homing, 62, 152, 154
  in nuclear vs. conventional wars, 160
  orbital bombardment systems and, 18, 22–23, 132–33
  performance requirements of, 173–74
  pop-up laser as, 177–78
  in post-attack assessment, 162
  realistic conditions for treaties on, 165–68
  role of, defined, 172–73
  satellite orbits and, 148–50
  SDI and, 93, 135, 271–72, 275–76
  sensors and, 67, 178–79
  Soviet, 23–24, 67, 75, 113, 134, 150–52, 174, 271–72
  test moratorium by USSR of, 135, 163
  upgrading problem of, 171, 176–77, 179–84
  U.S.-Soviet negotiations on, 135
  see also HKVs
"Assessing the President's Vision: The Fletcher, Miller, and Hoffman Panels" (Hafner), 91–107
Atlas D missile, 34
atom-smashers, 61–62, 335–37
"automated response," 121–22
automatic programming, 120–21

B-1 bomber, 221, 248, 300
B-52 bomber, 248
Backfire bomber, 158
ballistic missile boost intercept (BAMBI) system, 49
ballistic missile defense, *see* BMD
"Ballistic Missile Defense: Concepts and History" (Flax), 33–52
Ballistic Missile Defense Program (U.S. Army), 152
ballistic missiles, 19, 30, 33
  flights phases of, *see* flight, ballistic missile
  see also specific missiles
balloons, aluminized, 45–46, 57, 58, 66
Batitskii, P. F., 262
battle management systems, 65, 158, 316
  computers and, 84, 322–23
  real-time, 158–59
"beam splitting," 327, 347
Bertram, Christoph, 279–96
Bethe, Hans A., 53–71, 313–50
"Big Bird" satellite system, 136
birth-to-death tracking, 58, 82, 118–19, 349

blackout, radar, 39, 46, 66, 67, 225, 264
BMD (ballistic missile defense):
  ABM treaty and, *see* ABM treaty
  against accidental launches, 39, 232, 240, 251, 252, 264
  ASAT arms control and, *see* ASAT arms control
  ASAT overlap with, 115, 166–67, 174–84, 186, 188, 273
  ASATs vs., 172–73, 185–86
  command-and-control of, *see* command-and-control system
  cost-effectiveness of, 40, 51, 114, 124, 287, 300–301, 309
  cost-exchange ratios in, *see* cost-exchange ratios
  countermeasures to, *see* countermeasures to BMD
  decoys and, *see* decoys
  DEWs in, *see* DEWs
  homing kill (kinetic energy) vehicles in, *see* HKVs
  inception of, 33–34, 257
  legal issues and, 193–218
  modernization cycles in, 65–68, 77–78, 94, 100
  multilayered, *see* multilayered BMD systems
  new technologies in, 53–90, 127–45, 313–50
  perfect defense issue in, 53, 60, 117, 242, 243, 244–45, 288, 290, 298, 299–301
  performance requirements of, 114, 117, 122–25, 173–74, 242–47, 288, 290, 298, 299–301, 320–21
  platforms, 18, 63–64, 67, 115
  research and development of, 19, 34–39, 48–50, 58, 115, 131, 152, 165, 172, 174–75, 180–81, 184–85, 201–3, 207, 215, 298, 309
  SATKA systems in, 342, 346–48
  self-defense capability required in, 110
  sensors in, *see* sensors
  Soviet, 37–39, 46–48, 58, 113, 174, 194–95. 258, 262–65, 272, 306
  Soviet views on, 231–34, 240–42, 254–55, 263, 265–73, 276
  strategic instability and, 51, 239–55, 281, 306–7
  systems perspective on, 115–16, 123–25
  third-country attacks and, 36, 39, 41, 232, 240, 250–51, 252, 261, 264
  U.S. programs, reviewed, 34–37, 39–46, 48–52
  see also SDI
BMD, space-based, 51–52, 63–65, 172, 232–34, 348–49

"BMD and Strategic Instability" (Rathjens and Ruina), 239–55
"BMD Technologies and Concepts in the 1980s" (Bethe, Boutwell, Garwin), 53–71
bombers, manned, 53, 161, 221, 248, 300
  early warning and, 141, 226–27
  Stealth, 248, 300
  in strategic triad, 21–22
  see also aircraft
booster plumes, 56, 57, 204, 205, 320, 327
booster rockets, 56, 150–51, 286, 319, 332–34
(boost-phase intercept) BPI, 66, 121–22, 230, 244
  DEWs in, 78–79, 317
  duration of boost in, 55–56, 79, 314
  HKVs in, 338–41
  MIRV bus and, 44, 57, 58, 62, 66, 79, 317, 337
  pop-up weapons and, 316
  protection of missiles in, 66, 269, 314
  SATKA systems in, 346–47
  target acquisition and discrimination in, 81–82, 205
Boost Surveillance and Tracking System (BSTS), 205
Boutwell, Jeffrey, 53–71, 297–311
Brezhnev, Leonid, 265
Buchan, Alastair, 287
Bumpers, Dale, 308
Bundy, McGeorge, 194, 244
bus, MIRV, 44, 57, 58, 62, 66, 79, 317, 337

Canavan, Gregory, 323
Carter, Ashton B., 171–89, 206, 271–72
Carter, Jimmy, 29, 135
Carter administration, 23, 42, 48–50, 151
"Case for Ballistic Missile Defense, The" (Toomay), 219–37
Central Intelligence Agency (CIA), 263, 270
Chafee, John, 308
chaff, radar, 44, 45, 46, 57, 58
Chayes, Abram, 193–218
Chayes, Antonia Handler, 193–218
chemical lasers, 50, 60, 82–83, 176, 177, 204, 319, 321–25
Cherednichenko, M., 259
Cheysson, Claude, 289
China, People's Republic of:
  in Korean war, 159
  as nuclear threat, 24, 36, 39, 41, 250, 252
  Soviet monitoring of, 139
civil defense, 39–40, 221–22, 242
civilian space programs, 280, 285, 286
command-and-control system, 64–65
  computers in, *see* computers

(Command-and-control continued)
  defense of, 40–41, 48–49, 64–65, 67, 117–18, 292–93
  instability of, 162–63
  local area defense and, 226–27
  in multilayered BMD, 111, 117–18
  software for, 119–21
  testing of, 118
communication links, 65, 129, 142–43
communications, 40–41, 161–62
  EMP and, 40–41, 225–26
  encryption of, 157, 166
  geostationary satellites for, 137
  see also command-and-control system
communication satellites, 18–19, 129, 137, 140, 153–54, 160–61
communication transponders, 142
"Compliance of the Strategic Defense Initiative with the ABM treaty" (Compliance Appendix) (1985 Report to Congress), 202–3, 204–5
components (BMD), 179, 203–5, 207, 215, 309
Compton, Arthur Holly, 101
computers:
  automatic programming by, 120–21
  battle management and, 65, 84, 322–23
  data-processing, 119, 228, 231
  in decision-making roles, 122
  hardware of, *see* hardware, computer
  software of, 119–21
  space-based, 84
Congress, U.S., 24, 36–37, 50, 68, 95, 135, 200–203, 212, 302, 305
conventional warfare, 40, 115, 156, 158, 219, 229
  ASATs in, 155–60, 172
  satellites in, 139
cost-exchange ratios, 40, 41, 51, 78, 80, 87, 114, 242–44, 246–47, 248
countermeasures to BMD, 35, 43–46, 57–78, 65–68, 231–34, 240–42, 273, 303, 313–14
  ASATs as, 172, 185–86, 271–72, 275–76
  cost-effectiveness and, 51, 114, 124, 287, 300–301, 309
  decoys as, *see* decoys
  in defining BMD research and development, 77–78, 100
  infrared-based system, *see* infrared-based systems
  jammers as, 43, 45, 46, 118, 142, 154, 157–58, 166
  MaRVs as, 45, 60, 66, 67, 225
  MIRVs as, *see* MIRVs
  offense and defense conservative views of, 43

  radar cross-section reduction as, 46, 263–64
  see also BMD
cruise missiles, 20–22, 53, 68, 115, 221, 279
  air defenses and, 263
  air-launched, 21–22, 265, 300
  ground-launched, 22, 265, 274
  sea-launched, 265, 303–4
Cuban Missile Crisis (1962), 221, 279
Cunningham, Christopher, 323

data processing, 119, 228, 231
"decapitation," 162–63
deceptive missile basing, 224
decoys, 35, 44–46, 49, 56, 67, 79–80, 81–82, 119, 178, 180, 230, 270, 316
  aerosol clouds as, 57–58, 66
  anti-simulation technique and, 45
  balloons as, 45–46, 57, 58, 66
  chaff as, 44, 45, 46, 57, 58
  MIRV bus and, 44, 57, 58, 62, 66, 79, 317, 337
dedicated ASATs, 143, 164–168, 179
deep strike operations, 141
defense:
  air, *see* air defense
  allies and, 99, 117, 229, 279–96, 306
  "automated response" in, 121–22
  in boost phase, *see* BPI
  civil, *see* civil defense
  of command-and-control centers, 40–41, 48–49, 64–65, 67, 117–18, 292–93
  cost-effective, 40, 114, 124, 287, 300–301, 309
  cost-exchange ratios and, 40, 41, 51, 78, 80, 87, 114, 242–44, 246–47, 248
  fair-to-good, 246–47, 288
  hard-point, *see* hard-point defense
  light area, *see* light area defense
  local area, *see* local area defense
  partial, 300–301
  perfect, 53, 60, 99–100, 117, 242, 243, 244–45, 288, 290, 298, 299–301
  population, 40, 44, 105, 142, 226–27, 240
  preferential, *see* preferential defense
  terminal, 58–60, 66, 230, 249, 343–46
  third-party attacks and, 36, 39, 41, 232, 240, 250–51, 252, 262, 264
  very good, 242, 245–46, 250
  whole-country (area), 34, 230–31, 233–34
  see also BMD; SDI
Defense Department, U.S. (DOD), 27, 85–86, 92, 103, 150, 230, 235, 249, 263, 269, 297, 308

Index 379

SDI cost and, 25, 68–69, 86, 201, 234–35, 305, 308
  technology study teams and 76–77
defense-dominance, 101, 243, 303, 310
Defense Support Program (DSP) early-warning satellites, 137
defense suppression tactics, 66–67
defensive satellites (DSATs), 23, 67, 233
Defensive Technologies Study Team, see DTST
DeLauer, Richard, 76, 276
depressed trajectory, SLBMs in, 314
deterrence, 17–32, 98–100, 219–22, 293–95, 297–99
  ASATs and, 113, 147, 164
  defense dominance and, 101, 243, 303
  European attitudes toward, 281–86
  first-strike capability and, see first-strike capability
  hard-point defense and 223, 225, 233
  nuclear vs. conventional, 219, 229
  offense dominance and, 33, 243, 247–48, 303
  second-strike capability and, 139, 232–33, 245–46, 261, 265–66, 270–71, 291–92, 298, 301
  Soviet policy on, 257–62
  see also MAD
"development," in ABM treaty, 201–3, 215
DEWs (directed energy weapons), 60–63, 78–81, 231, 232, 314–38
  as ASATs, 142, 152–53, 160, 162, 173
  nuclear-driven, 60, 75, 82, 208, 315, 328–29
  performance requirements of, 320–21
  possible basing modes for, 317–19
  possible missions for, 315–17
  Soviet research on, 263
  see also lasers; particle beams
"diffraction limited," 320
Dougherty, Russell, 219
Drell, Sidney D., 26
DTST (Defensive Technologies Study Team) (Fletcher panel), 25–26, 31, 50–51, 76–77, 80–81, 85–86, 88–89, 92–96, 101–4, 110–12, 117–18, 121, 219, 222, 240, 269, 316, 320, 322, 328
dual-purpose technologies, in ABM treaty, 205–10
dummy warheads, see decoys
dynamic engagement models, 114

early-warning radars, 39, 48, 175, 213, 264
early-warning satellites, 21–22, 56, 129, 137, 139, 315–16
  dual nature of, 147
  Soviet, 140–41, 160–61

  Soviet vs. U.S. dependence on 140–41
Eisenhower, Dwight D., 31, 37, 131, 132
electromagnetic pulse (EMP), 41, 225–26
electromagnetic railguns, 62, 83, 338–39
electronic intelligence (ELINT) satellites, 136–37, 139, 148, 151, 152, 157
encryption, 157, 166
Energy Department, U.S. (DOE), 50, 85–86
EORSAT (ocean ELINT), 157–58
escalation dominance, 283
Europe:
  ABM treaty as viewed in, 43, 287–90
  ASATs and war in, 155–60
  attitudes toward deterrence in, 281–86
  civilian and commercial space programs in, 280, 285, 286
  indigenous nuclear forces in, 291–93
  light area defense in, 229
  reaction to SDI in, 287–90
  surveillance systems in, 156
  see also NATO
European Space Agency (ESA), 286
excimer (ultraviolet) lasers, 60, 64, 82–83, 177, 315, 316, 319, 326–28
extended deterrence, 921

F-15 anti-satellite weapon, 62, 165, 175, 339
Farley, Philip J., 26
fast-burn ICBMs, 269–70, 314
Federal Aviation Administration (FAA), 119–20
Fermi, Enrico, 94, 101
Field, George, 325
"field tests," in ABM treaty, 202
first-strike capability:
  defense against, 245–46, 267
  early-warning satellites and, 141
  retaliation against, 139, 232–33, 245–46, 261, 265–66, 270–71, 291–92, 298, 301
  Soviet reliance on, 270
Flax, Alexander, 33–52, 69–70
Fletcher, James C., 25–26, 50–51, 76, 93, 95
Fletcher panel, see DTST
flight, ballistic missile, 53–60, 179, 181, 207, 316
  boost phase of, 55–56, 66, 79, 81–82, 121–22, 230, 244, 269, 314, 316, 317, 338–41, 346–47
  mid-course phase of, 44, 46, 55, 57–58, 67, 79–80, 110, 317
  post-boost phase of, 55, 56–57, 79
  terminal phase of, see terminal phase
follow-on-forces attack (FOFA), 158
Ford, Gerald, 135

Index

fractional orbital bombardment system (FOBS), 18, 23, 133
France, 39, 229, 281–82, 283, 285–86, 287, 306
free-electron lasers (FELs), 60, 82–83, 315
Future Security Strategy Study team (Hoffman panel), 51, 76–77, 96–98, 102–5, 113, 222–23, 226, 240, 245

gallium arsenide (GaAs), 337
Galosh missile, 38–39, 47, 58, 67, 174, 262
gamma radiation, 337
Garthoff, Raymond, 199
Garwin, Richard L., 42, 53–71, 313–50
Geneva arms-control talks, 24, 188, 289, 302, 307
GEOs (geosynchronous orbits):
 ASAT attack in, 150, 153, 160, 166, 342–43
 BMD platforms in, 63
 DEW deployment in, 316, 326–28
 excimer (ultraviolet) lasers in, 326–28
 reconnaissance and, 18–19
 satellite vulnerability in, 150, 153, 160, 166, 342–43
 X-ray lasers in, 61
geodetic satellites, 20, 129
geostationary satellites, 137
 conventions for placement of, 193
geosynchronous orbits, *see* GEOs
Germany, 37, 159, 283–84
global positioning system (GPS) satellites, 18, 20, 22, 137, 141, 150
GLONASS navigation satellite system, 141, 150
Gorbachev, Mikhail, 275
Gottfried, Kurt, 147–70, 342
Graham, Daniel, 75
Great Britain, 39, 229, 275, 281–82, 283, 285–86, 287, 306
Grechko, A. A., 260
Griffon missile system, 38
Gromyko, Andrei, 273–75
ground-launched cruise missiles (GLCMs), 22, 265, 274

Hafner, Donald L., 91–107
hardened silo defense, 42, 211, 221, 223, 249
hard-point defense, 34, 42, 44, 45, 142, 222–26, 230
 cost-effectiveness of, 223, 225, 229–30, 298
 deceptive missile basing in, 224
 Hoffman Panel on, 222–23

interceptor missiles in, 223–24
 local area defense vs., 226
 offense dominance and, 243, 247–48, 303
 swarmjet, 42, 58, 345–46
 technical difficulties of, 225–26
 terminal defense and, 60, 249–50
hardware, computer, 65, 95, 119–21, 205
 see also computers
Healey, Denis, 294
high-altitude interceptor missiles, 34, 36, 38–39, 43, 47
high energy laser program, 50
High Frontier study, 49, 75, 339
high-speed data links, 65
HKVs (homing kill vehicles), 26, 135, 142, 151–52, 174–75, 207, 338–46
 for boost-phase intercept, 62–63, 78, 82–84, 338–41
 for mid-course intercept, 63, 341–42
 space- vs. ground-based, 63
 for terminal defense, 63, 343–46
 see also ASATs; infrared-based systems; tracking
Hoffman, Fred S., 50, 76, 349
Hoffman panel (Future Security Strategy Study team), 51, 76–77, 96–98, 102–5, 113, 222–23, 226, 240, 245
Holloway, David, 26, 257–78
homing kill vehicles, *see* HKVs
Homing Overlay Experiment, 49, 58, 152, 165, 174–75, 207
Howe, Geoffrey, 289
HSD (hard site defense), *see* hard-point defense
Hubble Space Telescope, 321
hydrogen fluoride (HF), 315, 321–22

ICBMs (intercontinental ballistic missiles), 45, 55, 74, 83, 206, 269, 276, 313, 316
 fast-burn, 269–70, 314
 flight phases of, *see* flight, ballistic missile
 flight trajectory of, 179, 181, 207
 tracking of, *see* tracking
 see also ballistic missiles; *specific ICBMs*
Ikle, Fred, 76
inertial guidance systems, 129
infrared-based systems, 49, 56, 57, 231
 ABM radars and, 44, 45–46
 sensors in, 178–79
 see also HKVs; tracking
infrared chemical lasers, 50, 60, 82–83, 176, 177, 204, 319, 321–25
infrared-homing interceptors, 62, 152, 154
infrared plumes, 56, 57, 204, 205, 320, 327
interceptor missiles, 34–39, 42, 47, 132, 229, 262–63

Index 381

as defined by ABM treaty, 199
  in hard-point defense, *see* hard-point defense
  see also ASATs; BMD, HKVs
intercontinental ballistic missiles (ICBMs), *see* ICBMs
intermediate-range nuclear forces (INF) negotiations, 221, 274, 284–85, 288
International Geophysical Year (IGY), 19, 131
internetting, 143
"intrawar deterrence," 283

Jackson, Henry, 202
jamming, 43, 45, 46, 118, 142, 154, 157–58, 166
Johnson, Lyndon B., 37, 252, 307
Jupiter rocket, 131

Kennan, George F., 194
Kennedy, John F., 22, 30, 37, 94, 132, 133, 221, 307
Kent, Glenn, 40
Keyworth, George A., II, 87, 269–71, 301
KH-9 (Big Bird) satellite system, 136
KH-11 satellite system, 136, 153
Khrushchev, Nikita, 131, 258
"kill radius," 322
kinetic energy weapons, *see* HKVs
"Kleenex radars," 345
Kokoshin, A. A., 266
Korean War, 19, 159, 279
Kosygin, Aleksei, 258
Krasnoyarsk phased-array radar, 179, 212–13, 309
Kwajalein Atoll, 184, 198, 207

"ladder down" tactics, 344
Laird, Melvin R., 200
LAMPF particle accelerator, 335
lasers, 26, 51, 53, 60–61, 304, 314–315, 320–23
  for ASAT purposes, 153
  for BPI, 56, 78–79, 317
  in conventional war, 115
  defined, 60
  ground-based, 24, 75, 82–84, 177, 271
  high energy, 50
  optical, 50, 60–61, 64, 82–84, 176–77, 204, 315, 316, 319, 320, 321–24
  power density of, 320–21
  space-based, 75, 82–84, 176–77, 271
  X-ray, *see* X-ray lasers
launch-on-warning, 21, 28, 223
Lawrence, Ernest Orlando, 101
Lebow, Richard Ned, 147–70, 342

light area defenses, 36, 41, 222, 228–30, 251
limited nuclear war, 139, 162, 291
Limited Test Ban Treaty (1963), 128, 208, 330
liquid-fuel ICBMs, 83, 150, 151, 316
local area defenses, 44, 60, 226–27, 229–30
LODE/LAMP, 204
Long, F. A., 7–12, 297–311
Long Range Naval Aviation, Soviet, 158
low-altitude interceptor missiles, 35, 36, 42, 45, 47
low earth orbit (LEO), 148, 317–19, 326–28
Lysenko, Trofim Denisovich, 273

McNamara, Robert S., 23, 194
MAD (mutual assured destruction):
  alternatives to, 27, 75, 239
  European attitude toward, 281–83
  evolution of, 33–34, 99
  see also deterrence
Malmstrom Air Force Base, Mont., 37, 211
maneuvering warheads (MaRVs), 45, 60, 66, 67, 225
Manhattan Project, 94, 101, 106, 239, 257
manned bombers, *see* bombers, manned
mapping satellites, 20, 21–22
Mathias, Charles, 308
MESFET circuits, 337
meteorological satellites, 129, 137–38
mid-course phase, 55, 57–58, 67, 317
  decoys in, 44, 317
  defensive advantages in, 79–80, 317
  DEWs for intercept in, 341–42
  multilayered BMD systems and, 46, 55, 57–58, 79–80, 110
Midgetman system, 224
Miller, Franklin C., 76, 96
Miller panel, 76, 96, 106
Millionshchikov, Mikhail, 267
miniature homing vehicles, 135, 151–52, 174–75
mini-radars, 345–46
Minuteman ICBMs, 44, 207, 221, 223, 228
Minuteman silos, 36–37, 42, 49, 211
MIRVs (multiple independently targetable reentry vehicles, 28, 30, 44, 55, 67–68, 248
  buses, 44, 57, 58, 62, 66, 79, 317, 337
  flight of, *see* flight, ballistic missile
  see also RVs
"missile gap," 21
missile site radar (MSR), 35, 36, 48
missile "skin," 61, 329

missile-tracking radar, 39
Mitterrand, François M., 289
mobile command centers, 41
Molniya orbits, 148, 160
Molniya strategic communication system, 137, 148
Moscow ABM system, 47–48, 75, 174, 262
multi-face phased-array radar, 48
multilayered BMD systems, 53, 55–60, 77–90
　boost phase in, 55–56, 79, 81–82, 110
　command-and-control of, 110–12, 117–18
　cost of, 298–99
　countermeasures limited by, 66
　elements of, 80–84
　"hand over" in, 65
　mid-course phase in, 46, 55, 57–58, 79–80, 110
　MIRV bus in, 44, 57, 58, 62, 66, 79, 317, 337
　operational complexities of, 59, 110–12, 117–18
　post-boost phase in, 55, 56–57, 79, 110
　software-dependent interactions in, 120–21
　survivability of, 84–90, 123–25
　terminal phase in, 58–60, 78, 80
multiple independently targetable reentry vehicles, *see* MIRVs
multi-spectral infrared sensor technologies, 231
mutual assured destruction, *see* MAD
MX missile, 42, 48–49, 56, 74, 105, 221, 265–66, 302, 316

NASA (National Aeronautics and Space Administration), 19–20
national command authority (NCA), 227
national technical means of verification (NTM), 197, 203, 206, 210–11
NATO (North Atlantic Treaty Organization), 115, 155–59, 208–9, 229, 279–84, 288–89, 292–95
navigation satellites, 18, 20, 22, 24, 128, 129, 137, 141, 150, 152, 157
Navstar GPS navigation satellites, 137, 141, 150
neutral particle beams, 62, 83, 153–54, 315, 321–22, 337
"New BMD Technologies" (Bethe and Garwin), 313–50
Nike Ajax, 37
Nike Hercules, 43
Nike-X, 35–36
Nike Zeus, 34–36, 133

Nitze, Paul H., 70, 275, 300, 303, 321
Nixon administration, 36
North Atlantic Treaty Organization, *see* NATO
Novaya Zemlya, nuclear test at, 23
nuclear blackout, 41, 225–26
"Nuclear Deterrence and the Military Uses of Space" (York), 17–32
nuclear explosions:
　EMP from, 41, 225–26
　X-ray lasers powered by, 60, 75, 82, 208, 315, 328–29
nuclear generators, 142
nuclear strategies, 103, 221, 283, 307, 344
　assured destruction vs. assured survival, 17
　luanch-on-warning, 21, 28, 223
　launch-under-attack, 272
　see also defense; MAD; SDI
nuclear war, 28–29, 228, 249, 291
　limited or protracted, 139, 162, 291
　post-attack assessment in, 139, 160, 161, 162

ocean reconnaissance satellites, 18, 20, 22, 24, 128, 129, 137, 141, 150, 152, 157
offense dominance, 33, 243, 247–48, 303
Ogarkov, N. V., 259, 261–62
Oppenheimer, Julius Robert, 101
optical/IR homing tests, 151
optical lasers, 50, 60–61, 64, 82–84, 176–77, 315, 316, 319, 320, 321–34
orbital bombardment systems, 22–23, 132–33
　fractional, 18, 23, 133
orbits:
　"absentee problem" in, 63, 318–19, 324–25, 341
　geosynchronous, *see* GEOs
　highly elliptical (Molniya), 148, 160
　low earth, 148, 317–19, 326–28
　semi-synchronous, 148–50, 160, 166
Outer Space Treaty (1967), 133, 193, 195–97, 208, 214

Palomar Observatory, 321
PAR (perimeter acquisition radar), 36
parity, deterrence and, 221, 259, 272
particle accelerators, 61–62, 335–37
particle beams, 26, 49, 51, 53, 56, 61–62, 334–38
　angular spread of, 335–36
　charged, 315, 321–22, 335–37
　critical parameters of, 82–83
　neutral, 62, 83, 153–54, 315, 321–22, 337

Soviet, 47
  space-based, 82–83, 176–77
passive tracking, 67
Pastoral Letter on War and Peace, (Catholic bishops), 74
Patriot, (SAM), 43
"Pave Paws," 48
"Peace and National Security" (Reagan's SDI speech, March 23, 1983) text of, 351–53
penetration aids, 35, 180, 185, 229–31, 263, 298
  see also decoys
perfect defense, 53, 60, 117, 242, 243, 244–45, 288, 290, 298, 299–301
perimeter acquisition radar (PAR), 36
peripheral early-warning radars, 48
Pershing II missiles, 209, 262, 264, 265, 274, 279, 302
phased-array radar, 39, 48, 262, 309
  at Krasnoyarsk, 179, 212–13, 309
photographic satellites, 128, 152–54, 156–57
  see also reconnaissance satellites
point defense, *see* hard-point defense
Polaris A-3 SLBM, 44
population defense, 40, 44, 105, 142, 226–27, 240
pop-up BMD systems, 316
  battle stations, 63, 142
  geometry of intercept, illustrated, 331
pop-up X-ray lasers, 61, 63–64, 142, 177–78, 316, 317, 319, 330–31, 333–34
Poseidon SLBM, 44, 333
post-attack assessment, 139, 160, 161, 162
post-boost phase, 55, 56–57, 79, 110
preemptive strike threat, *see* first-strike capability
preferential defense, 45, 225–26, 231–32
  in assuring retaliatory capability, 243–44
  defined, 223–24
  leverage from, 248–49, 344–45
preferential offense targeting, 243–44
programming, automatic, 120–21
Project Orbiter satellite, 19, 131
Project Vanguard satellite, 19, 131
Proxmire, William, 308
Pushkino radar, 48, 264

radar:
  of ABM-1 system, 39
  in ABM treaty, 47, 199
  Battle of Britain and, 30
  blackout of, 39, 46, 66, 67, 225, 264
  chaff, 44, 45, 46, 57, 58

  cross-section reduction, 46, 263–64
  decoys and, 35, 44–46, 49, 56, 57–58, 66, 67, 79–80, 81–82, 119, 178, 180, 230, 270, 316
  early-warning, 39, 48, 175, 213, 264
  earth-based, 21, 35, 39, 175, 228, 345
  jammers and, 43, 45
  phased-array, *see* phased-array radar
  Pushkino, 48, 264
  space-tracking, 178, 213
  Stealth techniques vs., 248
  see also sensors; tracking
radar corner reflectors, 44
radar ocean reconnaissance satellites (RORSATs), 22, 24, 157–58
radio communication links, 142
railguns, electromagnetic, 62, 207, 338–39
RAND Corporation, 18–19, 21
Rathjens, George, 69, 239–55
Reagan, Ronald, 25–26, 49, 91–107, 195, 198, 219, 246, 249, 251–52, 279, 290, 299, 301
  SDI speech of, 17–18, 50, 73, 76, 91, 219, 230, 239, 253, 265, 266, 297, 306, 351–53
Reagan administration:
  ABM treaty polity of, 194–95, 201–2
  ASAT arms control policy of, 135, 163, 275–76
  on BMD research, 184–85, 252, 275
Reagan SDI policy, 27–30, 91–107, 253–54, 287–88, 292–295, 297–303
  see also SDI
reconnaissance satellites, 19–22, 128, 132, 156–57, 196–97, 280
  ocean, 18, 20, 22, 24, 128, 129, 137, 141, 150, 152, 157–58
  photographic, 128, 152, 153–54, 156–57
  U.S. vs. Soviet, 134, 136–40
"red-out," 46
Redstone rocket, 131
reentry vehicles, *see* RVs
"Relationship of ASAT and BMD Systems, The" (Carter), 171–89
  in ABM treaty, 201–3, 215, 298
"research," in ABM treaty, 201–3, 215, 298
residual ASATs, 143, 164–65, 168
robotics, 123
Rogers, William Penn Adair, 200–201
Ruina, Jack, 69, 239–55
Rusk, Dean, 307
RVs (reentry vehicles), 98, 263–64, 313, 316–17
  see also MIRVs

SA-1 missile system, 37

Index

SA-2 missile system, 40
SA-3 missile system, 40
SA-5 missile system, 40, 43, 180
SA-10 missile system, 180
SA-12 missile system, 180, 263
Safeguard system, 36-37, 41-42, 58, 228, 252
Sagdeev, R. Z., 266
SAGE (semi-automatic ground environment) air defense, 119–20
Sakharov, Andrei, 267
SALT (Strategic Arms Limitation Talks), 44, 222, 259–62
SALT I, 134, 210, 211, 265
SALT II, 135, 206, 210–12, 270, 304
salvage fuzing, 58, 66, 80, 343
Salyut space station, 138
SAM radars, 46
SAMs (surface-to-air missiles), 34, 37–38
 aircraft and, 179–80
 intercepting tactical ballistic missiles with, 43, 79–80
 Soviet, 37, 40, 43, 179–81, 206–8, 263–64
 susceptibility to penetration aids of 180–81
 upgrading of, 179–81, 263–64
satellite-borne theater command-and-control, 158
Satellite Data System (SDS) satellites, 137
satellites:
 communication, *see* communication satellites
 in conventional war, 155–60
 crisis monitoring with, 134, 139
 defense of, 143–44, 153–54
 defensive (DSATs), 23, 67, 233
 deterrence and, 18–20, 22–25
 early-warning, *see* early-warning satellites
 electronic intelligence, 136–37, 139, 148, 151, 152, 157
 future military applications of, 141–42
 in geosynchronous orbits, *see* GEOs
 in LEO, 148, 317–19, 326–28
 multipurpose, 128, 129, 137–38
 navigation, 18, 20, 22, 24, 128, 129, 137, 141, 150, 152, 157
 in nuclear war, 160–63
 photographic, 128, 152, 153–54, 156–57
 for post-attack assessment, 139, 160, 161, 162
 reconnaissance, *see* reconnaissance satellites
 in semi-synchronous orbit, 160, 166
 survivability of, 67, 142–44, 173
 threats to, 127, 142–44
 U.S. vs. Soviet deployments of, 130, 134, 136–40, 148
 see also specific satellites
SA-X-12 missile system, 43
Schlesinger, James, 68, 305
Scowcroft Commission, 222, 249, 304
SDI (Strategic Defense Initiative), 50–51, 73–90, 257–78, 297–311
 ABM treaty and, 172, 194–95, 198–201, 214–16, 252–53, 308
 Apollo program and, 88, 106, 239, 257
 arms-control negotiations and, 101–3, 135, 172, 175–76, 185, 253, 273–77, 302–3, 307
 ASATs and, 93, 135, 271–72, 275–76
 budget for, 25, 68–69, 86, 201, 234–35, 305, 308
 crisis instability and, 241–42, 245–46, 247
 leverage on Soviets from, 97, 302–3, 307
 policy studies and, 87–90
 politics of, 251–55, 290–93, 299, 301–7
 principal ABM treaty rules and, 214–15
 research phase of, 79–81, 86–88, 201–3, 215, 241, 298, 304–5
 Soviet assessment of, 254–55, 263, 265–73, 276
 systems perspective on, 109–26
 see also BMD; defense; Reagan SDI policy
"SDI and U.S. Security, The," (Boutwell and Long), 297–311
SDI Program Office, U.S., 337–38
sea-launched cruise missiles (SLCMs), 265, 303–4
second-strike nuclear retaliation, 139, 232–33, 243–46, 261, 265–66, 270–71, 291–92, 298, 301
"semi-active homing," 342
semi-automatic ground environment (SAGE) air defenses, 119–20
semiconductor kill dose, 337
semi-synchronous orbits, 148–50, 160, 166
sensors, 67, 119, 204
 infrared, 45–46, 49, 56, 57, 62, 152, 154, 178–79, 231
 jamming of, 43, 45, 46, 118, 142, 154, 157–58, 166
 see also radar; tracking
Sentinel program, 36–37, 41–42, 58, 228, 252
SH-04 missile, 262–63
SH-08 missile, 262-63
shoot-look-shoot tactics, 225–26
Shultz, George, 274–75

Index 385

silos:
    hardening of, 42, 211, 221, 223, 249
    Minuteman, 36–37, 42, 49, 211
    swarmjet defense of, 42, 58, 345–46
SLBMs (submarine-launched ballistic missiles), 44, 177, 178, 206, 240, 333
    boost phase of, 55–56, 79
    in depressed trajectory, 314
    mid-course phase of, 57–58, 79–80
    in post-boost phase, 56–57
    in terminal phase, 58–60
Smith, Gerard C., 194, 200, 202–3
soft-area targets, 34, 36, 44, 60, 221, 222, 226–27, 242, 343–44
software, 119–21
Sokolovskii, V. D., 258–59
solar cells, 142
solid-fuel missiles, 269
Soviet Union, 127–45, 257–78
    ABM treaty and, 75, 194–95, 260–61
    air defense system of, 37–38, 40, 43, 180, 221–22, 248, 263, 300
    ASATs of, 23–24, 67, 75, 113, 133–34, 150–52, 174, 271–72
    ASAT test moratorium by, 135, 163
    ATBM development in, 209
    BMD as viewed by, 231–34, 240–42, 254–55, 263, 265–73, 276
    BMD programs in, 37–39, 46–48, 58, 113, 174, 194–95, 258, 262–65, 272, 306
    civil defense in, 221–22
    deterrence and defense in policy of, 34, 134, 221, 257–65
    nuclear attack doctrine of, 139–40
    satellite launch rate of, 136–38, 139
    upgrading of SAMS by, 180–81
space:
    European interests in, 285–86
    legal rules governing use of, 193–218
    as "sanctuary," 127, 164, 168–69
space militarization, 17–32, 128–30, 142, 144–45, 165, 168–69, 193–218, 285–86
space mines, 67, 85, 152–53, 271
space mirrors, 67, 204
space platforms, 18, 63–64, 67, 115
space shuttles, 123, 138, 154, 204
Space Surveillance and Tracking System (SSTS), 178–79, 205, 207
space-tracking radar, 178, 213
space transport, 123, 138
space trucks, 339, 341
"Space Weapons: The Legal Context" (Chayes, Chayes, and Spitzer), 193–218

Spartan missiles, 36
speed of light weapons, *see* DEWs
Spitzer, Eliot, 193–218
spoofing, 142, 157–58, 166
Sprint missiles, 35, 36, 47, 262
Sputnik, 19–20, 21, 127, 129, 131, 132
Stalin, Joseph, 27
Standing Consultative Commission (SCC), 210–14, 264, 309
Stares, Paul, 127–45
Stealth bomber techniques, 248, 300
Stimson, Henry, 101
"store dump" satellites, 137
Strategic and International Political Consequences of the Creation of a Space-Based Anti-Missile System Using Directed Energy Weapons (Committee of Soviet Scientists for Peace Against Nuclear Threat), 266–68
Strategic Arms Limitation Talks, *see* SALT
Strategic Arms Reduction Talks (START), 274
"Strategic Defense and the Western Alliance" (Bertram), 279–96
"Strategic Defense: A Systems Perspective" (Zraket), 109–26
Strategic Defense Initiative, *see* SDI
"Strategic Defense Initiate, The" (Yonas), 73–90
"Strategic Defense Initiative and the Soviet Union, The" (Holloway), 257–78
Strategic Rocket Forces, Soviet, 45, 131, 150, 259, 269–70
strategic triad, 21–22, 142
submarine-launched ballistic missiles, *see* SLBMs
submarines:
    navigation satellites and, 20, 129, 141–42, 150
    nuclear, 161, 221, 265–66, 285, 325
    pop-up BMD systems on, 64, 330–33
    see also SLBMs
super-hard silos, 42
surface-to-air missiles, *see* SAMs
surveillance, acquisition, tracking, kill assessment (SATKA) systems, 342, 346–48
swarm jet, 42, 58, 345–46
Szilard, Leo, 94

tactical nuclear forces (TNF), 291
targets:
    hard, *see* hard-point defense
    soft-'area, 34, 36, 44, 60, 221–22, 226–27, 242, 343–44
    see also defense

Teller, Edward, 75
terminal phase, 58–60, 343
  atmospheric drag in, 55, 230
  hard-point defense and, 230, 249
  MARVs in, 45, 60, 66, 67, 225
  in multilayered BMD, 58–60, 78, 80
terrain comparison ("tercom") system, 20
terrorism, 143, 228, 233, 251
test sites:
  Soviet, 23, 150, 179, 198, 212–13
  U.S., 184, 198, 207
Thatcher, Margaret, 283, 285
third country missile attacks, 36, 39, 41, 232, 240, 250–51, 252, 261, 264
Third World, 155
  nuclear threats from, 251
  Soviet Union and, 135
Thor booster, 23, 133
Toomay, John C., 78. 219–37
tracking, 67, 205, 304
  birth-to-death, 58, 82, 118–19, 349
  infrared, 45–46, 49, 56, 57, 62, 152, 154, 178–79, 231
  SATKA system, 347–48
  see also decoys; radar; sensors; *specific flight phases*
tracking satellites, 178–79, 205, 207
Treaty on the Limitation of Anti-Ballistic Missile Systems, *see* ABM treaty
TRIAD program, 204
Trident D-5 SLBM, 265–66, 285
Trident submarine, 161, 221, 325
TRIUMF particle accelerator, 335
Truman, Harry S, 29, 279
Tyuratam test range, Soviet, 150

ultraviolet (excimer) lasers, 60, 64, 82–83, 177, 315, 316, 319, 326–28
United Nations, 132, 135, 159, 195
"U.S. and Soviet Military Space Programs: A Comparative Assessment" (Stares), 127–45
Ustinov, Dmitri, 265

Vandenberg launch complex, 138
Vanguard launch vehicle, 131
Velikhov, E. P., 266-68, 271, 273

Wallop, Malcolm, 74–75
warheads, simulated as decoys, 45–46
Warsaw Pact, 155–58, 284, 288
weather satellites, 129, 137–38
Weinberger, Caspar, 92, 93, 239, 246, 255, 297, 307
West Germany, 283–84
White House Compliance Report to Congress (1985), 212
White Sands (N. Mex.), 198
whole-country defense (area), 34, 230–31, 233–34
"window of vulnerability," 249
Wohlstetter, Albert, 291
World War II, 30, 37, 131
Wörner, Manfred, 289

XB-70, 38
X-ray lasers, 50, 56, 60–61, 82–84, 177–78, 328–34
  performance requirements of, 320–21
  pop-up, 64, 177–78, 317, 319, 330, 333–34
  in post-boost vs. boost phase, 316–17
  powered by nuclear explosion, 60, 75, 82, 208, 315, 328–29

Yonas, Gerold, 73–90
York, Herbert F., 17–32

Zemskov, V. M., 260
Zraket, Charles A., 65, 109–26, 348

358.1

Long, Franklin A
AUTHOR
Weapons in space
TITLE

358.1

HARDEMAN COUNTY PUBLIC LIBRARY
TUESDAY, WEDNESDAY 9:30 TO 5:30
THURSDAY, FRIDAY 12:30 TO 5:30
SATURDAY 9:30 TO 3:30
PHONE: 663 - 8149